DICTIONARY OF BIOCHEMISTRY

PRINTED IN INDIA

Published by [illegible] Kumar for Anmol Publications [illegible] Delhi 110002 and Printed at Mehra Offset [illegible]

DICTIONARY OF BIOCHEMISTRY

Editors
SATISH ANAND
RAJ KUMAR

ANMOL PUBLICATIONS PVT. LTD.
NEW DELHI-110002 (INDIA)

ANMOL PUBLICATIONS PVT. LTD.
4374/4B, Ansari Road, Daryaganj
New Delhi-110 002

Dictionary of Biochemistry

First Edition 1991
Reprint 1992, 1993, 1994, 1995, 1996, 1997, 1998

Reprint 1999

ISBN 81-7041-272-2

PRINTED IN INDIA

Published by J. L. Kumar for Anmol Publications Pvt. Ltd., New Delhi-110 002 and Printed at Mehra Offset Press, Delhi.

Preface

The "Encyclopaedic Dictionary of Chemistry", the first of its kind in India, is an effort to keep pace with continuing rapid developments and expanding vocabularies in all areas of chemistry. This is a set of six volumes and includes terms which have been carefully selected from the various aspects of chemistry such as Inorganic Chemistry, Physical Chemistry, Organic Chemistry, Analytical Chemistry, Biochemistry and Drugs. This dictionary has been encyclopaedic in its contents and style of representation which is evident from the comprehensiveness of its coverage and contexual relevance of explanations with illustrations and mathematical expressions whenever necessary. Throughout the process of compilation and editing the volumes our sincere effort was to make the dictionary serve as a ready reference for undergraduate and post-graduate chemistry students and for those appearing for competitive examinations, research scholars, teachers, and authors.

Every effort has been carefully made to write the entries in a clear and lucid style to provide both straight forward definitions and invaluable background information. At certain appropriate places, the line diagrams have been included whenever the meaning of a word can be best understood by means of a diagram.

The presentation throughout has been aimed at sustaining the interest of readers while enriching their vocabulary and comprehension of technical terms and expressions.

The dictionary will be of value to students of chemistry and to scientists and engineers working on science projects.

Finally the editors express their sincere thanks to the publishers and printer for printing this book promptly.

All comments from users on omissions or shortcomings will be most welcome.

Editors

Preface

"The Encyclopaedic Dictionary of Chemistry", the first of its kind in India, is an effort to keep pace with continuing rapid developments and expanding vocabulary in all areas of chemistry. This is a set of six volumes and includes terms which have been carefully selected from the various aspects of chemistry such as Inorganic Chemistry, Physical Chemistry, Organic Chemistry, Analytical Chemistry, Biochemistry and Drugs. This dictionary has been encyclopaedic in its contents and style of representation which is evident from the comprehensiveness of its coverage and contextual relevance of contents with illustrations and mathematical expressions wherever necessary. Throughout the process of compilation and editing the volumes, our sincere effort was to make the dictionary serve as a ready reference for undergraduate and post graduate chemistry students and also those appearing for competitive examinations, research scholars, teachers and experts.

Every effort has been carefully made to write the entries in a clear and lucid style to provide both straightforward definitions and invaluable background information. At certain appropriate places, the line diagrams have been included wherever the meaning of a word can be best understood by means of a diagram.

The presentation throughout has been aimed at sustaining the interest of readers while enhancing their vocabulary and comprehension of technical terms and expressions.

The dictionary will be of use to students of chemistry and to scientists and engineers working on chemical problems.

Finally the editors express their sincere thanks to the publishers and printer for printing this book promptly.

All comments from users on omissions or shortcomings will be most welcome.

A

A. Abbreviation for adenine.

AAR. Abbreviation for Antigen-antibody reaction (see).

A Band. A transverse dark band, consisting of thick and thin filaments, see in electron microscope preparations of myofibrils from striated muscle.

Abietic Acid. A diterpene carboxylic acid, and the isomeric neoabietic acid, can easily be interconverted. These two resin acids are the main components of rosin (up to 90%), from which they could be obtained by treatment with heat or acids, possibly as products of the rearrangement of other diterpene carboxylic acids. Amber contains derivatives of A.a.

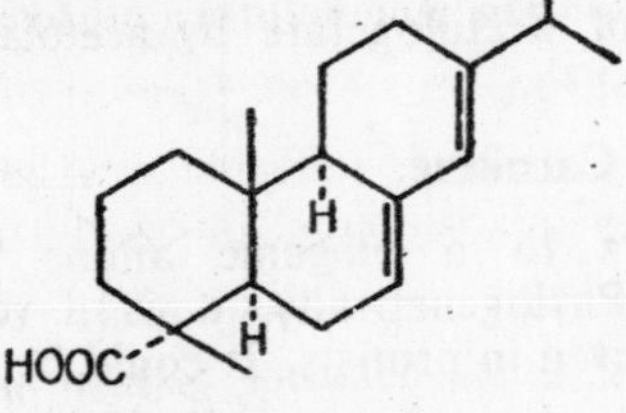

Abietic Acid

Abscisic Acid, Abb. ABA, abscisin, dormin: (S)-(+)-5(1′-hydroxy-4′-oxo-2′, 6′, 6′-trimethyl-2-cyclohexen-1-yl)-3-methyl-cis, trans-2,

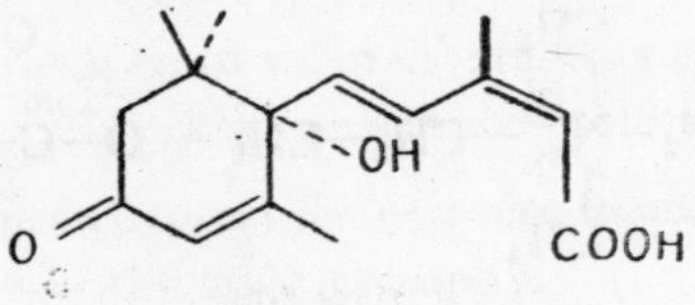

(S-)-(+)-Abscisic Acid

4-pentadienoic acid. It is a widely occurring, sesquiterpene plant hormone. Its action is mainly inhibitory. It inhibits growth and the germination of seeds. It induces dormancy in seeds and promotes the falling of leaves and fruits. The biosynthesis of ABA is still unknown.

Absolute Oils. See Essential oils.

Acceptor Site. The ribosomal binding site for the aminoacyl-tRNA during protein biosynthesis.

Accumulation of Metabolic Intermediates. See Mutant technique.

Acetaldehyde, Ethanal: CH_3-CHO_4. The important intermediate in the degradation of carbohydrats. m.p.—123°C, b.p. 20.1°C. In its activated form (Thiamine pyrophosphate), it gets involved in a number of reactions (Alcoholic fermentation). Two molecules A can undergo acyloin condensation to form Acetoin.

Acetic Acid, Ethanoic Acid: CH_3COOH. A very common monocarboxylic acid which occurs in the free form as the end product of fermentation and oxidation reactions in some organisms. Acetate is formed metabolically by dehydration of acetaldehyde, catalysed either by aldehyde oxidase (EC 1.2.3.1) or a $NAD(P)^+$-dependent aldehyde dehydrogenase (EC 1.2.1.3).

Acetoin, 3-hydroxy-2-butanone, Acetyl Methyl Carbinol CH_3-CO-CHOH-CH_3. A reduction product of diacetyl which arises under certain conditions as a side product of the pyruvate decarboxylase (EC 4.1.1.1) reaction. It is also formed by decarboxylation of acetologctate by acetolactate decarboxylase (EC 4.1.1.5).

Acetylcarnitine. See Carnitine.

Acetylcholine. Refers to a biogenic amine which is biologically highly active. Phylogenetically, it is a very ancient hormone which appears even in protists. It could be a predecessor of the neurohormones.

It acts as a cholinergic neurotransmitter in nerves and neuromuscular synapses; it induces a muscle contraction by changing the permeability of the sarcolemma. It is degraded by acetylcholinesterase (EC 3.1.1.7).

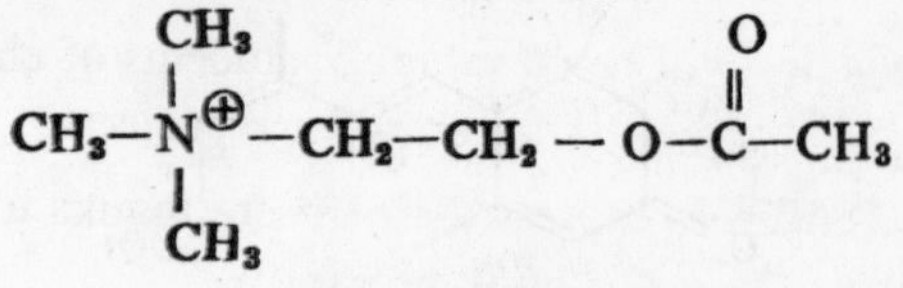

$$CH_3-\overset{\displaystyle CH_3}{\underset{\displaystyle CH_3}{\overset{|}{\underset{|}{N^{\oplus}}}}}-CH_2-CH_2-O-\overset{\displaystyle O}{\overset{\|}{C}}-CH_3$$

Acetycholine

Acetylcholinesterase (EC 3.1.1.7). Refers to the "true cholinesterase", which catalyses the hydrolysis of acetylcholine into choline and acetate. This enzyme is found in the central nervous system, particularly in the postsynaptic membranes of the striated muscles, the parasympathetic ganglia, the erythrocytes and the electric organs of fish.

Acetyl-coenzyme A, Acetyl-CoA, Active Acetate $CH_3CO \sim SCoA$. It is a derivative of acetic acid in which the acetyl residue is bound by a high-energy bond to the free SH-group of coenzyme A. The very reactive thioester has a high potential for transfer of the acetyl group, and is therefore a universal intermediate which provides the C_2 fragment for numerous syntheses.

N-Acetylglutamic Acid, N-acetylglutamate, abb. **Ac-glu,** HOOC-$CH(NHCOCH_3) \cdot CH_2 \cdot CH_2COOH$. The acetylated form of glutamic acid, is the cofactor of carbamoyl phosphate synthetase (ammonia) (EC 6.3.4.16) and allosterically activates this enzyme.

Acetyl Methyl Carbinol. See Acetoin

Acetyl Phosphate: CH_3-$COOPO(OH)_2$. An energy-rich acyl phosphate. It is the product of acetate activation in some organisms: Acetate+ATP=.Ap.+ADP; the reaction is catalysed by acetate kinase (EC 2.7.2.1). The back reaction can be used for ATP spythesis, for example in the phosphoroclastic cleavage of pyruvate.

Acid Plants, Ammonium Plants. Plants which accumulate organic acids in their leaf cells, which are neutralized by ammonium ions.

TABLE 1

Reactions in which Acetyl-coenzyme A is Synthesized

Enzyme	Reaction	Occurrence/ Significance
Acetyl-CoA synthetase (EC 6.2.1.1)	$CH_3COO^- + ATP + CoA \rightleftharpoons CH_3CO \cdot CoA + AMP + PP_i$	Yeasts, Animals, Higher plants
Acyl-CoA synthetase (GDP-forming) (EC 6.2.1.10)	$CH_3COO^- + GTP + CoA \rightleftharpoons CH_3CO\text{-}CoA + GDP + P_i$	Liver

Acetate kinase	$CH_3COO^- + ATP \rightleftharpoons CH_3CO\text{-}O\text{-}PO_3H_2 +$ ADP (Acetyl phosphate)	Microorganisms
Phosphate acetyl-transferase (EC 2.3.1.8)	$CH_3CO\text{-}O\text{-}PO_3H_2 + CoA \rightleftharpoons CH_3CO\text{-}CoA + P_i$	Microorganisms
ATP citrate (pro-3 S)-lyase (EC 4.1.3.8)	Citrate + ATP + CoA $\rightarrow CH_3CO$ CoA + oxaloacetate + ADP + P_i	Outside the mitochondria
Pyruvate dehydrogenase complex (EC 1.2.4.1, 2.3.1.12 and 1.6.4.3)	$CH_3COCOO^- + NAD^+$ + CoA (Pyruvate) $\xrightarrow{\text{TPP, LipS}_2}$ $CH_3\text{-}CO\text{-}CoA + CO_2 + NADH + H^+$	Mitochondrial particles
Acetyl-CoA transacetylase (EC 2.3.1.9)	$CH_3COCH_2CO\text{-}CoA$ + CoA (Acetoacetyl-CoA) $\rightleftharpoons 2CH_3CO$ CoA	Fatty acid degradation

Abb. TPP=thiamine pyrophosphate; $LipS_2$=Lipoamide

Aconitate Hydratase, Aconitase: (Ec 4.2.1.3). A hydratase which catalyses one stage of the tricorboxylic acid cycle, the reversible interconversion of citrate and isocitrate. The reaction proceeds via the enzyme-bound intermediate, cis-aconitate.

Aconitic Acid. An unsaturated tricarboxylic acid, usually occurring in the cis form, but sometimes in the trans. cis-A.a., m.p. 130°C., trans-A.a., m.p. 194 to 195°C. The anionic form of *cis*-aconitic acid (propen-cis-1, 2, 3-trioic acid) is important as an intermediate in the isomerization of citrate to isocitrate in the Tricarboxylic acid cycle (see).

Aconitine. An Aconitum alkaloid from the roots of aconite (Aconitum napellus) and other Aconitum and Delphinium species, m.p. 197 to 198°C, $[a]_D^{20} -36°$ (benzene). It is an esterified alkaloid. It is extremely poisonous and can cause death in adults at a dose of to 2 mg by paralysing the heart and respiration. In spite of useful physiological properties, it is rarely used in medicine, due to its toxicity. It is sometimes used internally as tincture for rheumatism and neuralgias and externally as a pain-killing salve.

Aconitum Alkaloids. A group of terpene alkaloids, some of them very poisonous, from various aconite (Aconitum) species. The best-known representative has been aconitine.

ACP. Abb. for acyl carrier protein.

ACTH. Abb. for adrenocorticotropic hormone.

Actin. See Muscle proteins.

Actinidine. A widely occurring terpene alkaloid. alkaloids.

Actinomycins. Refer to a large group of peptide lactone antibiotics which are produced by various strains of Streptomyces. These highly toxic red compounds are having a chromophore, 2-amino-4, 6-dimethyl-3-keto-phenoxazine-1, 9-dioic acid (actinocin), which is linked to two 5-membered peptide lactones by the amino groups of two threonine residues. The various actinomycins differ only in the amino acid sequence of the lactone rings. In vivo, actinomycins inhibit the DNA-dependent RNA synthesis at the level of transcription by interacting with the DNA.

Actinomycin D (Fig.) is one of the most widespread A. Its spatial structure has been elucidated by NMR studies, and the specificity of its interaction with deoxyguanosine was demonstrated by X-ray analysis. Actinomycin D is used as a cytostatic, *e.g.* in the treatment of Hodgkin's disease.

H_3C — CO—NH—Thr—D-Val—Pro—Sar—MeVal
O N
H_3C — CO—NH—Thr—D-Val—Pro—Sar—MeVal
O NH_2

Actinomycin D

Activated Carboxylic Acid. The derivatives of carboxylic acids which are very reactive, and thus capable of reactions which the free acids do not undergo.

Activated Fatty Acids. Fatty acyl coenzyme A thioesters which, as high energy compounds, are having a large potential for group transfer. They are formed during fatty acid biosynthesis, or by the activation of free fatty acids. Acyl CoA synthetases catalyse formation of the CoA derivatives according to the reaction :

$$CH_3(CH_2)_nCOO^- + ATP + HS\text{-}CoA \longrightarrow$$
$$CH_3(CH_2)_nCO{\sim}SCoA + AMP + PPi$$

Activated Glycol Aldehyde : 2-(1, 2-dihydroxyethyl)-thiamine pyrophosphate, Abb. DETPP, glycol aldehyde bound to the C-2 atom of the thiazole ring of thiamine pyrophosphate. It is formed in carbohydrate metabolism by cleavage of a ketose and is transferred as C-2 group to an aldose in a transketolation reaction.

Activated Fatty Acids. Derivatives of carboxylic acids which are very reactive, and thus capable of reactions that free acids do not undergo.

Activation Hormone. See Insect hormones.

Activator Protein. See Calmodulin.

Active Acetaldehyde. A-hydroxyethylthiamine pyrophosphate, abb. HETPP, the activated form of acetaldehyde formed by decarboxylation of active pyruvate. The aldehyde is bound to the C-2 atom of the thiazole ring of thiamine pyrophosphate. HETPP is an intermediate in alcoholic fermentation.

Active Center. That part of an enzyme or other protein which binds the specific substrate and converts it to product (enzymes) or otherwise interacts with it (heme proteins, various carrier and receptor proteins).

Active CO_2. See Biotin enzymes.

Active Formaldehyde. See Active one-carbon units; Thiamine pyrophosphate.

Active Formate. See Active one-carbon units.

Active Glycolaldehyde 2-(1, 2-dihydroxyethyl)-thiamine pyrophosphate, abb DETPP, glycolaldehyde bound to C-2 of the thiazole ring of thiamine pyrophosphate. It is formed in carbohydrate metabolism by cleavage of a ketose, and is transferred as a 2 C group to an aldose in a transketolation reaction.

Active One-carbon Units, abb. **C_1 Units.** C_1 fragments which are activated by binding to tetrahydrofolic acid, or less commonly, to thiamine pyrophosphate. The active ethylenediamine group of tetrahydrofolic acid serves as a carrier for the metabolic transfer of a formyl or methyl or group.

Active Pyruvate. α-lactyl-thiamine pyrophosphate. The lactyl is bound to the C-2 atom of the thiazole ring of the thiamine pyrophosphate. Active pyruvate is an intermediate in the oxidative decarboxylation of pyruvate to acetyl-coenzyme A and in its decarboxylation to acetaldehyde in alcoholic fermentation.

Active Succinate. Refers to the high-energy thioester succinyl-coenzyme A. It is important as an intermediate in the tricarboxylic acid cycle.

Active Transport. A process in which solute molecules or ions move across a biomembrane from lower to higher concentration, *i.e.* against the concentration gradient.

Acyl Carrier Protein, abb. ACP. A small, acidic, heat-stable globular protein which is part of the fatty-acid synthesizing complex in Escherichia coli and other bacteria, yeasts and plants. It is the carrier of the fatty acid chain during its biosynthesis.

Adaptor Hypothesis. A suggestion made by Crick to explain the translation of the genetic code. He proposed that there must be an adaptor between the information-carrying nucleic acid and the protein being synthesized which was able to "recognize" both kinds of molecules. The discovery of tRNA and the corresponding amino acyl-tRNA synthetases confirmed his hypothesis.

Addictive Drugs, Psychotropic Drugs. The drugs which create a sense of euphoria, and which have a strong potential for addiction.

Adenine, abb. A or Ade. 6-aminopurine. It is one of the common nucleic acid bases. It is also part of the adenosine phosphates and other physiologically active substances, including various nucleoside antibiotics. It is synthesized de novo from adenosine monophosphate, or is formed by degradation of nucleic acids. Adenine deaminase (EC 3.5 4.2) removes the 6-amino group to give hypoxanthine.

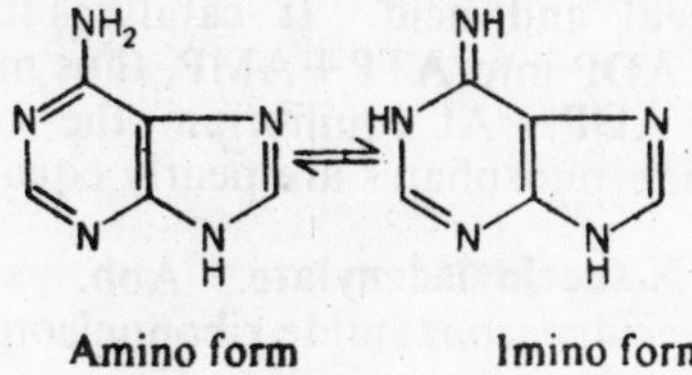

Amino form   Imino form

Tautomeric forms of adenino

Adenosine, abb. **Ado.** 9-β-D-ribofuranosyladenine. Phosphorylated derivatives of Ado are metabolically important.

Adenosine Phosphates, Adenine Ribonucleotides. Important as components of nucleic acids and as the major form in which chemical free energy is stored and transferred. They are also important metabolic regulators for example in glycolysis and the tricarboxylic acid cycle. The biologically significant derivatives, including cyclic adenosine 3′,5,-monophosphate, carry the phosphate ester on the C-5 of the ribose.

S-Adenosyl-L-methionine, S-(5′-deoxyadeno-sine-5′)-methionine, Active Methionine, Active Methyl. Abb. S-Ado-Met, SAM ; a reactive sulfonium compound which is the most important methylating agent in cellular metabolism (see transmethylation). The natural form is the L-(+)-isomer. $[\alpha]_D^{24}$ of SAM^+Cl^-=+48.5 (c=1.8 in 5N HCl). It is formed by activation of L-methionine with ATP: Met+ATP→SAM+PP_i+P_i. The adenosine residue of the ATP is transferred to the methionine.

S-Adenosyl-L-methionine

Adenylate Kinase, Myokinase (EC 2.7.4.3). A trimeric enzyme found in the mitochondria of muscles and other tissues. It is resistant to heat and acid. It catalyses the conversion of two molecules of ADP into ATP+AMP, thus making available the energy of the ADP. At equilibrium, the concentrations of the three adenosine phosphates are nearly equal.

Adenylosuccinate, N-succinyladenylate. Abb. sAMP: 5-aminoimidazole-4-N-succinocarboxamide ribonucleotide, an intermediate in purine biosynthesis.

Adenylylsulfate Reductases. Enzymes of sulfur metabolism which reduce either phosphoadenylylsulfate (APS reductase) or adenylylsulfate. Adenylylsulfate reductase, (EC 1.8.99.2) has been identical with one component of the sulfate reductase in sulfate assimilation since adenylylsulfate is the donor of the sulfate group.

Adermine. Vitamin B_6.

ADH. Abb. for Antidiuretic hormone.

Adjuvant. A mixture of oils, emulsifiers, killed bacteria and other components which serves to intensify unspecifically the immune response.

Ado. Abb. for Adenosine.

ADP. Abb. for Adenosine 5′-diphosphate.

ADP-ribosylation of Proteins. Attachment of monomeric or polymeric ADP-ribosyl groups to a protein by transfer from NAD^+ :

Adenine Nicotinamide
| |+
(ribose-P-P-ribose)$_n$ +Protein→
Adenine
|
Protein (ribose-P-P-ribose)$_n$+Nicotinamide+H^+, where n can vary from 1 to 50. Poly ADP-ribosyl groups represent a novel homopolymer of repeating ADP-ribose units linked 1′—2′ between respective ribose moieties :

↓2′ 1′
Adenine-ribose-P-P-ribose
↓2′ 1′
Adenine-ribose-P-P-ribose
↓

Phosphoadenylylsulfate reductase from Saccharomyces cerevisiae requires NADPH and has been partly purified and fractionated.

The free energy of hydrolysis of the β-N-glycosidic linkage of NAD^+ is—34.35 k Joules (−8.2 kcal)/mole at pH 7 and 25°C; it is therefore a so-called high energy bond, and NAD^+ can act as an ADP-ribosyl transferring agent. The transfer of one ADP-ribosyl group (n=1 in above equation) is catalysed by ADP-ribosyl transferase. Formation and concomitant transfer of poly ADP-ribose to an acceptor gets catalysed by poly (ADP-ribose) synthetase (n is greater than one in the above equation).

Adrenal Corticosteroids, Adrenocorticoids, Corticosteriods, Corticoids, Cortins. An important group of steroid hormones, formed in the adrenal cortex in response to adrenocorticotropic hormone. These are structurally related to pregnane (see Steriods); they contain a carboxyl group with a neighboring α, β-unsaturated bond in ring A, a ketol side chain in position 17, and other oxygen functions, particularly in positions 11, 17, 21 and 11.

More than 30 steroids have been found in the adrenal cortex.

These are biosynthesized from cholesterol via progesterone; the latter is converted into A.c. by stepwise hydroxylation in positions 17, 21 and 11.

Adrenal Gland, Suprarenal Gland, Glandula Suprarenalis. A heavily vasculated, vertebrate endocrine gland, weighing about 15 g in the adult human. There are two adrenal glands one just above each kidney. The adrenal gland consists of two developmentally and functionally distinct parts: the mesodermal adrenal cortex (AC) and the ectodermal adrenal medulla (AM). The AC, which contains three histologically distinct zones, produces and exports glucocorticoids (see Cortisol) and mineralocorticoids (see Aldosterone) in response to the action of the pituitary hormones, corticotropin and renin/angiotensin II, respectively. The AC also produces sex steriods.

Adrenalin, Epinephrine: 4-[1-hydroxy-2-(methylamino) ethyl]-1, 2-benzenediol. It is a catecholamine hormone and drug. The L-form is physiologically active, affecting carbohydrate metabolism and the circulatory system.

It is synthesized in the adrenal cortex and sympathetic nervous system from tyrosine (via dopa, dopamine and noradrenalin), stored in the chromaffin granules and released into the blood stream upon nervous stimulation by the nervus splanchnicus. It is an adrenergic neurotransmitter.

OH
HO–(benzene ring)–CH(OH)–CH_2–NH–CH_3

Adrenalin.

Adrenosterone androst-4-ene-3, 11, 17-trione. It is a steroid derived from androstane. It is synthesized in the adrenal cortex and is considered one of the male gonadal hormones, due to its weak androgenic effect.

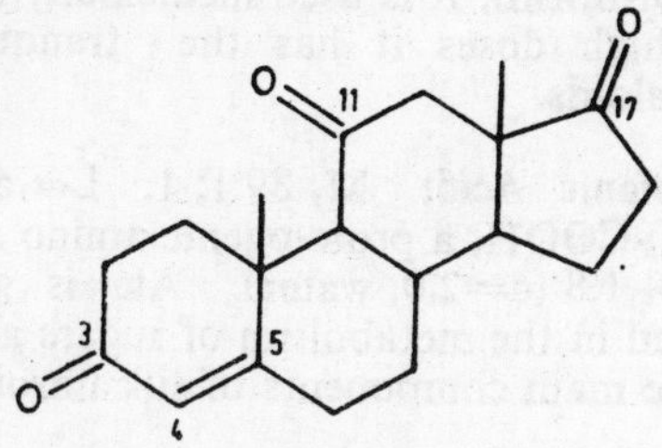

Adrenosterone

Aflatoxins. Refer to microbial products belonging to the group of mycotoxins. They are natural carcinogens, causing liver cancers, and they are 100 times as active as previously known liver carcinogens. They are produced by Aspergillus flavus, Aspergillus parasiticus and Aspergillus oryzae, as well as some Penicillium strains.

They are coumarin difuran derivatives.

AGA. Abb. for N-Acetylglutamate.

Agar-agar. A polysaccharide plant slime from various red algae. It consists of about 70% polygalactan, which is about 70% agarose and 30% agaropectin.

Agglutination. Means the clumping of insoluble antigens bound to particles, such as bacteria, viruses, erythrocytes, by the appropriate antibodies.

Agnosterol: 5α-lanosta-7, 9 (11), 24-trien-3β-ol, a tetracyclic triterpene alcohol derived structurally from 5α-lanostane (see Lanosterol. M_r 424.7, m.p. 165°C, $[\alpha]_D+66°$. It is a zoosterol.

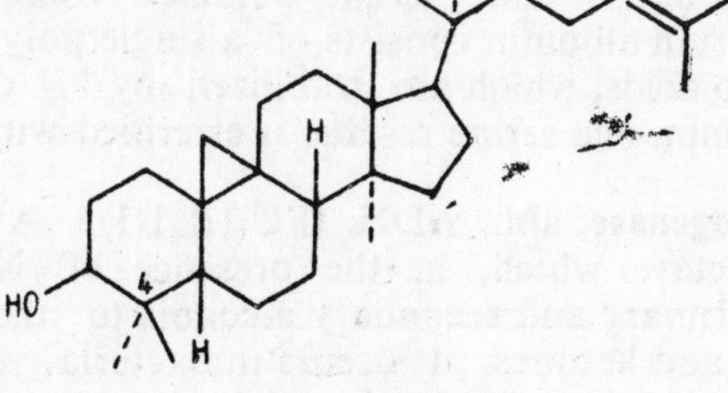

Agnosterol

AICAR. Abb. for 5(4)-Aminoimidazole-4(5)-carboxamide ribotide.

AIR. Abb. for 5-Aminoimidazole ribotide.

Ajmaline. A Rauwolfia alkaloid, m.p. 205 to 207°C, $[\alpha]_D^{20}=+144°$ (c=0.8 in chloroform). It is used medicinally to normalize heart rhythm. In high doses it has the tranquilizing effect of Rauwolfia alkaloids.

Alanine, Aminopropionic Acid: M_r 89.1; 1. L-α-alanine, abb. Ala, CH_3-CH(NH_2)-COOH, a prote-ogenic amino acid. m.p. 297°C (d.), $[\alpha]_D^{25}=+1.8$ (c=2.0, water). Ala is glucogenic and is closely involved in the metabolism of sugars and organic acids. It is one of the main components of silk fibroin.

2. **β-Alanine,** H_2N-CH_2-CH_2-COOH. A nonproteogenic amino acid, m.p. 196°C (d.). It occurs in free form, for example in the human brain, and is a component of the dipeptides carnosine and anserine, and of coenzyme A.

Alar 85. See N-Dimethylsuccinamide.

Albizziin, 2-amino-3-ureidopropionic Acid: $H_2NCONHCH_2CH(NH_2)$-COOH. A nonproteogenic amino acid which is occurring primarily in species of the genus Albizzia. It is presumably formed from carbamylphosphate and 2, 3-diaminopropionic acid by transcarbamylation. It is an antagonist of glutamine.

Albomycin. An antibiotic which is synthesized by Actinomyces subtropicus. It is a cyclic polypeptide with a pyrimidine base (cytosine) (Fig.). It is effective against both Gram-positive and Gram-negative bacteria, and inhibits the aerobic metabolism of Staphylococcus aureus and Escherichia coli.

Albumins. Refers to a group of simple proteins. They are found in body fluids and tissues and in some plant seeds. Serum albumin (plasma albumin) makes up 55 to 62% of the serum protein, and is one of the few carbohydrate-free proteins in blood plasma, or the serum obtained from it by clotting. Human serum albumin consists of a single polypeptide chain of 548 amino acids, which are stabilized by 17 disulfide bridges. In ovalbumin, one serine residue is esterified with phosphate.

Alcohol Dehydrogenase, abb. ADH (EC 1.1.1.1). A zinc-containing oxidoreductase which, in the presence of NAD^+, reversibly oxidizes primary and secondary alcohols to the corresponding aldehydes and ketones. It occurs in bacteria, yeasts, plants and the liver and retina of animals.

Alcoholic Fermentation. The anaerobic (occurring in the absence of oxygen) formation of ethanol and carbon dioxide from glucose. The most important feimenting organism are yeasts and other microorganism. For lack of pyruvate decarboxylase, there is no alcohalic fermentation in animals. The process produces energy under anaerobic conditions : the fermentation of 1 mol glucose yields 1 mols ATP.

The starting point for A. f. has been glucose 6-phosphate, which is converted by the glycolysis reactions to pyruvate. Pyruvate is the branching point for the last step of carbohydrate degradation and gets decarboxylated by pyruvate decarboxylase (EC 4.1.1.1) to acetaldehyde, which is then reduced to ethanol by alcohol dehydrogenase (EC 1.1.1.1. Balance :

$C_6H_{12}O_6 + 2Pi + 2\ ADP \rightarrow 2\ CH_3CH_2OH + 2\ CO_2 + 2ATP$ (glucose) $+ 2H_2O$.

Alcohols. Hydrocarbon derivatives carrying one or more hydroxyl (-OH) group. In nature alcohols in the form of esters are important components of the essential oils, fats and waxes. A number of lower A., ethanol for example, are formed by fermentative processes from carbohydrates and proteins.

Aldonic Acids. The monocarboxylic acids which are (derived from aldoses by oxidation of the aldehyde group. Some important aldonic acids are L-arabonic acid, xylonic acid, D-gluconic acid D-mannonic acid and galactonic acid.

Aldoses. The polyhydroxyaldehydes, one of the two main subdivision of monosaccharides. These are characterized by their terminal aldehyde group —CHO, which is always given the number 1 in systematic nomenclature.

Aldosterone. 11β,21-dihydroxy-3, 20-dioxopregn-4-en18-al-11 →11-hemiacetal, a highly active mineralocorticoid hormone from the

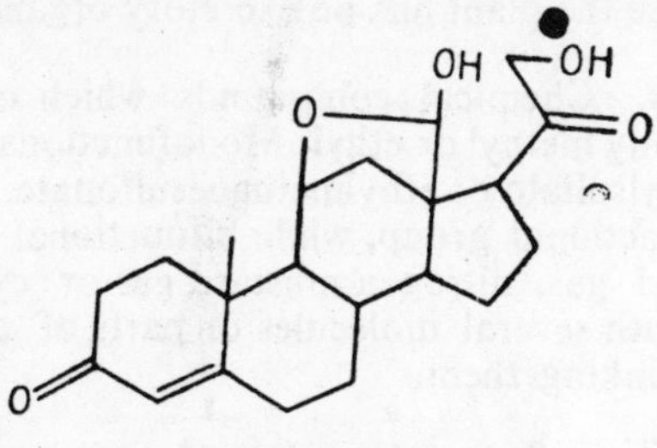

OH OH O O O

Aldosterone

adrenal cortex. It is the most important mineralocorticoid, regulating NaCl resorption and potassium excretion. It also has a certain degree of glucocorticoid activity. It is synthesized in the liver from progesterone via cortexone and corticosterone, which is oxidized at C-18.

ALG. Abb. for Antilymphocyte globuline.

Alginic Acid. A polyuronic acid extracted from seaweeds. It is composed of varying proportions of D-mannuronic and L-guluronic acids, linked β-1,4.

Alizarin. 1,2-dihydroxyanthraquinone, a red dye. m.p. 290 °C. It occurs in the root of the madder plant (Rubia tinctorum L.) and other Rubiaceae in combination with 2 moles glucose, forming the compound ruberythric acid.

Alkaloids. Basic natural products occurring primarily in plants. They contain one or more heterocyclic nitrogen atoms and are generally found in the form of salts with organic acids.

Probably 10 to 20% of all higher plants contain alkaloids; they are particularly frequent in some families of dicotyledons.

The free alkaloids are displaced from the genuine salts by treatment with alkaline solutions and can be extracted with organic solvents. They can be further separated and purified by means of salt formation (picrates, Reinecke salts) and by chromatography. The Dragendorff reagent can detect alkaloids in concentrations of a few μg. In recent years, the classical techniques of structure determination have been largely replaced by physical methods such as UV, IR NMR and mass spectroscopy. Most A. are optically active, and almost all are levorotory.

Alkaloids are the end products of secondary metabolism, and are not subject to significant degradation. They are accumulated because the plant has no excretory organs.

Alkylating Agents. Chemical compounds which can donate aikalyl groups,usually methyl or ethyl. Monofunctional alkylating agents like dimethylsulfate or ethylmethanesulfonate, can transfer only a single functional group, while bifunctional alkylating agents, like mustard gas, nitrogen mustard gas or cyclophos-phamide, can react with several molecules or parts of a macromolecule, thus cross-linking them.

Allantoic Acid. Diureidoacetate, a degradation product of allantoin in aerobic purine degradation and in anaerobic allantion degradation.

Allantoin. 5-ureidohydantoin, glyoxyldiureid, an intermediate in aerobic purine degradation. It is the end product of purine metabolism in most mammals and some reptiles, and is excreted in their urine.

Allergy. A hypersensitivity of the immune apparatus, a pathological immune reaction induced either by antibodies immediate hypersensitivity) or by living lymphoid cells (delayed type A.)

Allocholane. The outdated term for 5α-cholane, Steroids.

Allodeoxycholic Acid. The 3α, 12α-dihydroxy-5α-cholan-24-oic acid, one of the bile acids. A dihydroxy steroid carboxylic acid. It was isolated from the bile and feces of rabbits.

Allopregnane. Outdated term for 5α-pregnane.

Allostery. The phenomenon of changes in conformation of proteins with quaternary structure upon binding to certain low-molecular weight ligands. It has an important role in enzyme regulation and in the uptake of oxygen by hemoglobin.

ALS. Abbreviation for Antilymphocyte.

Amaranthin. A red dye belonging to the betacyanin group. It is found in Amaranthus species, for example the foxtail Celosia argentea.

Amaryllidacease Alkaloids. A group of complicated alkaloids found only in the plant family Amarylliaceae. Their biosynthes is is similar to that of the isoquinoline alkaloids, beginning with phenylamine or tyramine and a carbonyl compound. The main alkaloid galanthamine is isolated from Caucasians nowdrops, Galanthus woronowii, and is used therapeutically as an inhibitor of acetylcholinesterase.

Amatoxins. A group of bicyclic octapeptides which, together with the phallatoxins, are the most important poisons in the death cap fungus, Amanita phalloids. These poisons inhibit the nucleoplasmic RNA polymerase II (EC 2.7.7.6) of eukaryotic cells, which leads to necrosis of liver and kidney cells.

Amber Codon, Nonsense Codon. The sequence UAG in a mRNA. It does not code for any of the 20 proteogenic amino acids, and it results in the premature termination of protein synthesis. It may be formed by mutation of a sense codon : potential precursors are the codons UCG (serine), UAU and UAC (tyrosine) and CAG (glutamine).

Amber Mutants. The mutant bacteria in which the mRNA contains the codon UNT because of a point mutation.

Amicetins. Pyrimidine antibiotic (Nucleoside antibiotics) synthesized by various Streptomyces species. A micetin A is formed by Streptomyces jasciculatus and St. vinaceusdrappus. It is primarily bacteriostatic, especially against Gram-positive bacteria 0.5 μg amicetin A/ml inhibits the growth of Mycobacterim tuberculosis.

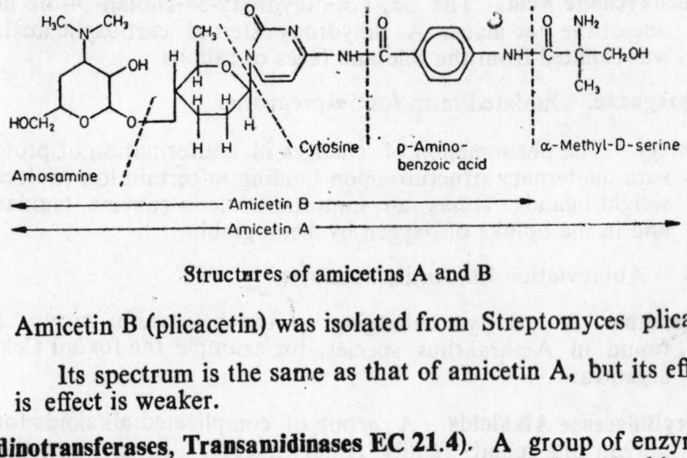

Amosamine — Cytosine — p-Amino-benzoic acid — α-Methyl-D-serine

Amicetin B

Amicetin A

Structures of amicetins A and B

Amicetin B (plicacetin) was isolated from Streptomyces plicatus

Its spectrum is the same as that of amicetin A, but its effect is effect is weaker.

Amidinotransferases, Transamidinases EC 21.4). A group of enzymes catalysing transamidination. They are involved in the biosynthesis of creatine.

Amination. The introduction of the amino group (—NH_2) inio an organic carbon compound.

Aminoacetic Acid. Glycine.

Amino Acids. The aminocarboxylic acids, organic acids carrying amino groups, usually not more than two. The α-amino groups, as the components of proteins and peptides, but also in their free form are one of the most important classes of organic substances in the cell.

There are about 20 amino acids which are normally components proteins. They are called proteogenic or proteinogenic amino acids (Table 1).

TABLE 1

Proteogenic Amino Acids

Amino acid	abbreviation
L-Alanine	Ala
L-Arginine	Arg
L-Asparagine	Asn
L-Aspartic acid	Asp
L-Cysteine	Cys
L-Glutamic acid	Glu
L-Glutamine	Gln
Glycine	Gly
L-Histidine	His
L-Isoleucine	Ile
L-Leucine	Leu
L-Lysine	Lys
L-Methionine	Met
L-Phenylalanine	Phe
L-Proline	Pro
L-Serine	Ser
L-Threonine	Thr
L-Aryptophan	Trp
L-Tyrosine	Tyr
L-Valine	Val

Amino Acids Occurring only in Special Proteins

Amino Acid	Occurrence
δ-Hydroxy.L-lysine	Fish collagen
L-3,5-Dibromotyrosine	Skeleton of Primnoa lepadifera (coral)
L.3,5-Diiodotyrosine	Skeleton of Gorgonic cavolinii (coral)
L-3,5,3′-triiodothyronine	Thyreoglobulin (tissue protein in the thyroid gland)

L-Thyroxin	Thyreoglobulin
Hydroxy-L-proline	Collagen, gelatins
α-Aminoadipic acid	Maize protein

$$\underset{\text{(variable)}}{\text{Residue}}\quad R-\overset{H}{\underset{NH_2}{C}}-COOH \quad \alpha\text{-Carboxyl group}$$

α-Amino group

Structure of an α-amino acid

The amino acids are classified as acidic or basic, depending on their isoelectric points, or, depending on the nature or their side chain, they are divided into four groups :

1. Amino acids with neutral, hydrophobic (non-polar) side chains, glycine, alanine, valine, leucine, isoleucine, phenylalanine, tryptophan, proline and methionine;
2. Amino acids with neutral and hydrophilic (polar) side chains, serine, threonine, tyrosine cysteine, asparagine and glutamine;
3. Amino acids with acid and hydrophilic (polar) side chains, aspartic acid and glutamic acid;
4. Amino acids with basic and hydrophilic (polar) side chains, lysine, arginine and histidine.

TABLE 2

Minimal Requirements of Human beings for Essential Aminoacids in mg per kg and Day

	Ile	Leu	Lys	Phe	Met	Cys	Thr	Trp	Val
Child	90		90	90[1]	85		60	30	85
Man	10.4	9.9	8.8	4.3[2]	1.5	11.6	6.5	2.9	8.8
				13.3[3]	13.2	0			
Woman	5.2	7.1	3.3	3.1[4]	4.7	0.5	3.5	2.1	9.2
WHO Norms	3.0	3.4	3.0	2 0[5]	1.6	1.4	2.0	3.0	

Properties. Amino acids are amphoteric, since they carry both an—NH_2 and at —COOH group and their solutions are ampholytes. In the solid state and in strongly polar solvents, they are zwitterions, $H_3^+N-CHR-COO^-$.

Amino Acid Reagents. Reagents for the colorimetric identification and quantitation of amino acids. One of the most important is the ninhydrin reaction, in which a blue-violet dye, called Ruhemann's Purple (absorbance maximum at 570 nm, for proline at 440 nm) is formed by the reaction of 2, 2-dihydroxy-1 H-indene-1; 3 (2H)-dione (ninhydrin) with the amino acid.

In the flourescamine technique, the amino acids have been converted by reaction with 4-phenyl [furan-2 H (3 H)-1'-phthalane]-3, 3'-dione (fluorescamine) into strongly fluorescing compounds which can be detected even in nanomole quantities at 336 nm. The reagent itself is not fluorescent, and in contract to ninhydrin, it is not sensitive to ammonia.

Other highly sensitive reagents have been 2, 4, 6-trinitrobenzosulfonic acid, 1, 2-naphthoquinone-4-salfonic acid (Folin's reagent) and 4, 4'-tetramethyldiaminodiphenylmethane (TDM). Intensely fluorescing amino acid derivatives are formed by reaction with o-phthaladehyde in the presence of reducing agents; with pyridoxal and zinc^{2+} ions; and with dansyl chloride (5-dimethylaminonphthalene sulfonyl chloride.

Aminoacyladenylate, Activated Amino Acid. The product of the first enzymatic reaction of protein biosynthesis. It consists of an amino acid linked by an acid anhydride bond to the phosphate of AMP.

Aminoacyl-tRNA. A transfer RNA charged with a specific amino acid; the transport form in which the amino acid is brought to the specific acceptor site on the ribosome. The carboxyl group of the amino acid is esteritied to either the 2' or the 3' OH-group of the ribose of the terminal adenosine of the tRNA.

Aminoacyl-tRNA Synthetases, Amino-acid Activating Enzymes (EC 6.1.1). A group of enzymes which activate amino acids and transfer them to specific tRNA molecules as the first step in protein biosynthesis. The process consists of two steps, illustrated here with leucine :

(1) Leu + ATP + leucyl-tRNA synthetase $\rightleftharpoons$ [Leu AMP Enzyme]+pP_i

(2) [LeuAMP Enzyme] + $tRNA^{Leu}$ $\rightarrow$ leucyl-$tRNA^{Leu}$ + AMP+enzyme.

(1) and (2) Leu + ATP + $tRNA^{Leu}$ $\rightarrow$ Leucyl-$tRNA^{Leu}$ + AMP + PP_i.

These are highly specific with respect to the amino acid they activate, and they also recognize the tRNA with great precision. The mechanism by which the enzyme recognizes the appropriate tRNA is still unclear.

2-Aminoadipic Acid, Abb. Aad. $HOOC\text{-}CH_2\text{-}CH_2\text{-}CH_2\text{-}CH(NH_2)\text{-}COOH$. An amino acid which is proteogenic only in maize. M_r 161.1, m.p. 206°C(d). Amino adipic acid is an intermediate in the biosynthesis of L-lysine by the Aad-pathway. The free acid cyclizes in boiling water to piperidone carboxylic acid.

2-Aminoadipic Acid Pathway. See L-Lysine.

4-Aminobutyric Acid, abb. 4-Abu, γ-Aminobutyric Acid, abb. GABA. $H_2N\text{-}CH_2\text{-}CH_2\text{-}CH_2\text{-}COOH$. A nonproteogenic amino acid. M_r 103.12, m.p. 202°C (d.). Formation of GABA from L-glutamic acid, by the action of glutamate decarboxylase (EC 4.1.1.15), has been demonstrated in brain, various microorganisms (*e.g.* Clostridium welchii, Escherichia coli), higher plants (*e.g.* spinach, barley) and other animal tissues (liver and muscle). It can also be formed from 4-guanidobutyric acid by removal of urea in higher fungi (Basidiomycetes) and Streptomycetes.

Amino Citric Acid.

$$\begin{array}{lll} CH(NH_2)\text{-} & C(OH)\text{-} & CH_2 \\ | & | & | \\ COOH & COOH & COOH, \end{array}$$

An amino acid identified in acid hydrolysates of ribonucleoproteins from calf thymus, bovine and human spleen, Escherichia coli and Salmonella typhi.

3′-Amino-3′-deoxyadenosine. A purine antibiotic synthesized by Cordyceps militaris and Helminthosporium species. It has antitumor activity.

5 (4)-Aminoimidazole-4 (5)-carboxyamide Ribonucleotide, abb. AICAR. 5-amino-1-ribofuranosylimidazole-4-carboxamide 5′-phosphate, an intermediate in purine biosynthesis.

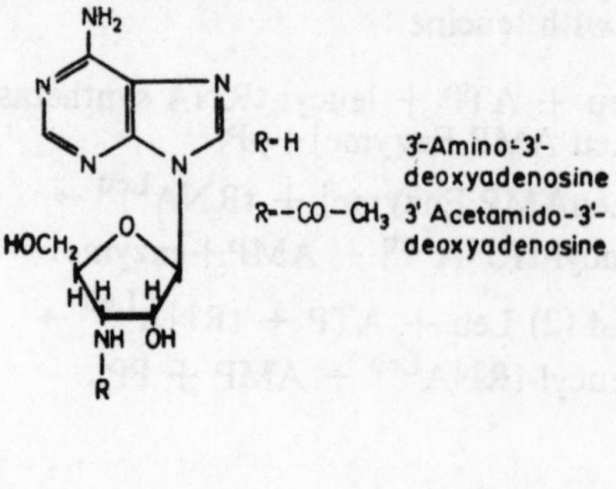

NH_2; N; N; N; N; $HOCH_2$; O; H; H; H; H; NH; OH; R

R-H 3′-Amino-3′-deoxyadenosine

$R\text{-}CO\text{-}CH_3$ 3′ Acetamido-3′-deoxyadenosine

5 (4)-Aminoimidazole-4(5)-carboxyribonucleotide. An intermediate in purine biosynthesis.

5-Aminoimidazole Ribonucleotide, abb. AIR. An intermediate in purine biosynthesis, and in the formation of thiamine.

5-Aminoimidazole-4-N-succinocarboxamide Ribonucleotide. An intermediate in purine biosynthesis.

Aminoisobutyric acid. In the β-form (2-methyl-β-alanine), $H_2N-CH_2-CH(CH_3)-COOH$, a product of the reductive degradation of thymine. The α-form, (2-methylalanine) $H_2N-C(CH_3)_2-COOH$, does not occur in nature.

5-Aminolevulinic Acid, δ-aminolevulinic Acid. $HOOC-CH_2-CH_2-CO-CO_2-NH_2$. An amino acid. It is an intermediate in the biosynthesis of the Shemin Cycle. Schemin cycle.

Aminopeptidases. Expopeptidases, EC 3.4.11, usually containing metal ions, which shorten proteins and peptides from the N-terminal end of the chain, removing one amino acid residue per step.

Amino Sugars. Monosaccharides in which an hydroxyl group has been replaced by an amino group ($-NH_2$). The amino group is often acetylated. Examples are D-galactosamine, D-glucosamine, D-mannosamine, neuraminic acid and muramic acid.

Ammonia, NH_3. A colorless gas with a sharp smell.

NH_3 is the product of nitrate reduction, biological nitrogen fixation, and deamination of amino acids and various catabolic pathways, for example oxidation purine degradation and reductive pyrimidine degradation.

Ammonia Assimilation. The utilization of ammonia in the net synthesis of the nitrogen-containing groups of nitrogenous cell constituents, *e.g.* amino acids, amides, carbamyl and guanido compounds. Incorporation of ammonia into the amide group of glutamine, catalysed by glutamine synthetase (EC 6.3.1.2), is of central importance : L-glutamate + NH_3 + ATP → L-glutamine + ADP + P_r. The amide nitrogen of L-glutamine is then used in various syntheses :

1. L-glutamine + α-ketoglutarate + 2 H^+ + 2 e^- → L-glutamate (Glutamate synthase.
2. L-glutamine + HCO^-_3 + 2 ATP + H_2O → carbamyl phosphate + L-glutamate + 2 ADP + P_r (see carbamoyl phosphate).

3. The amide group of glutamine is used in purine synthesis, where it provides N-3 and N-9 of purine ring, and the 2-NH_2 group of guanine.
4. In several syntheses, nitrogen is derived directly from the amide group of L-glutamine, *e.g.* histidine synthesis; conversion of chorismate into anthranilate (see tryptophan synthesis); the synthesis of amino sugars; amination of UTP to CTP.
5. In some organisms, the amide group of glutamine is transferred to aspartate by the action of asparagine synthetase (glutamine hydrolysing) (EC 6.3.5 4):

L-glutamine + L-aspartate + ATP → L-glutamate
+L-asparagine + AMP + PP_i.

Ammonia Detoxification. The detoxification of the toxic, non-dissociated compound by formation of ammonium salts and nitrogen-excretion compounds. The ammonia produced by catabolism in animals is either excreted directly, or it is converted into other nitrogenous compounds for excretion; it is not reassimilated.

AMO 1618. (2 isopropyl-5-methyl -4-trimethylammonium chloride)-phenyl-1-piperidinocarboxylate, a synthetic growth retardant of plants. It is an antagonist of gibberellin and inhibits the biosynthesis of gibbrrellin A_8.

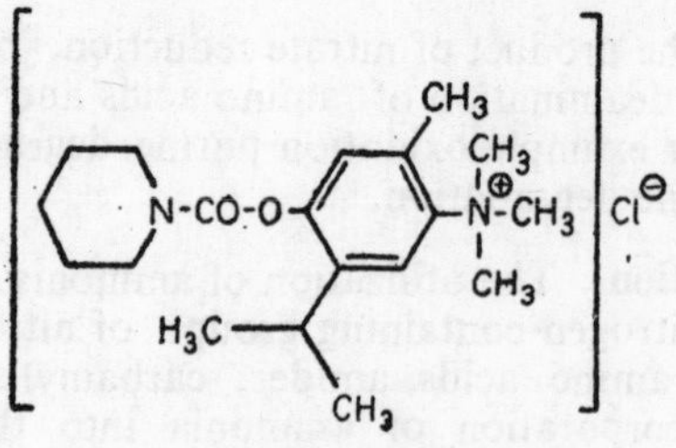

AMO 1618

AMP. Abb. for Adenosine 5′-monophosphate. 3′, 5′-AMP. Abb. for Cyclic adenosine-3′, 5′-monphosphate.

Amphibian Toxins. A group of chemically very heterogeneous toxins (biogenic amines, peptides, alkaloids, steroids) produced by toads, salamanders, frogs and newts. Pharmacologically, they include heart, muscle and nerve toxins, sympathicomimetica, local anaesthetics and hallucinogens.

The amphibian toxins serve as a protection against both natural enemies and bacteria and fungi which might attack the animal's skin.

Amphibolic Pathways. The metabolic pathways serving both breakdown and synthesis.

Amylases Diastases (EC 3.2.1.1, 3 2.1.2 and 3.2 1.3). A widely occurring group of hydrolases which cleave the α-1, 4 glycosidic bonds in oligosaccharides (from trisaccharides upwards) add polysaccharides like starch, glycogen, and dextrins. α-Amylase is in endoamylase, while β-amylase (saccharogenic A.) and γ-amylase (glucoamylase) are exoamylases.

Amylo-1, 6-glucosidase, Debranching Enzyme (E.C. 3.2.1.33). An endoglucosidase which lyses the 1→6 glycosidic bonds at the branching points in glycogen and amylopectin.

Amyloid Protein. A pathological, fibrillar, low-molecular-weight protein. In amyloidosis, it is deposited together with glycoproteins and proteoglycans, primarily in the spleen, liver and kidneys.

Amylopectin. A component of starch (the other is amylose). It is a branched, water-insoluble polysaccharide consisting of a main chain of D-glucose units linked α-1,4 with side chain, attached to every 8th or 9th glucose by ah α-1, 6 bond.

Amylose. A component of starch (the other is amylopectin). It is an unbranched; water-soluble polysaccharide consisting of 100 to 300 D-glucopyranoside residues.

Amyrin. A pentacyclic triterpene alcohol with one double bond; α-A. M_r 426.73, m.p. 186°C, $[\alpha]_D^{10}+94.6$ (c = 1.9 in $CHCl_3$). α-A. is found free, esterified and as the aglycone of triterpene saponins in many plants and has been isolated from many latexes, for example that of the dandelion, Taraxncum officinale.

Anabasine. A Nicotiana alkaloid. It is isolated in the L-form, primarily from Nicotiana glauca and the asiatic Anabasis aphylla, in which it is the main alkaloid. Its physiological effects are similar to those of nicotine, and like nicotine, it is used as an insecticide.

Anabolic Steroids. A group of synthetic steroids which stimulate the production of body protein. Male gonadal hormones have this effect.

Anabolism. The sum of synthetic metabolic reactions.

Androgens. A group of male gonadal hormones, including testosterone, androsterone and androstenolone, which are formed in the intermediary cells of the testes tissue, and a number of less active androgens produced in the adrenal cortex, *e.g.* androstenedione and adrenosterone.

The androgens are biosynthesized primarily from progesterone via 17 α-hydroxyprogesterone, by degradation of the pregnane side chain.

Androstanazole. A synthetic and highly active anabolic steroid. It is used, in the therapy of inflammation and tumors and in reconvalescence.

Androstenedione. Androst-4-ene-3, 17-dione, M_r 286.42, m.p. 174°C, $[\alpha]_D^{30}$ +191° (alcohol). It is synthesized in the adrenal cortex as an intermediate in the biosynthesis of testosterone from progesterone via androsterone. Due to its weakly androgenic effects, it is counted as one of the male gonadal hormones.

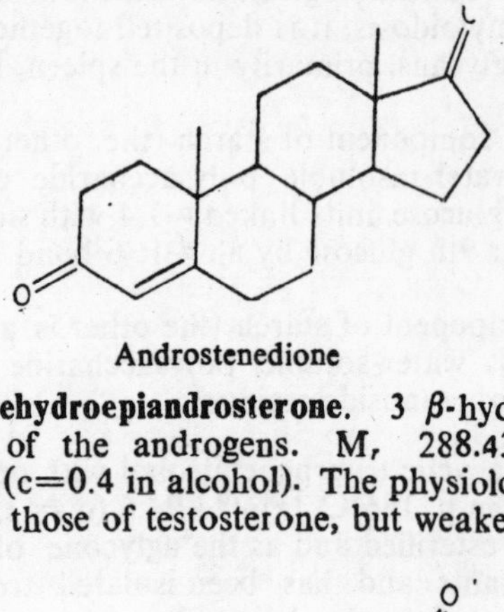

Androstenedione

Androstenolone, Dehydroepiandrosterone. 3 β-hydroxyandrost-5-en-17-one, one of the androgens. M_r 288.43, m.p. 142 °C, $[\alpha]_D^{18}$ +10 9° (c=0.4 in alcohol). The physiological effects of A. are similar to those of testosterone, but weaker.

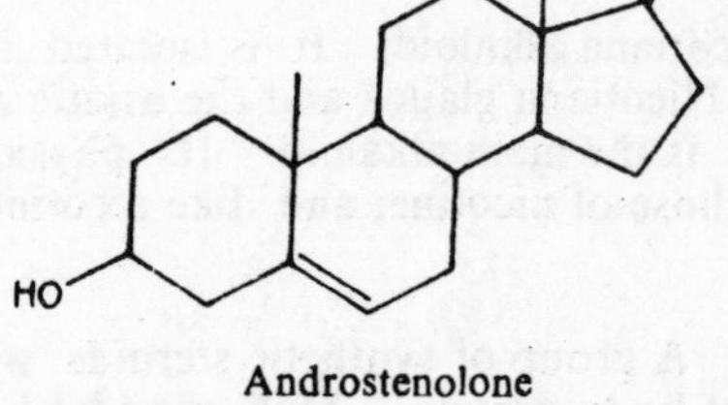

Androstenolone

Androsterone. 3 α-hydroxy-5α-androstan-17-one, an androgen. M_r 290.54, m.p. 183°C, $[\alpha]_D^{25}$ +86° (c = 2 in ethanol). It is formed in the interstital cells of the testes. It is similar to testosterone, but is androgenic activity is 7 times weaker. It is biosynthesized from progesterone via 17 α-hydroxyprogesterone and androstenedione.

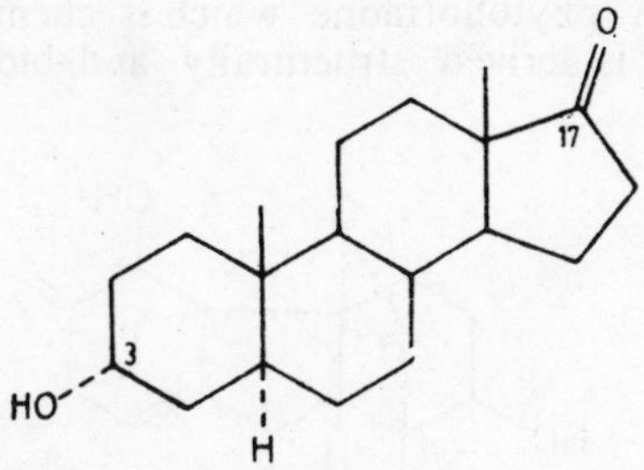

Androsterone

Angiotensin, Angiotonin, Hypertensin. A tissue peptide hormone affecting the blood pressure.

Angustmycins. The purine antibiotics (see Nucleoside antibiotics) synthesized by various species of Streptomyces. It is specific for mycobacteria. Gram positive and Gram negative bacteria and fungi are not sensitive to angustmycin A. It acts by inhibiting the formation of 5-phosphoribosy l-1-pyrophosphate in purine biosynthesis.

Anhalonium Alkaloids. Cactus alkaloids, isoquinoline alkaloids found primarily in cacti.

Both the biosynthesis and the chemical synthesis are based on a Mannich condensation of a β-phenylethylamine derivative with a carbonyl component.

These are weak narcotics and paralysing agents. Pellotine has a cramp-relaxing effect similar to that of acetylcholine.

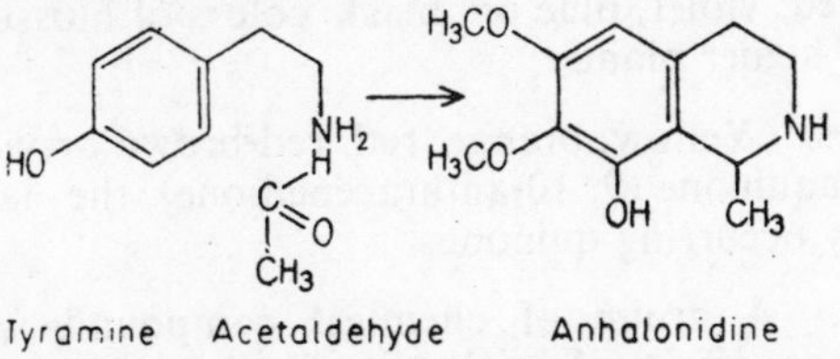

Tyramine Acetaldehyde Anhalonidine

Biosynthesis of the anhalonism base, anhalonidine, by a Mannich conden sation

Ankyrin, Anchorin Syndein. An erythrocyte membrane protein which binds Spetrin.

Antamanide. Cyclo-(Pro-Phe-Phe-Val Pro-Pro-Ala-Phe-Phe-Pro), a cyclic decapeptide (all the amino acid residues have the L-configuration) isolated from the death cap fungus, Amanita phalloides.

Antheridiogen. A phytohormone which is chemically a diterpenoid. M_r 330. It is derived structurally and biogenetically from the gibberellins.

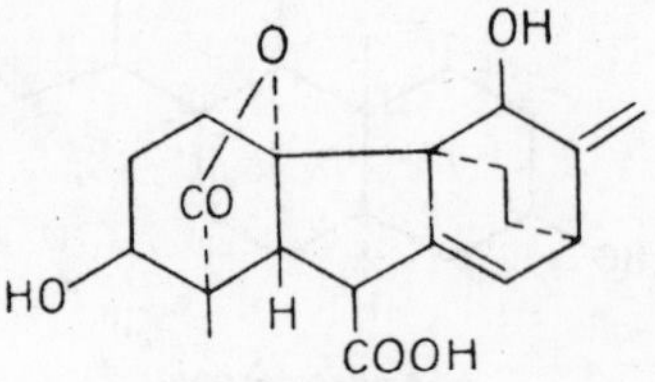

Antheridiogen

Antheridiol. A steroid plant hormone, the first plant sex hormone discovered. It is derived structurally from the hydrocarbon stigmastane (see steroids) and contains a lactone group.

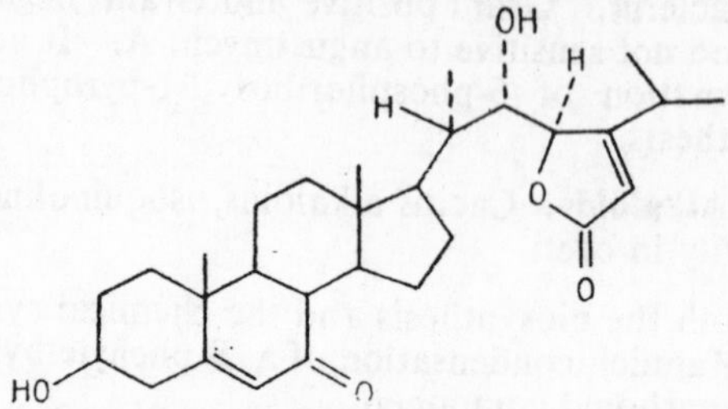

Antheridiol

Anthocyanins. Widely occurring flavonoid plant pigments responsible for the red, violet, blue or black colors of blossoms, leaves and fruits of higher plants.

Anthraquinones. Yellow, orange, red, red-brown or violet derivatives of anthraquinone (9, 10-anthracenedione), the largest group of naturally occurring quinones.

Antiandrogens. A group of chemical compounds which reversibly inhibit the effect of male sex hormones by competing with them for their receptors.

Antibiotics. Substances produced by microbes which kill or inhibit the growth of other microorganisms. In contrast to general cell poisons, the antibiotics are selective.

Anticodon. Refers to a sequence of three nucleotides in one loop of a transfer RNA molecule which recognize the nucleotides of the codon in messenger RNA by forming H bonded base pairs with them. This anticodon loop thus insures that the correct amino acid is built into the polypeptide sequence.

Anticytokinins. Antagonists of the cytokinins which can partially negate the physiological effects of the hormones.

Antidepressants. The drugs with stimulatory and antidepressant action. They decrease fatigue, reduce appetite and decrease sleeping time. These effects are the result of reduced autonomic activity, especially of cholinergic systems, and of central sympathetic activation.

Antienzymes. The polypeptides or proteins which act as enzyme inhibitors or antibodies. The inhibitors include the many animal and plant protease inhibitors such as soybean trypsin inhibitor and serum antitrypsin.

Antigen-antibody Reaction. Abb. AAR: together with phagocytosis, the most important protective mechanism of the animal organism against invading foreign substances.

Antigens. The substances which induce an immune response. They are normally foreign to the body, and may be natural or synthetic macromolecules, particularly proteins and polysaccharides ($M_r > 2000$), or surface structures on foreign particles which can be phagocytosed.

Anti-gray-hair Factor. A member of the B_2 complex.

Antitemophilic Factor, Factor VIII. An oligomeric β_2-glycoprotein (6% carbohydrate) which activates factor X in the process of blood clotting, and is thereby completely consumed.

Antilymphocyte Serum. Abb. ALS : An immune serum which has an opsonizing (see Opsonization) and cytotoxic effect on lymphocytes, thus producing immune suppression.

Antimetabolite. A compound so similar in structure to a metabolite that it can occupy the enzyme binding sites specific for the latter.

Antipyrine. 1, 2-dihydro-1, 5-dimethyl-2-phenyl-3H-pyrazol-3-one, a weak base, pK_a 1.4. After oral administration it gets is rapidly absorbed and becomes distributed throughout the total body water within 2 hours.

Antiserum, Immune Serum. The serum of an animal (including humans) which has been immunized against an antigen. It can be monovalent or polyvalent, *i.e.* contain one or more specific antibodies, depending on whether the animal was immunized with a purified antigen or a mixture of antigens.

Antitumor Enzymes. Enzymes which stimulate either the irreversible degradation of amino acids which cannot be synthesized by tumor cells, or the inhibition of tumor-specific DNA, leading to a stopping of tumor growth.

Antitumor Proteins. The proteinaceous antibiotics isolated from culture filtrates of various Streptomyces strains which inhibit tumor growth.

Antivitamins. Antimetabolites of the vitamins which inhibit the growth of vitamin-dependent microorganisms and cause symptoms of vitamin deficiency in animals.

d-Apiose. A monosaccharide with a branched carbon chain. M_r 150.13, $[\alpha]_D^{19} +9°$ (pure syrup). A is found in various glycosides and as a component of various polysaccharides. The bio-synthesis starts from D-glucuronic acid.

Apoprotein. The protein component of a conjugated protein.

APP. Abb. for Aneurine pyrophosphate.

APS. Abb. for Adenosine 5′-phosphosulfate.

Apurine Acids. Polynucleotides which have been subjected to short treatment with mild acid, which removes the purines and leaves the phosphate, pentose and pyrimidines.

Apyrimidine Acids. The nucleic acids from which the pyrimidines have been removed by chemical treatment, such as exposure to hydrazine.

Arabans. The high-molecular-weight, branched polysaccharides composed of L-arabinose linked 1, 5 and 1, 3 in furanose form.

Arabinose. A pentose occurring naturally in both the D-and the L-forms.

```
     CHO                CHO
      |                  |
  HO—C—H             H—C—OH
      |                  |
   H—C—OH           HO—C—H
      |                  |
   H—C—OH           HO—C—H
      |                  |
    CH₂OH              CH₂OH
  D-Arabinose        L-Arabinose
```

Arabinosides, Arabinonucleosides. Structural analogs of the ribonucleosides in which the hydroxyl group on the C-2 atom of the sugar is cis to the glycoside bond. The sugar is arabinofuranose (Fig.). The base is the same as in the corresponding ribonucleoside.

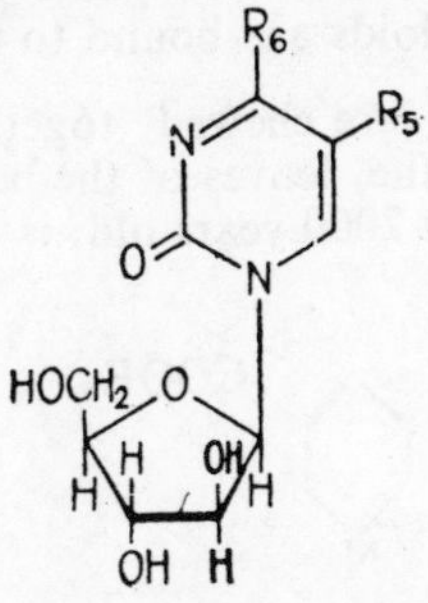

Structures of known pyrimidine arabinosides.

Arabinoside	R_5	R_6
1-β-D-Arabinofuranosylcytosine	NH_2	H
1-β-D-Arabinofuranosyluracil	H	OH
1-β-D-Arabinofuranosylthymine	OH	CH_3

Arachidic Acid. Icosanoic acid, CH_3-$(CH_2)_{18}$-COOH, a fatty acid, M_r 312.5, m.p. 75.3°C, b.p. 205°C. It occurs widely as a component of glycerides, but is usually present only in low concentrations. In sunflower oil, soybean oil, milk fat and peanut oil, this acid may represent up to 3% of the fatty acids.

Arachidonic Acid: 5, 8, 11, 14-icosatetraenoic acid, CH_3-$(CH_2)_4$-CH=CH-CH_2-CH=CH-CH_2-CH=CH-CH_2-CH=CH-$(CH_2)_3$-COOH. An essential fatty acid. M_r 30.45, m.p. 49.5°C. This is an esterified component of fish liver oil and of animal phosphatides.

Arachin. The peanut protein composed of 6 sub-units, M_r 345000. Each subunit is composed of 2 equal-sized, covalently bound polypeptide chains. A. is very similar to edistin.

Arcain. 1, 4-diguanidobutane, H_2N—C(=NH)-NH-$(CH_2)_4$-NH-C(=NH)-NH_2, a strongly basic guanidine derivative first isolated from a mussel (Arca noae), but also found in higher fungi, for example Panus tigrinus.

Areca Alkaloids. The pyridine alkaloids in which the pyridine ring is partially hydrated (Fig.). These alkaloids are obtained from betel nuts, the seeds of the betel palm (Areca catechu). In the plant, the these alkaloids are bound to tannins.

The betel nuts are chewed together with lime (to release the alkaloids) and the leaves of the betel pepper (Piper betle). The practice, at least 2000 years old, is common in East Africa, India and Oceania.

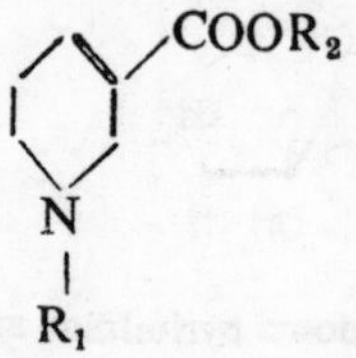

$COOR_2$

N

R_1

	$R_1=H$	$R_1=CH_3$
$R_2=H$	Guvacine	Arecaidine
$R_2=CH_3$	Guavacoline	Arecoline

Areca alkaloids

Arecoline. 1, 2, 5, 6-tetrahydro-1-methyl-3-pyridinecarboxylic acid methyl ester, the most important of the areca alkaloids. The compound is the methyl ester of the alkaloid arecaidine (M_r 141.19, m.p. 232°C.) It is responsible for the physiological effect of betel nuts It acts as a parasympatheticomimetic, but due to its high toxicity, it is used only in veterinary medicine.

Arginase (EC 3.5 3.1). A highly active and specific liver enzyme which catalyses the last reaction in the urea cycle (L-arginine $+H_2O \rightarrow$ L-ornithine+urea) in ureotelic animals. In land ureoteles such as mammals, frogs and swamp turtles, it is found practically only in the liver, with traces in the pancreas, mammary glands, testes and kidneys.

l-Arginine. Abb. Arg. 2-amino-5-guanidovaleric acid, the most strongly basic amino acid. It is unstable in hot alkaline solutions and forms nearly insoluble nitrates, picrates and picrolonates, and a particularly insoluble salt with flavianic acid. The Sakaguchi reaction is used for the detection and quantitative determination of Arg.

It is found in particularly large amounts in protamines and histones.

Aristolochic Acids. A group of related aromatic nitro compounds from Aristolochia spp. The most important is aristolochic acid I, m.p. 173°C.

Biosynthesis these acids are formed from isoquinoline alkaloids of the norlaudanosine type by oxidation of the nitrogen-containing ring (Fig.).

HO HO HO HO NH → H_2C O O COOH NO_2 OCH_3

Norlaudanosoline Aristolochic acid I

Biosynthesis of Aristolochic acid I

Aromatic Biosynthesis, Aromatization. The most important mechanisms are 1. the shikimic acid/chorismic acid pathway, in which the aromatic amino acids L-phenylalanine, L-tyrosine and L-tryptophan, 4-hydroxybenzoic acid (precursor of ubiquinone), 4-aminobenzoic acid (precursor of folic acid), and the phenylpropanes $C_6 \cdot C_3$, including the components of lignin, cinnamic acid derivatives and flavonoids are synthesized, and 2. the polyketide pathway in which acetate molecules are condensed and aromatic compounds (*e.g* 6-methysalicylic acid) are synthesized via poly-β-keto acids. The biosynthesis of flavonoids (*e g.* the anthocyanidins) can occur by either pathway.

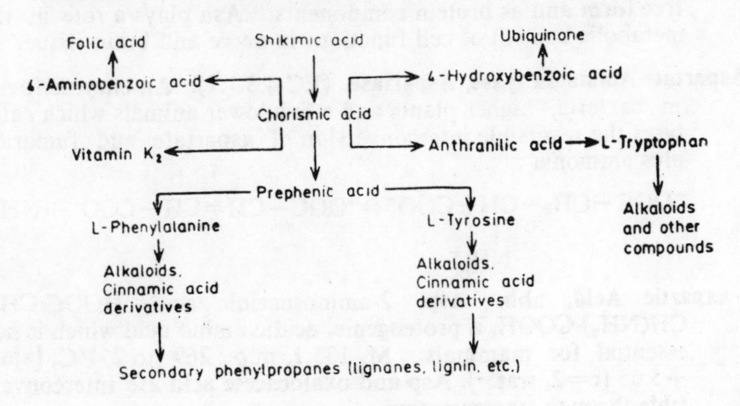

Key role of chorismic acid

Arrow Poisons. The natural toxins used for coating arrows, spears or blow pipe darts. In ancient Greece, extracts of aconitin were used. Ouabain and similar cardiac glycosides were (are) used in Africa. The jungle Indians of South America use various kinds of Curare alkaloids and tribes in Columbia use Batrachotoxin from the Columbian arrow poison frog.

Arylamidases. An ubiquitous group of aminopeptidases which preferentially cleave amino acid arylamides, *e.g.* alanine β-naphthylamide. Determination of arylamidases in the serum and urine is considered to have a certain diagnostic value in some liver diseases.

Arylsulfatases. The best-studied group of sulfatases. These hydrolyse aromatic sulfate esters at the O—S bond in the reaction

$$R\text{-}O\text{-}SO_3^- + H_2O \xrightarrow{A} R\text{-}OH + H^+ + SO_4^{2-}$$, where R-OH is a phenol.

Asn, Asp·NH₂. Abb. for L-asparagine.

l-Asparaginase (EC 3.5.1.1). A widely occurring enzyme which hydrolyses L-asparagine to L-aspartate and ammonia. The enzyme is used as an antitumor agent, particularly against lymphosarcomas and lymphatic leukemia. The cells of these cancers, unlike normal cells, cannot compensate for the lack of asparagine.

L-Asparagine, Abb. **Asn or Asp-NH₂.** $H_2N\text{-}OC\text{-}CH_2\text{-}CH(NH_2)\text{-}COOH$ the β-half amide of L-aspartic acid. M_r 132.1, m.p. (hydrate) 236°C (d.). Asn and aspartic acid occur ubiquitously, both in free form and as protein components. Asn plays a role in the metabolic control of cell functions in nerve and brain tissue.

Aspartate Ammonia-lyase, Aspartase. (EC 4.3.1.1). An enzyme found in bacteria, higher plants and a few lower animals which catalyses the reversible interconversion of aspartate and fumarate plus ammonia :

$$^-OOC-CH_2-\underset{\substack{|\\NH_3^+}}{CH}-COO^- \rightleftharpoons {}^-OOC-CH=CH-COO^- + NH^+_4$$

l-Aspartic Acid, abb. **Asp.** 2-aminosuccinic acid, $HOOC\cdot CH_2\text{-}CH(NH_2)\cdot COOH$, a proteogenic, acidic amino acid which is not essential for mammals. M_r 133.1, m.p. 269 to 271°C, $[\alpha]_D^{25}$ +5.05 (c=2, water). Asp and oxalocacete acid are interconvertible through transmination.

Aspergillic Acid. An antibiotic, M_r 224.3, produced by Aspergillus flavus.

Assimilate. In the wider sense, a product of assimilation. In the narrower sense, a stabilized end product of Photosynthesis.

Assimilation. Refers to the incorporation of nutrients into an organism.

Astaxanthine. 3,3′-dihydroxy-β,β-carotene-4,′4-dione, M_r 596 82, m.p. 216°C. The compound is found widely as a red animal pigment, especially in crustaceans, echinoderms and tunicates. It also

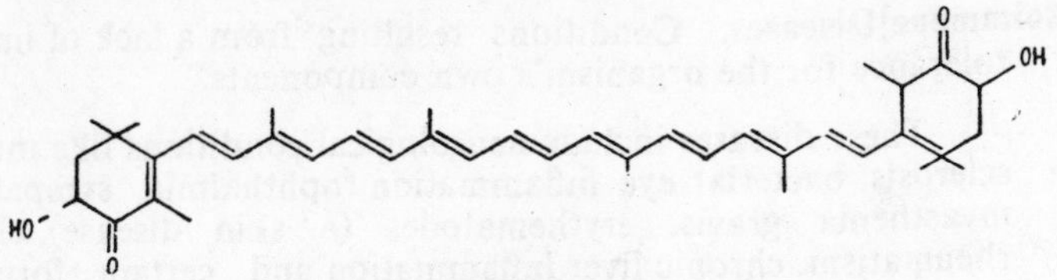

Astaxanthine

occurs in the feathers and the skin of flamingos and other birds, but is not found in many plants.

Asterosaponin A. A steroid saponin (see Saponins). It was first isolated from the starfish Asterias amurensis.

ATP. Abb. for Adnosine 5′-triphosphate.

ATP Citrate (pro-3S)-lyase, Citrate Cleavage Enzyme (EC 4.3.1.8). A microsomal enzyme converting citrate to acetyl-CoA and oxaloacetate and simultaneously cleaving one ATP to ADP : Citrate+ATP+CoA=Acetyl-CoA+oxaloacetate+ADP+P_r.

Atractyloside. A glucoside from the Mediterranean thistle Atractylis gummitera. It is a competitive inhibitor of adenine nucleotide binding and transport across the inner mitochondrial membrane.

Atropine, DL-hyoscyamine. A racemate formed during alkaline treatment of L-hyoscyamine. M_r 289.48. It is the ester of DL-tropic acid with tropine, m.p. 114–116°C It specifically inhibits those cholinergie neurons which are activated by muscarine.

Attenuation. A regulatory mechanism employed by the bacterial cell. Whereas Enzyme repression allows the cell to respond to extreme concentrations of metabolites, it probably represents a means of "fine tuning" to relatively mild fluctuations in the concentrations of metabolites.

Aurones, 2-benzylidene-3-cumarones. The yellow flavonoid plant pigments. These are very common in the petals of Compositae, Leguminosae and Scrophulariaceae.

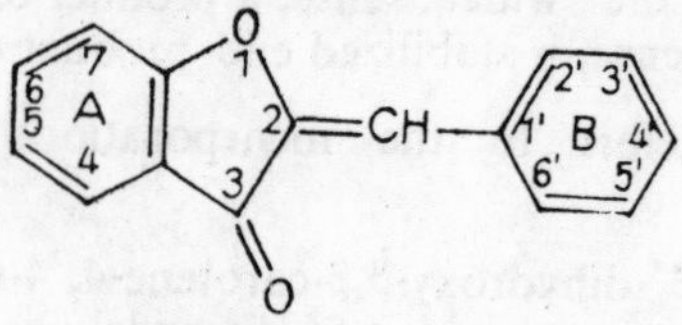

Aurone ring system

Autoimmune Diseases. Conditions resulting from a lack of immune tolerance for the organism's own components.

These diseases include neurological conditions like multiple sclerosis, bacterial eye inflammation (ophthalmia sympathica), myasthenia gravis, erythematodes (a skin disease), chronic rheumatism, chronic liver inflammation and certain forms of chronic kidney inflammation.

Autolysis. The self digestion of dying cells which occurs when the cathepsins get released from lysosomes and attack cytoplasmic proteins.

Autotrophy. Refers to the ability to synthesize all organic components from simple inorganic compounds like carbon dioxide, ammonia, nitrate and sulfate.

Auxin Antagonists. Inhibitors of auxins whose effects can be at least party reversed by auxins.

Auxins. A group of plant hormones which regulate growth. They stimulate extension growth and cell division in the cambium and root; they influence certain enzyme activities. Natural auxins are indole derivatives biosynthesized from tryptophan. The most important auxin is indole-3-acetic acid standard for comparison of the activity of other growth stimulants.

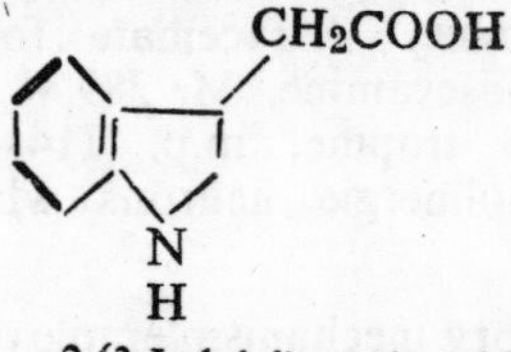

2-(3-Indolyl) acetic acid

Auxotrophy. Refers to the condition of requiring nutritional supplements for growth. It can arise through mutation.

Avenasterol, 28-isofucosterol. An isomer of Fucosterol found in green marine algae.

Avidin. A basic glycoprotein found in the egg whites of many birds and amphibia. Like the unrelated lysozyme and conalbumin, avidin protects the egg white against bacterial invasion.

5-Azacytidine, 1-β-D-ribofuranosyl-5-azacytosine. A pyrimidine antibiotic synthesized by streptoverticillius lakadamus var. 5-A. is effective against Gram-negative bacteria.

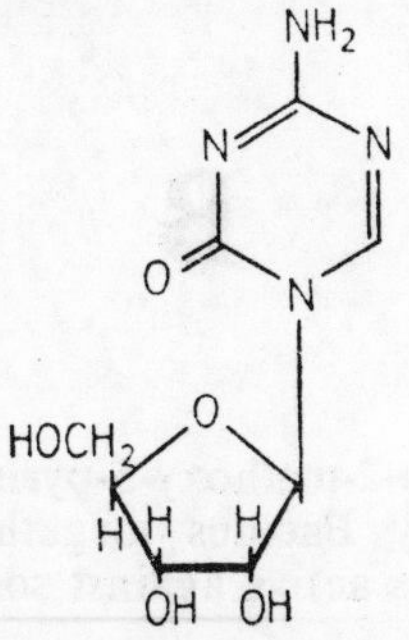

NH_2 N N O N HOCH$_2$ O H H H H OH OH

5-Azacytidine

8-Azaguanine, Pathocidin. A purine antagonist which is shown to be identical with pathocidin, in antibiotic from Streptomyces spectabalis. 8-Azaguanine affects many different enzymes of purine metabolism and synthesis. In particular it inhibits translation and causes errors of translation, due to its incorporation into mRNA.

L-Azaserine, O-diazoacetyl-L-Serine:

$\overset{-}{N}=\overset{+}{N}=CH\cdot CO\cdot OCH_2\cdot CH(NH_2)\cdot COOH$. A glutamine analog. It inhibits the transfer of the amide group (Transmidination) from L-glutamine to formylglycinamide ribotide.

Azomycin, 2-nitroimidazole. An antibiotic synthesized by Nocardia mesenterica and Streptomyces eurocidicus. m.p. 281 to 283°C. Both Gram-negative and Gram-positive bacteria are highly sensitive to the compound.

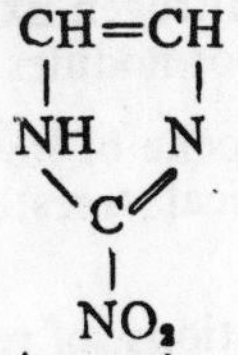

CH=CH | | NH N \ // C | NO_2

Azomycin

A-Z Solution. See Nutrient medium.

Azulene, Cyclopentacycloheptene. The parent compound of a group of blue to violet, nonbenzoid aromatics. These compounds are artefacts, however, produced from colorless sesquiterpenes, the product lines. The compounds have antiinflammatory properties.

B

Bacimethrin. It is 4-amino-2-methoxy-5-pyrinidine-methanol, an antibiotic synthesized by Bacillus megatherium. M_r 155.16, m p. 174°C. B. was, and is active against some yeasts and bacteria.

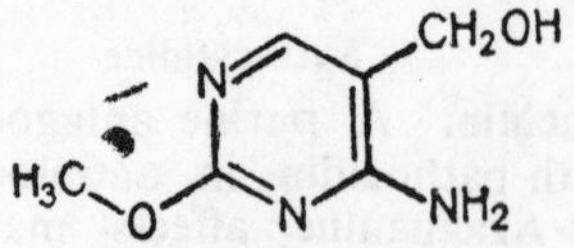

Bacimethrin

Bacitracins. Branched, cyclic peptides produced by various strains of Bacillus licheniformis. They are effective against Gram-positive bacteria. It has little antimicrobial activity.

Bacterial Photosynthesis. A primitive form of photosynthesis using H_2S, thiosulfate, fatty acids or other organic reducing agents rather than water as the source of hydrogen for carbon fixation.

Bacteriochlorin. 7, 8, 17, 18-tetrahydroporphyrin.

Bacteroids. The symbiotic, nitrogen-fixing, intracellular forms of Rhizobium spp. in the root nodules of leguminous plants.

Balata. A rubber-like polyterpene of low molecular weight obtained from the latex of tropical trees, especially from Mimusops balata.

Balsams, Oleoresins. The solutions of resins in volatile oils. B. are produced by plants, either normally or pathologically. The most important is turpentine.

Bamicetins. A pyrimidine antibiotic (see Nucleoside antibiotics) synthesized by Streptomyces plicatus. M_r 604.65, m.p. 240 to 241 °C, $[\alpha]_D^{26}$ + 123° (c=0.5 in 0.1 N HCl). It is effective in low concentrations against Mycobacterium tuberculosis.

Base Pairing. Refers to the specific hydrogen bonding between adenine and uracil (RNA) or thymine (DNA), and cytosine and guanine in a double-stranded molecule of RNA or DNA (Fig.).

Basic Protein. Refer to a group of small proteins rich in arginine and lysine found in the cell nuclous and in sperm. The exact function of these complexes is unknown.

Batrachotoxin. Refers to a neurotoxin from the skin of the Columbian arrow poison frog, genus Phyllobates. It brings about a selective and irreversible increase in the permeability of nerve membranes to sodium ions.

Bdellins. A group of protease inhibitors from the leech. They inhibit trypsin and plasmin, and they show a strong inhibitory activity towards the trypsin-like protease, acrosin, present in the acrosomes of spermatozoa.

Bee Toxin. A defense secretion which is produced in an abdominal gland or queen and worker bees and delivered by the sting. It contains three types of active principle :

(1) biogenic amines, including histamine, which cause pain and dilate the blood vessels, allowing wider penetration;

(2) biologically active peptides such as Mellitin and apamin, and

(3) enzymes like hyaluronidase and phospholipase A. B.t. is applied intracutaneously or percutaneously for treatment of neuralgias, rheumatism and allergies.

Beeswax. The solid secretion from which honeycomb is made. m.p. 62 to 65°C. It is a mixture or esters of straight chain fatty alcohols (even-numbered carbon chains from C_{24} to C_{36}) with straight-chain fatty acids (even numbers of carbon atoms up to C_{36}), *e.g.* myricyl palmitate (myricin), myricyl cerotate. The corresponding unesterified acids and alcohols are also present.

Behenic Acid, n-docosanoic Acid. CH_3-$(CH_2)_{20}$-COOH, a fatty acid. M_r 340.6, m.p. 82 °C, b.p. 262 °C. It is a component of glycerides in seed oils, for example rape seed oil, wool fat and cerebrosides.

Bence-jones Proteins. The proteins excreted in the urine of patients with multiple myeloma, a malignancy of antibody-producing cells.

Benzoquinones The compounds derived from p-benzoquinone. p-Benzoquinone, and its monodi, and trimethy, ethyl, methoxy and 2-methoxy-3-methyl derivatives are found in the defense secretions of certain arthropods. More than 90 different benzoquinones have been isolated from higher plants.

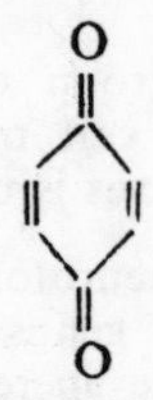

p-Benzoquinone

6-Benzylaminopurine, N-benzuladenine. A frequently and synthetic cytokinin.

Benzylisoquinoline Alkaloids. A group of alkaloids found mainly in poppy plants (Papaveraceae). The benzyl substituent on the Cl atom of the isoquinoline radical can enter various secondary cyclizations through phenol oxidation.

Beri-beri. The deficiency disease resulting from lack of vitamin B_1.

Betacyanins. The members of the betalain group of plant pigments with absorption maxima between 534 and 552 nm.

Betaines. The widely occurring biogenic amines. The simplest is betaine (glycine betaine, trimethylglycocoll betaine), $(CH_3)_3N+-CH_2COO-$M, 117.2, m p. of the hydrochloride, 227–228°C (d). They are synthesized in the ethanolamine cycle and are metabolically related to mono – and dimethylglycine. They can serve as methyl donor in methylation.

Betalains. A group of nitrogen-containing plant pigments found almost exclusively in the family Centrospermae.

Betalaminic Acid. The skeleton of the betalains. It is biosynthesized from tyrosine via dopa. Bonding to various amino acids, such as cyclodopa or proline, gives rise to the Betacyanins and Betaxanthins.

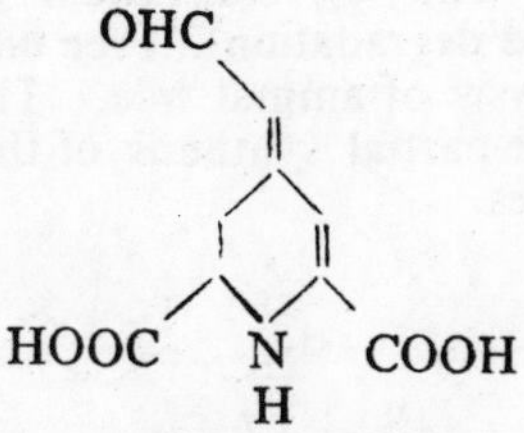

OHC

HOOC N COOH
H

Betalaminic acid

Betamethasone. The 9-fluoro-11β, 17, 21-trihydroxy-16β-methyl-pregna-1, 4-diene,-3,20 dione a synthetic pregnane derivative.

Betanin. The red pigment in beets (Beta vulgaris). It is a member of the Betalain group, and is highly water-soluble.

Betananthins. Yellow Betalain plant pigments with absorption maxima between 474 and 486 nm.

Betulenols. Isomeric bicyclic sesquiterpene alcohols from birch bud oil.

Betulin. A pentacyclic triterpene diol. M_r 442.73, m.p. 261 °C, $[\alpha]_D + 15°$ (chloroform). It is found in birch and hazelnut bark, in rose hips and in the cactus Lemaireocereus griseus.

Betulinic Acid. A pentacyclic triterpene carboxylic acid found in many plants, *e g.* Gratiola officinalis, Melaleuca spp., various cacti, the bark of Platanus spp. and the bark of the pomegranate tree, Punica granatum.

Bicuculline. An alkaloid from Dicentra cucullaria, Adlumia fungosa and several spp. of Corydalis.

Biflavonyls. The dimers of flavonyl units, very often apigenin, which are very widespread in plants.

Bile Acids. The components of bile which serve as emulsifying agents for fats. They are steroid carboxylic acids bound in peptide linkage to tauring or glycine.

The salts of bile acids, reduce the surface tension and emulsify fats, so that they can be absorbed in the intestine. The lipases are also activated by bile acids. In humans, the daily production is 20 to 40g. 90% of this is resorbed in the intestine and returned to the liver in the enterohepatic circulation. 1 liter bile contains 40 bile acids.

B.a. are biosynthesized from cholesterol by 7α hydroxylation, reduction of the double bond at position 5, and epimerization at position 4. The C_{27} side chain gets shortened by β-oxidation (Fatty acid degradation). Free bile acids are obtained by alkaline hydrolysis of animal bile. They are important as starting materials for partial synthesis of therapeutically important steroid hormones.

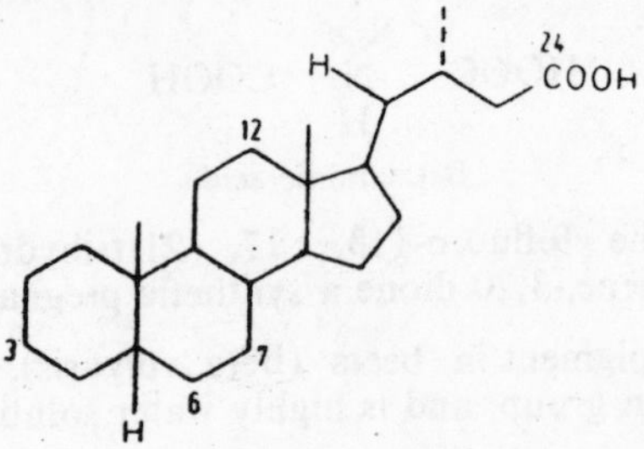

5β-Cholan.24 oic acid

Bile Alcohols. A group of polyhydroxylated steroids which are derived structurally from cholestane. They occur as sulfuric acid esters in the bile of lower vertebrotes.

Bile Pigments. Refers to the degradation products of prophyrins, especially heme. The α-methine bridge between rings A and B gets oxidatively cleaved to form CO and biliverdin IX microsomal hydroxylases. (In meso-biliervdin, the two vinyl substituents are reduced to ethyl). Biliverdin gets reduced to bilirubin) and transported to the liver as a complex with serum albumin. Bilirubin is produced mainly from the hemoglobin of aged erythrocytes in the reticuloendothelial system (spleen), bone marrow and liver.

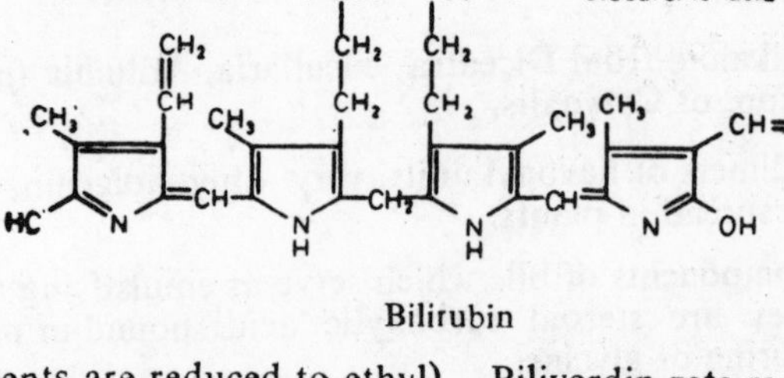

Bilirubin

Biliproteins, Phycobilins. Chromophores which serve as photosynthetic pigments in the blue green bacteria (Cyanophyceae and red algae (Rhodophyta) and are responsible for the colors of these organisms.

Biochemical Oxygen Demand, BOD. Refers to the rate at which the oxygen dissolved in water is consumed for the oxidation of organic compounds in the water by microorganisms.

Bioelements Refers to those chemical elements required by organisms. The elemental composition of organisms is considerably different from that of the earth's crust (Table 1); only abour 40 of the 90-odd crust elements are represented in living matter, and the six elements C,O,H,N,S and P together account for about 90% of it. The six main elements are present both as constituents of biomolecules and in inorganic matrix substances and in water, the medium of organic processes. Table 2 shows the elemental composition of the human body.

Carbon (C) forms the skeleton of all organic molecules.

Oxygen (O) is a component of nearly all biomolecules, providing a reactive "handle" for metabolic transformations (hydrocarbons are not generally biodegradable) of acids, aldehydes and ketones, alcohols and ethers.

Hydrogen (H) is present in all biomolecules, attached to C,N,O, and S.

Nitrogen (N) is a component of many biomolecules, especially of proteins and nuclei acids.

Sulfur (S) is present in two amino acids, cysteine and methionine, and in certain coenzymes.

Phosphorus (P) is present in both inorganic and organic compounds as phosphate. The nucleotides in nucleic acids are linked by phosphate bonds, and energy gets transferred from one molecule to another in the form of a high-energy phosphate bond, very often in ATP. Many coenzymes also certain phosphate.

TABLE 1

Relative Abundance of the Chemical Elements in the Earth's Crust and in the Human Body (adapted from Report)

Element	Earth's crust %	Human body %	Concentration (-fold)
Oxygen	50	63	—
Silicon	28	0	—
Aluminium	9	0	—
Iron	5	0.004	—

Calcium	3.6	1.5	—
Potassium	2.6	0.25	—
Magnesium	2.1	0.04	—
Hydrogen	0.9	10	10
Carbon	0.09	20	200
Phosphorus	0.08	1	10
Sulfur	0.05	0.2	4
Nitrogen	0.03	3	100

Table 2

Elemental Composition of the Human Body, Based on Dry Weight

Element	Percent
Carbon	50
Oxygen	20
Hydrogen	10
Nitrogen	8.5
Sulfur	0.8
Phosphorus	2.5
Calcium	4.0
Potassium	1.0
Sodium	0.4
Chlorine	0.4
Magnesium	0.1
Iron	0 01
Manganese	0.001
Iodine	0.000 05

Biogenic Amines Refer to a biologically and pharmacologically important class of compounds characterized by the presence of an amine group. They occur widely in plants and animals.

They can be precursors of alkalods (which is why they are also called protoalkaloids) and hormones. In addition, some are Neurotransmitters or components of phospholipids, vitamins, ribosomes and bacteria.

The biosynthesis and metabolism of the biogenic amines are generally similar. Some B.a. are hallucinogens, *e.g.*, mescaline and psilocybin. Many are toxic, *e.g.*, cadaverine and putrescene.

Bioluminescence. The emission of visible light by an organism. The light is emitted as a result of a redox reaction catalysed by the enzyme luciferase (Luciferin). The energy obtained from this reaction is used to excite the electrons of an oxidation product of luciferin to a higher electronic state. Light gets emitted as the oxyluciferin returns to the ground state.

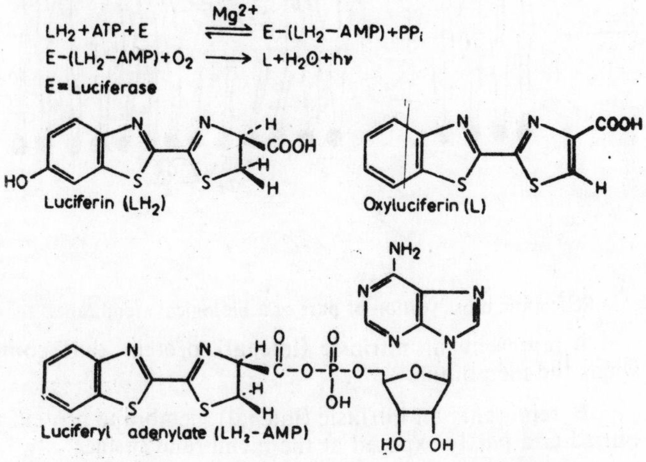

$$LH_2 + ATP + E \xrightleftharpoons{Mg^{2+}} E{-}(LH_2{-}AMP) + PP_i$$

$$E{-}(LH_2{-}AMP) + O_2 \longrightarrow L + H_2O + h\nu$$

E = Luciferase

Luciferin (LH_2)

Oxyluciferin (L)

Luciferyl adenylate (LH_2-AMP)

Mechanism of light emission by Photinus luciferin.

Biomass. The amount of organic substance in living organisms on a given area. The term is used in ecology.

Biomembrane. A structure having lipids, glycolipids, proteins and glycoproteins, bounding the cell (cell membrane) or subdividing it into compartments. It is a sheet- like structure about 60 to 100 Å thick. The lipids have hydrophilic "head" groups (indicated by spheres in the Fig.) and hydrophobic "tail" regions.

The currently accepted model of the structure of biomembrane is based on the fluid mosaic concept.

Physically, the surface of a biomembrane is an interface between a polar solvent (water) and a nonpolar solvent (the hydrophobic tails of the lipid molecules). Hydrophobic molecules may cross biomembrane because they can dissolve in the hydrophobic milieu of the membrane interior.

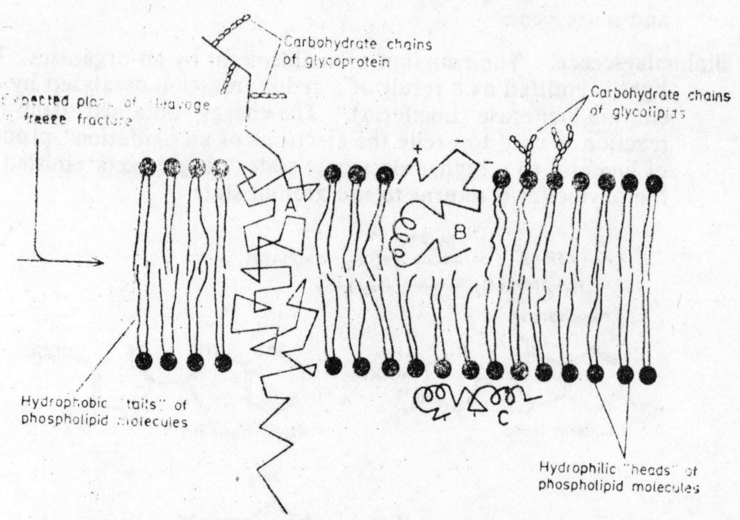

Schematic cross section of part of a biological membrane.

A represents an nitrinsic (integal) protein that compelely spans the membrane.

B represents an intrinsic (integal) membrane protein partly buried and partly exposed at the membrane surface.

C is representative of many extrinsic (peripheral) proteins that are more or less firmly associated with membranes, but

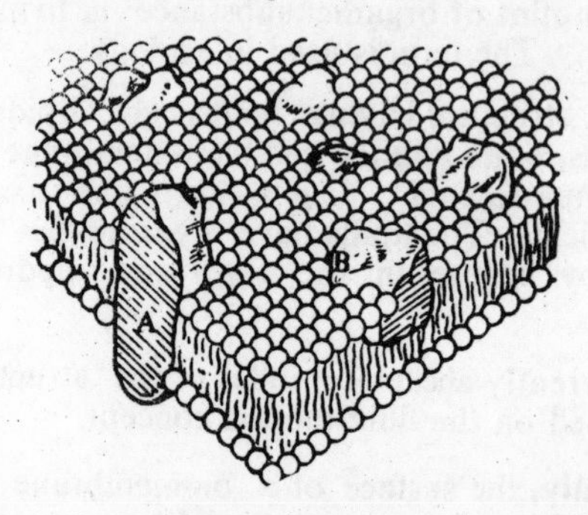

Three dimensional artistic impression of a phospholipid bilayer with intrinsic proteins.

appear not to be integrated into the phospholipid layer, *e.g.*, spectrin of the erythrocyte membrane.

Cytochrome c may represent an association intermediate between B and C; it is an important functional constituent of the inner mitochondrial membrane, but extremely easily removed.

Bionics The term used for the application of structural and functional principles discovered in organisms to man-made objects, especially in electronics. The goals of the discipline are the optimization of technological processes and the development of new ones.

Biophysics. The study of the physics of biological systems. It can be subdivided into three main areas : molecular and cellular biophics, which is concerned with membrane properties, bio-electric phenomena and the conformation and conformational changes in biopolymers; radiation B., which is the study of the changes in cells and macromolecules caused by radiation; and medical B., in which B. is applied to diagnosis and therapy and medical technology.

Biopolymer, biomacromolecule. A naturally occurring substance which can be regarded as a chain (polymer) of identical or similar subunits (monomers).

Synthesis of biopolymers begins with the transfer of an activated monomer to a primer molecule (initiation). It continues with the addition of more activated monomers, one at a time (elongation). At some point the growth terminotes, for reasons which have not been entirely elucidated. The biopolymer is not necessarily complete at this point (Post-translational modification); it may be further modified.

Biosynthesis, Biogenesis. The biological synthesis of a compound, *i.e.*, an enzymatic synthesis.

It can occur either in vivo, that is, in a living organism or in a cultivated part of it (such as a perfused organ), or in vitro, in cell homogenates, extracts, enzyme preparations and reconstituted systems.

The sum of all biosynthesis, in an organism is referred to as anabolism.

Biotechnology. Refers to an engineering discipline which develops the methods for carrying out microbial and biochemical processes on an industrial scale.

Biotest, Biological Assay. Refers to a procedure in which the biological activity of a compound is tested by observing its effect on an appropriate organism.

These are frequently used for hormones, inhibtiors and antibiotics The test objects are usually whole organisms, or isolated organs or tissues, sometimes in tissue culture.

Biotin Enzymes. Refers to the enzymes which catalyse carboxylation reactions, using biotin as a cofactor.

Biobenzyllsoquinoline Alkaloids. A group of isoquinoline alkaloids with a skeleton formed by fusion of two isoquinoline units by a phenol oxidation. The curare alkaloids, for example tubocurarine, belong to his class.

Bitter Acids. Refer to a mixture of substance in the resin fraction of hops. They are monoacylphloro-glucides. The compounds are chemically very labile. During the brewing of beer, they undergo a reduction of ring size to produce the isoacids, which also taste bitter.

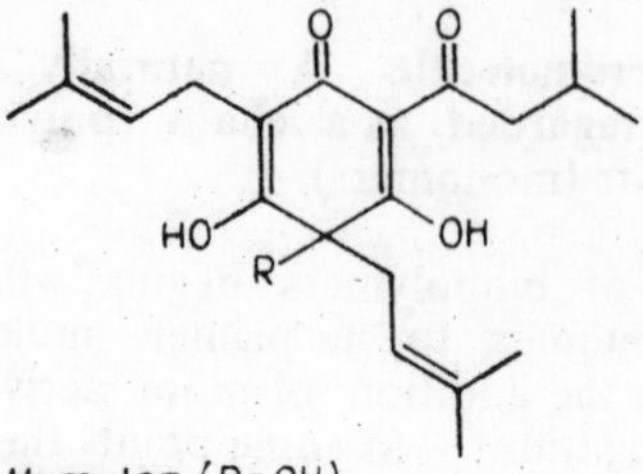

Humulon (R = OH)
Lupulon (R = $-CH_2-CH(CH_3)_2$)

Bitter Peptides. Refers to the bitter tasting peptides that may spoil the palatability of certain foodstuffs, *e.g.* These sometimes arise during the ripening of certain types of cheese, and they have also been found in fermented soyabean products.

Bitter Principles. Bitter-tasting substances which occur found especially in Compositate, Gentianaceae and Labiatae, which produce a reflexive increase in the secretion of saliva and digestive juice. Extracts from such plants have been used as bitter spices to increse the appetite and promote digestion, and they are used in the preparation of stomachic bitters.

Bixin. The monomethyl ester of a C_{24} dicarboxolic acid called norbixin, M_r of bixin, 394, m.p. 198°C.

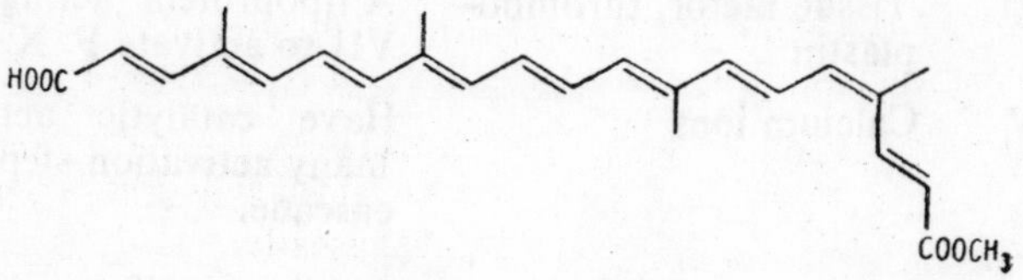

HOOC COOCH₃

Bixin

It is a yellow to redorange pigment found in the seeds of Bixa orellana. The extract of these seeds is used as a food color.

Blasticidiness. The pyrimidine antibiotics synthesized by Streptomyces griseochromogenes. An important representative of the group is blasticidine S., m.p. 235 to 236°C, $[\alpha]_D^{11}+108.4°$ (c=1 in water). They inhibit the growth of fungi, for example the rice fungus Piricularia oryzae, and a few bacteria. The antibiotic effect is due to the suppression of the elongation of polypetide chains during protein biosynthesis.

Blood Coagulation, Blood Clottong. Refers to a process in which blood is solidified to stop bleeding from a wound.

The clotting factors are given in Table 1.

TABLE 1

Coagulation Factors

No.	Name	Properties and functions
I	Fibrinogen	M_r 340000, compound of 6 chains : $(A)_2$ $(B)_2$ γ_2. Converted to fibrin by removal of 2A and 2B peptides.
II	Prothrombin II_a is thrombin	M_r 70000, 582 amino acids with 12 disulfide bridges. Glycoprotein. Converted to thrombin by F. X; thrombin inhibited by antithrombin III, α_2-macroblobulin, α_1-antitrypsin and hirudin.

III	Tissue factor, thromboplastin	A lipoprotein acting with F. VII to activate F. X.
IV	Calcium ions	Have catalytic activity in many activation steps in the cascade.
V	Proacceleıin V_a is accelerin	Labile modifier protein. V_a accelerates conversion of F. II to II_a.
VII	Proconvertin	Labile, activated by tissue trauma. Part of the extrinsic system.
VIII	Antithemophilic factor	Labile modifier protein, $M_r > 10^6$, acting together with F. IX_a to active F.X. Its absence causes hemophilia A.
IX	Christmas factor	M_r 55000 (bovine), single-chain glycoprotein. Its absence causes hemophilia B.
X	Stuart factor	M_r 55000, glycoprotein composed of a light and a heavy chain. Activated by a mixture of F. IX_a and VIII_a+Ca^{2+} or VII_a+tissue factor+Ca^{2+}.
XI	Plasma thromboplastin antecedent	M_r 124000 a glycoprotein composed of two similar or identical polypeptides joined by a disulfied bond(s). Activates F.IX. XI_a is inhibited by antithrombin III, trypsin inhibitors, α_1-trypsin inhibitor and CI inhibitor.
XII	Hageman factor	M_r 76000 (human), single-chained glycoprotein. Activated by plasmin, kallikrein and XII_a. Inhibited by antithrombin III, Cl esterase inhibitor and lima bean trypsin inhibitor.

		Inhibition by antithrombin III accelerated by heparin. Activation of F. XII initiated by contract with abnormal surfaces. It is the first factor in the intrinsic pathway.
XIII	Fibrin-stabilizing factor (Laki-Lorand factor)	M_r 35000, 4-chained α_2-globulin. XIII$_a$ is the transpeptidase responsible for cross-linking precipitated fibrin monomers.
—	Prekallikrein (Fletcher factor)	Activated to kallikrein, a serine protease which activates F.XII.
—	HMW kininogen (high molecular weight kininogen, contact activation cofactor, Fitzgerald factor, Williams factor, Flaujeac factor)	Activated to a kinin involved in activation of F. XII, at least in vitro.

Blood Group Antigens. Refers to the specific oligosaccharide structures attached to glycoproteins in the membranes of blood cells which are recognized as antigens by the immune systems of other individuals or organisms. The antigens get attached to the protein glycophorin in erythrocytes and to both proteins and lipids in other parts of the body. In humans, five systems of antigens could be identified, the ABO system, the MN, P, rhesus and Lutheran systems. Only the ABO and rhesus systems affect blood transfusions between humans; the other systems have been identified using antimal antibodies against human blood.

Blue-green Bacteria, Cyanobacteria, Blue-green Algae, Cyanophyta, Cyanophyceae. Refer to a group of photosynthetic prokaryotic organisms using H_2O as hydrogen donor (Photosynthesis). Many are also able to fix atmospheric nitrogen (Nitrogen fixation). Since they are prokaryotes, they have no nucleus, mitochondria or chloroplasts.

There are two major classes of blue-green the chroococcals, in which the cells are solitary (*e.g.* Anacystis) or colonial, held together by mucoid hulls; and the hormogonals, which grow in filaments (trichomas), often enclosed in a sheath.

Bombykol, 10-trans-12-cis-hexadecadienol-(1). A pheromone exuded by female silk months (Bombyx mori) to attract males. It is an oil, n_D^{20} 1.4835.

Bombykol

Bongkrekic Acid. The 3-carboxymethyl-17-methoxy-6, 18, 21-trimethyldocosa-2, 4, 8, 12, 14, 18, 20-heptaenedioic acid, M_r 486.61. One of two toxic antibiotics produced by Pseudomonas cocovenenans in spoiled bongkrek (a coconut product consumed in Indonesia). It is an inhibitor of adenine nucleoside translocation and affects carbohydrate metabolism.

Boron. An element essential for growth of higher plants. Lack of boron causes heart rot of sugar beet and other root.

Bradykinin, Kallidin I, Kinin 9. Arg-Pro-Pro-Gly-Phe-Ser-Pro-Phe, Arg, one of a group of plasma hormones called kinins. It causes dilation of blood vessels, and thus a reduction of blood pressure, causes the smooth mucsles of the bronchia, intestines and uterus to contract, and is a potent pain-producing agent. Lysylbradykinin has similar activity.

Brassicasterol, Ergosta-5, 22-dien-3β-ol. A plant sterol (Sterols). M_r 398.69, m.p. 148°C, $[\alpha]_D$—64°. It was first isolated from rapeseed (Brassica campestris) oil.

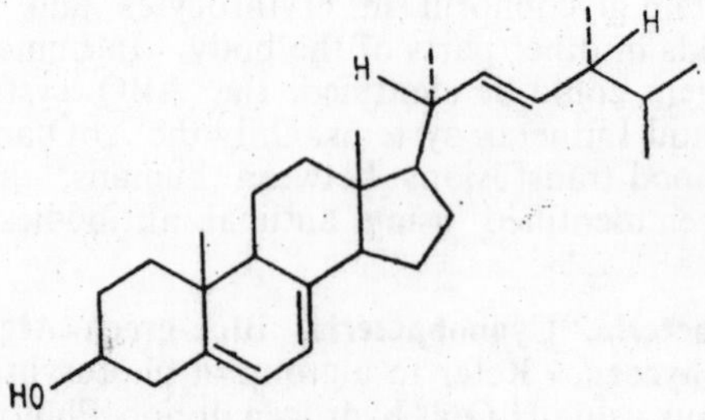

Brassicasterol

Bromelain. A thiol enzyme (EC 3.4.22.4) from the stems and fruits of the pineapple plant. The stem enzyme is a basic glycoprotein (M_r 33000; IP 9.55) which is structurally and catalytically similar to papain% It is an endopeptidase and is used in protein chemistry to hydrolyse polypeptide chains into large fragments.

Brucine: 2, 3-dimethoxystrychnine. M_r 394.47, m.p. 105°C (tetrahydrate), m.p. 178°C (anhydrous). The main physiological effect is a paralysis of the smooth musculature.

Bufotenine, 5-hydroxy-N-dimethyltryptamine. A toad poison which is also found in toadstools from which it was first isolated by Wieland.

Bufotoxin. The main toxin in the venom of the European toad Bufo vulgaris. M_r 757. It is a steroid derivative.

n-Butyric Acid, Butanoic Acid: CH_3-$(CH_2)_2$-COOH, the simplest fatty acid M_r 88 1, m p.—5 C. B a accounts for 3 to 5% of the fatty acids esterified to glycerol in butterfat. When butter becomes rancid, it is the free n-butyric acid produced by hydrolysis which is responsible for the unpleasant odor.

Buxus Alkaloids. A group of steroid alkaloids which are characteristic of plants in the boxwood genus (Buxus).

C

C_4-acid cycle. See Hatch-Slack-Kortschack cycle.

Cadaverine, 1, 5-diaminopentane. A biogenic amine produced enzymatically by decarboxylation of lysine. It is a precursor of a few alkaloids.

β-Cadinene. An optically active sesquiterpene found in the essential oils of junipers and cedars.

Caffeine, 1, 3, 7-trimethylxanthine. A purine derivative which is found in coffee beans and leaves, tea leaves and cola nuts. It is usually produced from tea leaves (1.5 to 3.5% caffeine content) and as a byproduct from the production of caffeine-free coffee.

Calciferol. Same as vitamin D.

Calcitonin, Thyreocalcitonin. A polypeptide hormone containing 32 amino acids. M_r (human) 3420. It is formed in the para-follicular cells of the thyroid in mammals, and in the ultimobranchial gland of nonmammalian species.

Calmodulin. A ubiquitous calcium-binding protein, which mediates the function of Ca^{2+} in eukaryotes. Many effects of Ca^{2+} are exerted through calmodulin-regulated enzymes:

$$(\text{calmodulin})_{inactive} + Ca^{2+} \rightleftharpoons (\text{calmodulin}.Ca^{2+})_{active}$$
$$(E)_{low\ activity} + (\text{calmodulin}.Ca^{2+}) \rightleftharpoons (E.\text{calmodulin}.Ca^{2+})_{high\ activity}$$

Calvin Cycle, Photosynthesis Cycle, Reductive Pentose Phosphate Cycle. A series of at least 15 enzymatic reactions which, taken together, generate 1 molecule of hexose phosphate from 6 molecules of CO_2. In the process, which does not require light (dark reactions), 12 molecules NADPH and 18 molecules of ATP are consumed per molecule of hexose phosphate.

TABLE

Reactions regenerating ribulose 1, 5-bisphosphate in the Calvin cycle

Reaction	Enzyme
Glyceraldehyde-P $\rightleftharpoons$ Dihydroxyacetone-P	Triosephosphate isomerase (EC 5.3.1.1)
Glyceraldehyde-P + Dihydroxyacetone-P $\rightleftharpoons$ Fructose-1, 6-P_2	Fructose-bisphosphate aldolase (EC 4.1.2.13)
Fructose-1, 6-P_2 $\rightleftharpoons$ Fructose-6-P + P_i	Fructose-bisphosphatase (EC 3.1.3.11)
Fructose-6-P + Glyceraldehyde-P $\rightleftharpoons$ Xylulose-5-P + Erythrose-4-P	Transketolase (EC 2.2.1.1)
Erythrose-4-P + Dihydroxyacetose-P $\rightleftharpoons$ Sedoheptulose-1, 7-P_2	Aldolase (EC 4.1.2.13)
Sedoheptulose-1, 7-P_2 $\rightleftharpoons$ Sedoheptulose-7-P + P_i	Sedoheptulose-bisphosphatase (EC 3.1.3.37)
Sedoheptulose-7-P + Glyceraldehyde-P $\rightleftharpoons$ Ribose-5-P + Xylulose-5-P	Transketolase (EC 2.2.1.1)

Ribose-5-P⇌Ribulose-5-P	Ribosephosphate isomerase (EC 5.3.1.6)
Xylulose-5-P⇌ Ribulose-5-P	Ribulosephosphate 3-epimerase (EC 5.1.3.1)
Ribulose-5-P+ATP⇌ Ribulose-1, 5-P_2+ADP	Phosphoribulokinase (EC 2.7.1.19)
Sum: 5 Triose-P→3 Pentose-P	

Calvin Plants. See C_3 Plants.

cAMP. Abb. for cyclic adenosine 3′,5′ monophosphate.

Campesterol: **(24R)-ergost-5-en-3β-ol.** A plant sterol which is found in the oils of rapeseed (Brassica campestris), soybean and wheat germ and in some molluscs.

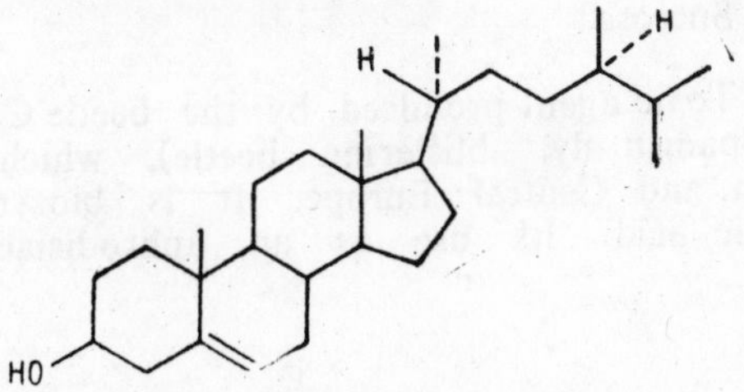

Campesterol

Camphor. A bicyclic monoterpene ketone found widely in plants. It is obtained commercially from camphor trees (Cinnamomum camphora) native to the coastal areas of Eastern Asia. The partial synthesis from pinene is also important, although the product is a racemic mixture used mostly in plastics.

Camptothecin. The main alkaloid from the wood and bark of the Chinese tree Camptotheca acuminata. Its total it is one of the most active natural substances against leukemia and tumors.

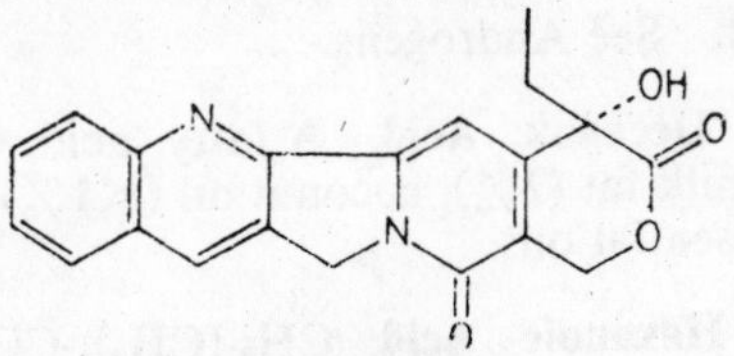

Camptothecin

l-Canalin. 2-amino-4-aminoxy-butyric acid. It is a hydrolysis product of L-Canavanine and is found in a few canavanine-containing legumes.

l-Canavanine, 2-amino-4-guanidinohydroxybutyric Acid: $H_2N\text{-}C(NH)_2\text{-}O\text{-}(CH_2)_2\text{-}CH(NH_2)\text{-}COOH$. It is structural analog of Arginine. It is found only in certain legumes. The presence or absence of C. is a useful trait in the chemical taxonomy of the legumes.

Cancer Research. Scientific study of :

(1) the factors which lead to the formation of the various malignant tumors.

(2) the continuous growth of the malignant tumors, and

(3) of their inhibition by anti-cancer agents, in order to find effective methods of therapy.

Candicine, N, N, N-trimethyltyramine. A biogenic amine found especially in grasses and cacti.

Cane Sugar. Sucrose.

Cantharidin. Toxic agent produced by the beetle Cantharis vesicatoria (Spanish fly, blistering beetle), which is native to Southern and Central Europe. It is biosynthesized from mevalonic acid. Its use as an aphrodisiac has caused fatalities.

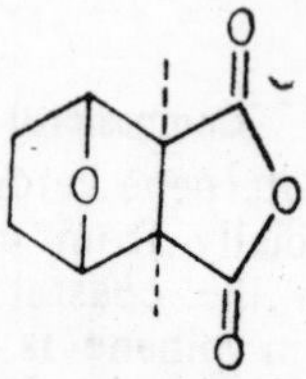

Cantharidin

Caoutchouc. Elastic, high-polymer hydrocarbons (elastomers) which become rubber on vulcanization.

Capon-comb Unit. See Androgens.

N-caproic Acid, Decanoic Acid. A fatty acid, $CH_3\text{-}(CH_2)_8\text{-}COOH$. Occurs in milk fat (2%), coconut oil (<1%) and various other seed and essential oils.

N-capronic Acid, Hexanoic Acid: $CH_3\text{-}(CH_2)_4\text{-}COOH$. A fatty acid. Found in milk fats (2%) and in small amounts in coconut oil and other palm oils. It also occurs in essential oils from plants and plant fats.

n-Caprylic Acid, Octanoic Acid: CH_3-$(CH_2)_6$-COOH. A fatty acid. It is found in various glycerides, *e.g.* 1 to 2% in milk fat, and as 6 to 8% of the coconut oil fats. It is also found in other plant fats.

Capsaicin. A pungent principle in the fruits of some peppers. The aromatic part is biosynthesized from phenylalanine. It is occasionally used as a counter-irritant.

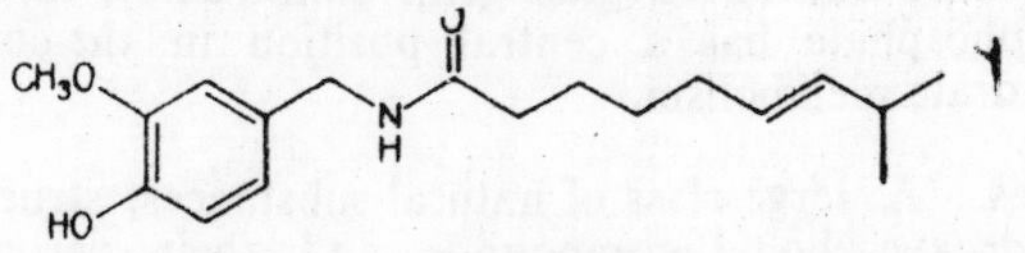

Capsaicin

Capsanthin. A carotenoid pigment isolated from paprika (Capsicum annuum). It is characterized by a terminal five-membered ring.

Capsid. See Viral coat protein.

Capsorubin. A carotenoid pigment found in paprika (Capsicum annuum) Contains two identical cyclopentanol rings. The hydroxyl groups have the S-configuration on C-atom 3 and the

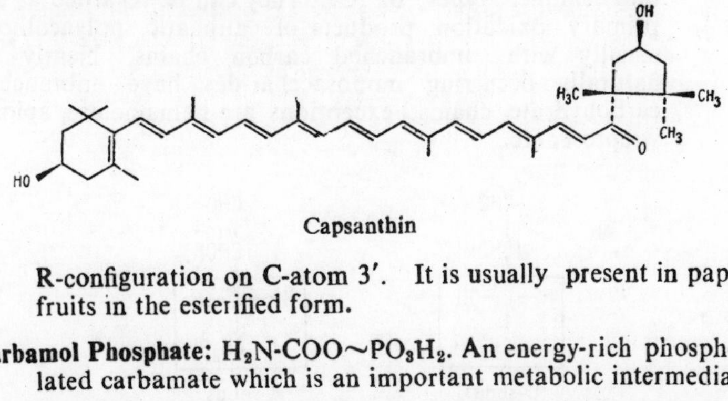

Capsanthin

R-configuration on C-atom 3′. It is usually present in paprika fruits in the esterified form.

Carbamol Phosphate: H_2N-COO~PO_3H_2. An energy-rich phosphorylated carbamate which is an important metabolic intermediate.

6-Carbamoylthreonyl Purine Nucleoside, N-(Nebularin-6-ylcarbamoyl)-threonine. Refers to one of the rare nucleic acid bases. It has so far been detected in six specific transfer RNAs. The analogous compound 6-carbamoylglycyl purine nucleoside has been isolated from yeast tRNA.

Carbohydrates Metabolism. The constant formation, transformation and degradation of the carbohydrates in the organism. The most important reactions in carbohydrate metabolism are :

(1) Interconversions of the polymeric storage forms (glycogen and starch) and the monomeric transport and substrate form (glucose),

(2) reactions of carbohydrate degradation and interconversion, and

(3) reactions for the synthesis of glucose from noncarbohydrate substances (glucogenic amino acids, fats). Glucose 6-phosphate has a central position in the entire carbohydrate metabolism.

Carbohydrates. A large class of natural substances, structurally the polyhydroxycarbonyl compounds and their derivatives. In general they correspond to the composition $(C)_n(H_2O)_n$. They are present in every plant or animal cell and make up the largest portion, in terms of mass, of organic compounds present on earth. They are formed in plants in the course of assimilation processes. Together with fats and proteins, they are the organic nutrients for humans and animals. The carbohydrates are subdivided on the basis of their molecular size.

1. Monosaccharides (simple C.) cannot be further hydrolysed into simpler types of C. They can be regarded as the primary oxidation products of aliphatic polyacolhols, usually with unbranched carbon chains. Nearly all naturally occurring monosaccharides have unbranched carbohydrate chains; exceptions are hamamelose, apiose, streptose, etc.

CHO — $(CHOH)_n$ — H—C—OH — CH_2OH (D-series)

CHO — $(CHOH)_n$ — HO—C—H — CH_2OH (L-series)

Fischer projections of D- and L-carbohydrates

2. Oligosaccharides are made up of 2 to 10 monosaccharides α- or β-glycosidically linked. They are classified as di-, tri-, tetrasaccharides, etc. depending on the number of subunits. They can be hydrolysed by acids or enzymes into their subunits, which they resemble in physical and chemical properties. Oligosaccharides are widespread in the plant and animal kingdoms, and occur in free and bound forms.

They are synthesized via the nucleoside diphosphate sugars.

3. Polysaccharides. In this quantitatively very large group of C., 10 or more monosaccharide units are bound, according to the same structural principles as in oligosaccarides, α- or β-glycosidically to branched or unbranched chains. The chains may be arranged linearly, spirally or spherically, The most common components are the hexoses D-glucose, D-fructose, D-galactose and D-mannose, the pentoses D-arabinose and D-xylose, and the amino sugar D-glucosamine.

Carbon Dioxide Assimilation. In the narrower sense, the fixation of atmospheric carbon dioxide in photosynthesis. It is part of photosynthesis; it is often incorrectly used as a synonym for photosynthesis.

Carbonic Acid Anhydrase, Carboanhydrase, Carbonate Dehydrates (EC 4.2.1.1). A widely occurring, zinc-containing enzyme, usually monomeric. It catalyses the reversible hydration of carbon dioxide ($CO_2+H_2O \rightleftharpoons H^++HCO_3^-$) with one of the largest known turnover rates.

Carbon Monoxide. A colorless and practically odorless, combustible gas. It is extremely (abb. Hb) is 300 times that of oxygen. Otherwise, CO is completely inert. The reaction between Hb and CO is reversible : $HbO_2+CO \rightleftharpoons HbCO+O_2$. The CO displaces O_2 from hemoglobin, so that the red blood cells cannot perform their normal function, oxygen transport to the tissue. O_2 can only displace CO from Hb when it is present in large excess. Death by CO poisoning occurs after the following series of events :

1. the transport capacity of blood for oxygen is reduced by formation of HbCO;

2. intoxication of oxygen-sensitive tissues, especially in the brain, occurs (symptom, headache);

3. the respiratory center in the brain is incapacitated (symptom, unconsciousness);

4. the heart steps being for lack of adequate oxygen supply. CO can only enter the body through the alveoli of the lungs. Concentrations$>0.01\%$ are considered toxic. The maximal allowable concentration at a place of work is 55 mg CO/m^3 air.

Carbonyl Cyanide-p-trifluoromethoxyphenylhydrazone, FCCP. An uncoupler of oxidative phosphorylation. For structure and mode of action.

4-Carboxyglutamic acid, γ-carboxyglutamic Acid, Gla. An amino acid residue found in certain proteins, and formed by post-translation carboxylation of glutamate residues (4-Glutamyl carboxylase). It is present in the blood clothing factors, prothrombin, factor VII, factor X and factor IX; in the low M_r protein isolated from the bones of several vertebrates.

Carboxylase

1. See Carboxylation.
2. Old term for pyruvate decarboxylase.

Carboxylation. The transfer of carbon dioxide, frequently in activated form. The most important form of carboxylation is photosynthetic fixation of CO_2. but carboxylation is also required for fatty acid and purine metabolism.

Carboxyl Carrier Protein. See Biotin enzymes.

Carboxylic Acid Esterases. See Esterases.

Carboxylic Acids. Organic compounds containing the carboxyl group, COOH. Alkane and alkene monocarboxylic acids are also called fatty acids. They are synthesized from activated acid derivatives, such as anhydrides or thioesters (for d le, acetyl-coenzyme A).

Carboxylic Esterases

1. A sub-group of esterases acting on carboxylic esters;
2. Carboxylesterase, (EC 3.1.1.1), a class of enzymes with wide specificity, usually for a short-chain acid and an alcohol with only one hydroxyl group.

Carboxypeptidases. Single-chain exopeptidases which contain zinc. They remove successive amino acids from the C-terminal ends of proteins. The animal carboxypeptidases play a role in the digestion of protein in the small intestine, into which they are secreted as inactive precursors (zymogens) from the pancreas. They are converted to the active form by trypsin.

Cardiac Glycosides. Cardiotoxins in the narrower sense, a group of poisonous vegetable glycosides with specific carniac effects and related chemical structures.

Cardiotoxins. In the widest sense, substances which, in toxic doses, cause heart damage and may lead to heart stoppage. They may interfere with the generation or conduction of stimuli or with the heart's own blood supply, or they may directly attack the heart muscle.

Carminic Acid. A red glucoside pigment from the scale insect Coccus cacti L. which lives on Central American cacti of the genus Opunita. It is the principle component of cochineal, which was formerly one of the most prized dyes for wool and silk.

Carnitine, Vitamin B_1, (3-carboxy-2-hydroxyproply) trimethyl-ammonium hydroxide, $(CH_3)_3N^+\text{-}CH_2\text{-}CH(OH)\text{-}CH_2\text{-}COO^-$, internal. It serves as a carrier of acetyl and acyl groups through the mitochondrial membrane. These are bound by a high-energy bond by transfer from acetyl-(or acyl)-coezyme A.

Carotenes. A group of isomeric unsaturated hydrocarbons with nine conjugated trans double bonds, four methyl branches and a β-ionone ring at one end.

Carorenoids. A large class of yellow and red pigments which are highly unsaturated aliphatic and alicyclic hydrocarbons and their oxidation products. These arise biogenetically from isoprene units (C_5H_8), and therefore have the methyl branches typical of isoprenoid compounds.

Cartageenan, Carrageen. A mixture of polysaccharides, similar to agar-agar, obtained from red algae by hot water extraction. It contains about 45% corragenin, a polysaccharide containing galactose and galactose sulfate.

Caryophyllenes. Isomeric cyclic sesquiterpene hydrocarbons found in many essential oils.

Catabolism, Dissimilation. The sum of all degradative metabolic processes. Amino acids, purines, pyrimidines, etc. formed in the turnover of cell components can be degraded, in some cases to inorganic compounds (water, carbon dioxide, ammonia, etc.). Catabolic processes are linked with the formation of ATP by substrate phosphorylation and respiratory chain phosphorylation. C. is therefore, in a certain sense, identical to energy metabolism.

Catabolite Repression. The inhibition of enzyme synthesis by increased concentration of certain metabolic products.

Catalase. A tetrameric heme enzyme which catalyses the removal of the highly poisonous hydrogen peroxide from the cell: $H_2O_2 \rightarrow H_2O + \frac{1}{2}O_2$.

Catecholamines. The alkylamino derivatives of pyrocatechol (o-dihydroxybenzene), a group of substances including the hormones Adrenalin, Noradrenalin and Dopamine. These are derived from tyrosine. They affect the blood vessels, the intermediary metabolism and nerve transmission. Their degradation products are excreted in the urine and can be detected there.

Catechol Estrogens. 2-Hydroxylated derivatives of estrogens, are hydroxylated by the action of the microsomal cytochrome P450 system of the silver. The 2-hydroxyl group then becomes methylated by the action of a methyl transferase and S-adenosyl-L-methionine in the liver cytosol. The corresponding methoxy derivatives are excreted in the urine.

Causal Analysis, Induction. The scientific method of reasoning in which a generalization (subsumptive generalization) is postulated on the basis of previous experiments and verified or falsified through the experimental analysis of a single case derived from it strictly in accordance with the rules of logic.

CCC. Abb. for chlorocholine chloride.

CDNA. Abb. for complementary DNA.

CDP. Abb. for cytidine 5′-diphosphate.

CDP-choline. Abb. for cytidine diphosphocholine.

CDP-glyceride. Abb. for cytidine diphosphoglyceride.

CDP-sugars. A metabolically activated form of sugars and sugar derivatives. The sugar alcohol CDP-ribitol plays a role in the synthesis of bacterial cell walls.

Cell. The smallest living unit of an organisms. Living organisms are classified as prokaryotic (prokaryotes) or eukaryotic eukaryotes), depending on their type.

Cell Cycle, Mitotic Cycle. The sequence of phases during nuclear and cell division in a mitotically dividing eukaryotic cell. The following phases are recognized : the G_1 (G stands for "gap") or postmitotic phase, without DNA synthesis; the S phase, when DNA is synthesized the G_2 or premiotic phase without DNA synthesis; the M or mitosis phase, culminating in cell division (Fig.). The duration of each phase is variable, depending on the organisms, *e.g.* in the root meristem of the broad, bean, G_1 = 12 h; S = 6 h; G_2 = 8 h; M = 4 h. Sometimes G_1 or G_2 may be too brief to measure.

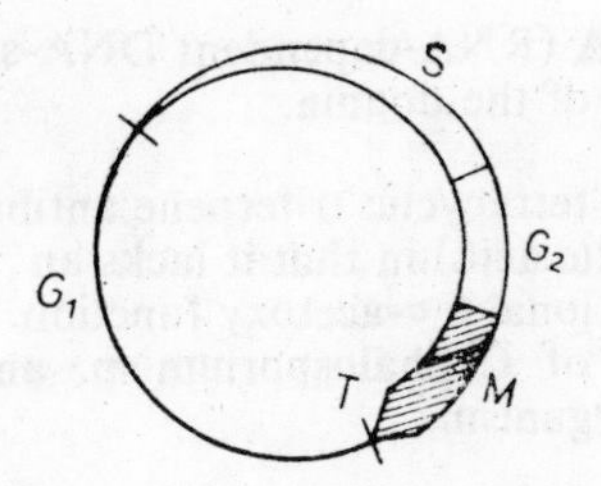

Schematic representation of the cell cycle

Cell Membrane, Plasmalemma. A biomembrane, which serves as the outer boundary of the cell. In addition to a C.m., plant and bacterial cells also have a Cell wall.

Cellobiose. A reducing disaccharide consisting of two molecules of D-glucose linked β-1,4.

Cellulases. Enzymes found in plants, microbes and fungi which hydrolyse cellulose to cellobiose.

Cellulose. A plant polysaccharide built of glucose units liked β-1, 4. The chains are not branched and have M_r between 300 000 and 500 000, corresponding to 3000 to 5000 glucose units.

Cell Wall. A rigid structure external to cell membrane and synthesized by the protoplasm. Animal cells do not possess a cell wall only a cell membrane. Cell wall enclose prokaryotic cells bacteria, blue-green bacteria), and plant cells.

Cembranes, Cembranoids. Monocyclic diterpenes isolated from several plants, especially from the gum resins of pines. They have also been found in marine coelenterates and insects.

Central Dogma of Molecular Biology. The fact that genetic information can be transferred from DNA to protein, but not in the reverse direction. The discovery that RNA can code for the

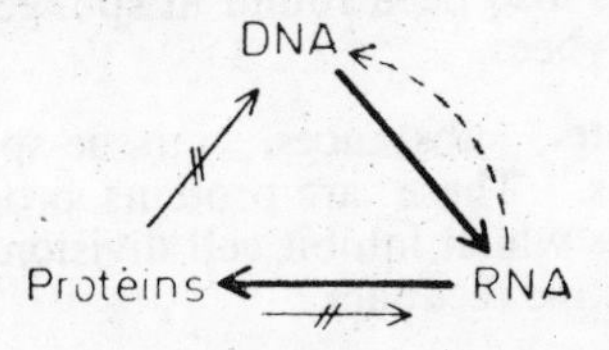

Schematic representation of the central dogma of molecular biology

synthesis of DNA (RNA-dependent DNA-synthetase), does not alter the validity of the dogma.

Cephalosporin P_1. A tetracyclic triterpene antibiotic. It differs from the related Fusidic acid) in that it lacks an 11 α-hydroxyl group and has an additional 6 α-acetoxy function. It is obtained from culture filtrates of Cephalosporium sp. and is effective against Gram-positive organisms.

Cerebronic Acid, 2-hydroxytetracosanoic Acid, α-hydroxylignoceric Acid. $CH_3-(CH_2)_{21}-CHOH-COOH$. An hydroxy fatty acid. M_r 384.63, m.p. 101 °C. It is a component of various glycolipids.

Cerotinic Acid, n-hexacosanoic Acid. $CH_3-(CH_2)_{24}-COOH$, A fatty acid. It is a component of beeswax, wool grease, carnuba wax and montan wax, and occurs in traces in plant fats.

Ceruloplasmin A blue, copper-containing glycoprotein found in mammalian blood plasma. It consists of 4 subunits (α_2/β_2) and contains 8 cu^{2+} ions per molecule.

It is phylogenetically related to lacease (EC 1.10.3.2) and ascorbate oxidase (EC 1.10.3.3), which are both tetrameric, blue, copper oxidases. C. is both a storage and a transport protein for copper (II) It has a central role in the metabolism of copper-containing enzymes, such as phenol oxidases, and in copper-dependent reactions, for example hemoglobin synthesis. When the organism has no C., as in congenital Wilson's disease, copper is deposited in the tissues, which causes death.

cGMP. Abb. for cyclic guanosine 3′, 5′-monophosphate.

Chalcones. A group of yellow to orange flavonoids (derived from the parent compound chalcone. Some examples are butein, coreopsin and pedicin.

Chalinasterol, Ostreasterol, 24-methylenyl Cholesterol, ergosta-5, 24 (28)-dien-3 β-ol. An animal sterol. It is a characteristic sterol in pollen, and has also been found in sponges, oysters and mussels, and in honeybees.

Chalones. Antitemplate substances, tissue-specific; endogenous mitosis inhibitors. These are proteins produced by mature or differentiated cells which inhibit cell division in primordial cells by a kind of negative feedback.

Charge-relay System. A network of hydrogen bridges in the catalytic center of chymotrypsin and other serine proteases which is responsible for the high degree of nucleophilicity of the Ser_{195} hydroxyl group.

Chemical Taxonomy. The deduction of taxonomic relatiohships from the distribution of certain natural products.

Chemosynthesis

1. Chemical synthesis.
2. The utilization of inorganic compounds or ions (ammonia, nitrite, hydrogen sulfide, thiosulfate, sulfide, iron(II) or manganese(II) ions) and of hydrogen or elemental sulfur to obtain reducing equivalents and ATP.

Chemotherapy. C. was defined by its founder, Paul Ehrlich, as "the use of a drug to combat an invading parasite without damaging the host". The term is also used for treatment of cancer with chemicals which are more damaging to dividing sells than to differentiated ones.

Chenodeoxycholic Acid. 3 α, 7 α-dihydroxy-5 β-cholan-24-oic acid. It is the main component of the bile of hens, geese and other fowl, and occurs in small amounts in the bile of ox, guinea pig, bear, pig and human.

Chicken Pancreas Hormone. A polypeptide hormone which occurs in the amide form. It consists of 36 amino acid residues.

Chirality. The necessary and sufficient condition for optical activity (rotation of the plane of rotation of polarized light). It means "handedness" (from Greek, KEtρ=hand). Chiral molecules have no second-order symmetry element (center, plane or axis of symmetry) and exist in two mirror-image forms enantiomers) which cannot be rotated in such a way as to coincide.

Chitin. A nitrogen-containing polysaccharide which is a major component of the exoskeletons of arthropods. It consists of straight chains of N-acetyl-D-glucosamine residues linked β-1, 4.

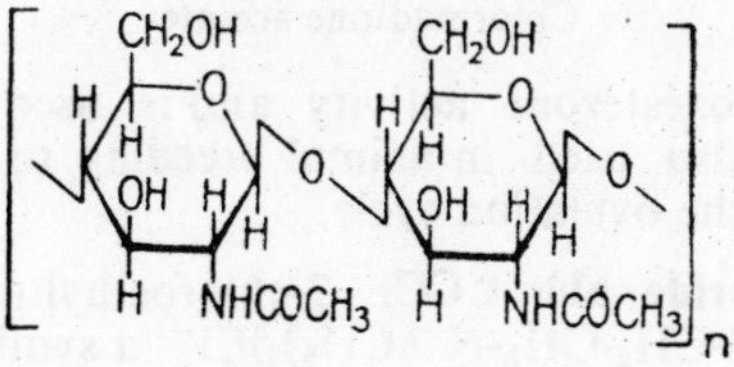

Chitin

Chloramphenicol. An antibiotic isolated from Streptomyces venezuelae. M_r 323. There are four different stereoisomers, of which only the D(-)-threo-C. shown here is antibiotic. It inhibits

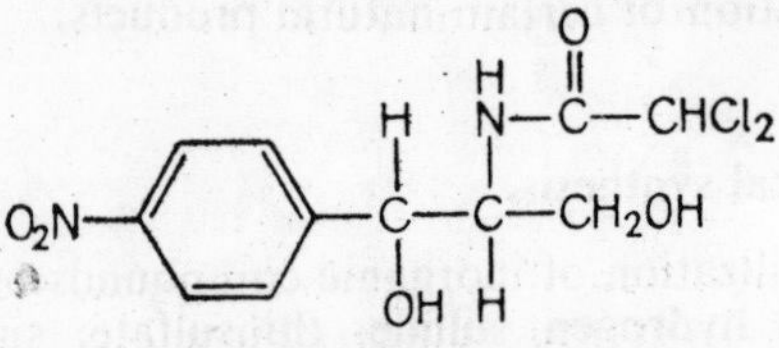

O
H ‖
H N—C—CHCl$_2$
O_2N—⟨ring⟩—C—C—CH_2OH
OH H

Chloramphenicol

protein synthesis on the 70S ribosomes of prokaryotes and also on the mitochondrial ribosomes of eukaryotic cells. Protein synthesis on the 80S ribosomes of eukaryotes is not affected. It inhibits the formation of peptide bonds and translocalization on the 50S subunit of the ribosomes, possibly through specific bonding to one of the ribosomal proteins involved in these reactions.

Chlorin. One of the basic ring structures of the porphyrins. Chlorin is 17, 18-dinphyoporphyrin.

Chlormadinone Acetate, 6-chloro-17 α-hydroxypregna-4, 6-diene-3, 20-dione Acetate. A synthetic gestagen. Administered orally, it

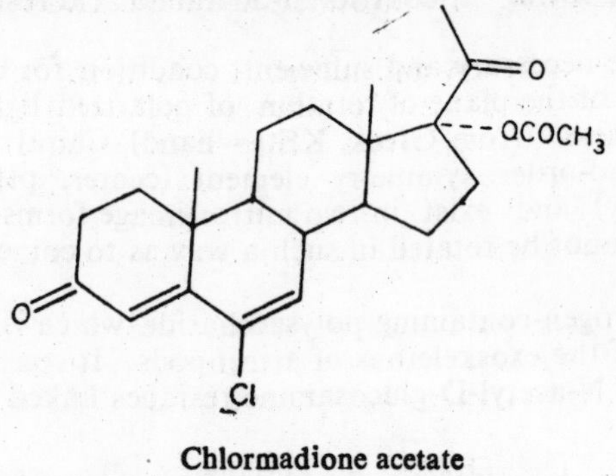

O
OCOCH$_3$
O
Cl

Chlormadione acetate

has high progesterone activity and is used in oral-contraceptives. It is also used in animal breeding to bring on heat and synchronize the ovulation cycle.

Colorocholine Chloride, abb. CCC. 2-chloroethyltrimethylammonium chloride, $(Cl\text{-}CH_2\text{-}CH_2\text{-}N^{-}(CH_3)_3)Cl^{-}$ a synthetic plant growth retardant. It causes a shortening and thickening of the straw in grains, and in used especially with wheat to prevent lodging. It is a gibberellin antagonist, and as such, it blocks the biosynthesis of gibberellins.

Chlorocruorin. A green respiratory proteins containing home(II) iron found in the hemolymphs of some marine annelids.

Chlorocruoroheme. The heme prosthetic group of the chlorocruorins, green respiratory pigments of certain invertebrates, such as the annelid Spirographs.

Chloroethyl Choline Chloride, abb. CCC, **2-chloroethyl Trimethyl Ammonium Chloride.** $[Cl\text{-}CH_2\text{-}CH_2\text{-}N^+(CH_3)_3]\ Cl^-$. A synthetic growth retardant used on grain to produce shorter, thicker straw and to prevent lodging.

2-Chloroethylphosphonic Acid, Ethrel. $Cl\text{-}CH_2\text{-}CH_2\text{-}PO(OH)_2$. A synthetic growth regulator. It is an ethylene generator; ethylene is a ripening hormone for fruit. It is used on Prums fruits to loosen the fruits, and on pineapple to stimulate blossoming.

Chlorophyll. The green photosynthetic pigments found in all higher plants within the Chloroplasts. They are magnesium complexes of tetrapyrroles and can be considered derivatives of protoporpyrin, a Porphyrin with two free or esterified carboxyl groups C. differ from other porphyrins in that 1) they have a saturated rather than a double bond between C atoms 7 and 8; 2) they have a pentanone ring carrying a methylated carboxyl group fused to ring III of the pyrrole; 3) atom 7 carries an esterfied propionic acid residue; in bacteriochlorophyll a and C. a, this is esterified to phytol $C_{20}H_{39}OH$ (Fig. 1), and in other bacteriochlorophylls, it is esterified to farnesol.

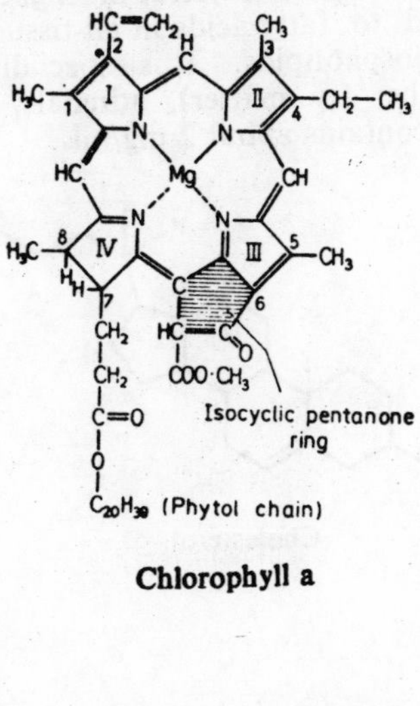

Chlorophyll a

Chlorophyllase (EC 3.1.1.4). A plant carboxy-esterase which catalyses the reversible transformation of chlorophyllides to chlorophyll in the last step of chlorophyll biosynthesis.

Chloroplasts. The photosynthesis organelles of higher plants. The whole chloroplast is lens-shaped and contains stacks of membranes (thylakoids) embedded in an aqueous phase (stroma) The chlorophyll is embedded in the thylakoids, and here the transformation of light energy to ATP and reductive equivalents (NADPH) occurs (both are required for reduction of CO_2.)

4-Chlorotestosterone : 4-chloro-17β-hydroxy·androst-4-en-3-one. An anabolic steroid obtained by partial synthesis from testosterone. It is used therapeutically as an ester, for example in convalescence, inflammation and tumors.

Chlorotriazine Dyes. The dyes containing chlorotriazinyl groups. They react readily with polysaccharide matrices under alkaline conditions, forming stable dyed products.

Cholanoic Acid: 5β-cholan-24-oic acid. See Bile acids.

Cholecalciferol. Vitamin D_3

Cholecystokinin, Pancreozymin. A tissue hormone consisting of 33 amino acids (porcine) with a M_r of 3838. It is formed in the mucosa of the upper intestine.

Cholestanol: 5-α-cholestan-3β-ol. A zoosterol. It is a 5, 6-dihydro-derivative of cholesterol and occurs in small amounts with cholesterol in animal cells.

Cholesterol. The most important sterol in higher animals. Found free or esterified to fatty acids in all tissues of the body, often together with phospholipids. It is especially abundant in brain (about 10% of the dry matter), adrenals, egg yolk and wool grease. Blood contains about 2 mg/ml.

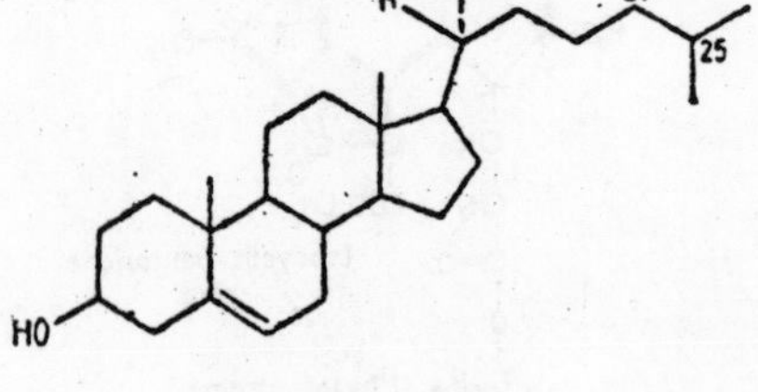

Cholesterol

It is synthesized ultimately from acetyl-coenzyme A, via the triterpene lanosterol and zymosterol. In turn, it is a key intermediate in the biosynthesis of many other steroids, including steroid hormones, steroid sapogenins and steroid alkaloids.

Cholic Acid: 3α, 7α, 12α-trihydroxy-5β-cholan2·4-oic acid. A bile acid found in conjugation with lysine or taurine in the bile of most vertebrates. It is used as starting material for the partial synthesis of therapeutically important steriod hormones.

Cholic Acid

Choline : $[(CH_3)_3N^+\text{-}CH_2\text{-}CH_2\text{-}OH]OH$—m.p. 180°C (d). As the natural hydrolysis product of lecithin, it is found in many plants and animals, in the brain, egg yolk, hops, Belladonna and Strophanthus. It is a methylating agent in metabolic processes. It reduces the deposition of body fat, lowers the blood pressure and causes the uterus to contract. It is required for the resynthesis of methionine from homocysteine.

Cholinesterase, Pseudocholinesterase, (EC 3.1.1.8). An unspecific acylcholinesterase which hydrolyses butyroyl and propionoyl choline much faster than acetylcholine. C is found primarily in the serum the liver and pancreas. It is also found in cobra venom.

Chondroitin Sulfate. A water-soluble mucopoly-saccharide found in animals. Chondroitin A and C consist of equimolar amounts of D-glucuronic acid and N-acetyl-D galactosamine linked in alternating β-1, 3 and β-1, 4 bonds; they differ in the position of the sulfate ester.

Chondrome. The genetic information contained in the mitochondria of a cell.

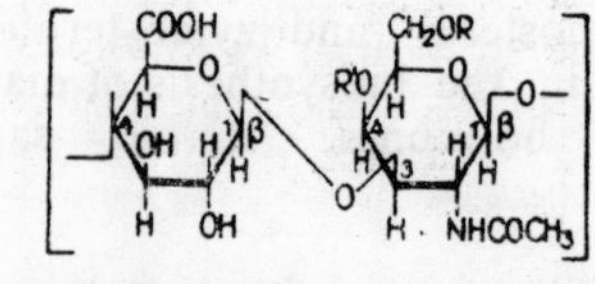

Chondroitin sulfate A R = H, R' = SO_3H
C R = SO_3H, R' = H

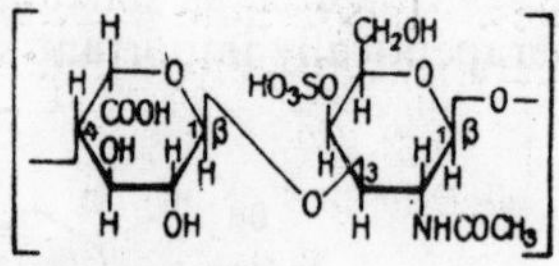

Chondroitin sulfate B

Choriogonadotropin, Placental Gonadotropin, Human Chorionic Gonadotropin, abb. **hCG.** The most important hormone formed in the placenta during pregnancy. It is a glycoprotein containing about 30% carbohydrate and an α-and a β-polypeptide chain of 92 and 139 amino acids, respectively.

Choriomammotropin Placentalactogen, Abb. PL, Human Lactogen. A single chain polypeptide hormone of known primary structure (191 amino acid residues, M_r 22308). It is synthesized in increasing amounts by the placenta during pregnancy, and secreted into the maternal circulation. Its action is similar to that of Somatotropin.

Chromatin. The stainable material of the interphase nucleus, consisting of the uncoiled chromosomes (DNA).

Chromatography. A type of method used for analytical or preparative separation of mixtures of compounds. It has revolutionized organic and biological chemistry by making possible separations of closely related compounds which could not be resolved by any other method.

Definition of terms used in chromatography

Term	Definition
Equilibration	Saturation of the paper to be used with the vapor of the elution solvent.
Detection	Visualization of the separated material : staining of spots or bands, UV or light absorption, radioactivity, etc.

Elution	Washing out of the components
Eluant	The material which has been eluted
Eluted	A solution emerging from a chromatography column
Developer, Eluent, or Elutant	The mobile (usually liquid) phase
Development	The process of chromatography, or, treatment with the reagent(s) used for detection
Solvent	Either a pure liquid or a mixture of solvents
Front	The position of the leading edge of the solvent
Running Time	The time during which C occurs
Standard substance	A pure, authentic substance used for identification

Chromatophores

(1) In botany, plastids: Chloroplasts Chromoplasts and Leuçoplasts.

(2) The photosynthetic organelle of the Photosynthetic bacteria.

Chromium, Cr. An essential dietary constituent for animals. Cr is known to be a constituent of glucose tolerance factor (GTF), a water-soluble, relatively stable organic complex of Cr, with M_r about 500, which is essential in animals and humans for normal glucose tolerance. The earliest symptom of Cr deficiency is impaired glucose tolerance. More severe deficiency leads to glycosuria, fasting hyperglycemia, impaired growth, shortening of life span, cases of diabetes refractory to insulin may be the result of Cr deficiency.

Chromoplast. A chromatophore filled with carotenoids, and therefore red-orange to yellow in color.

Chromoproteins. The proteins which contain a colored prosthetic group bound either covalently or noncovalently.

Chromosomes. Carriers of the genetic information in the cell. All chromosomes contain the cell's genes in linear array on a single DNA molecule. They can be exactly duplicated, and the information can be made available for use in the cell (transcription).

Chymopapain (EC 3.4.22 6). An enzyme from the latex of the papaya tree. It is made up of several equal sized segments (C.A and B, M_r 35000).

Chymotropic Pigment. A pigment dissolved in the vacuole of a plant cell.

Chymotrypsin. A family of structurally and catalytically homologous serine proteases formed and stored in the pancreas in the form of precursors.

CLMP. Abb for cyclic inosine 3′, 5′-monophosphate

Cinchona Alkaloids. A group of about 30 alkaloids from the bark of tropical trees, especially Cinchona succiruba. They are included among the indole alkaloids on account of their precursors, although the main representatives have a quinoline structure. Cinchona, the dried bark, contains 7 to 10% alkaloids. The main alkaloid is Quinine which makes up 5 to 7%, secondary alkaloids include quinidine, epiquinine and epiquinidine Other important C a. are cinchonine, and its stereoisomers, and cinchonamine.

Cinchonamine. See Cinchona alkaloids.

Cinchonine. See Cinchona alkaloids.

1, 8-Cineole, Eucalyptol. The main constituent oil of oil of eucalyptus. It finds use in cough syrup.

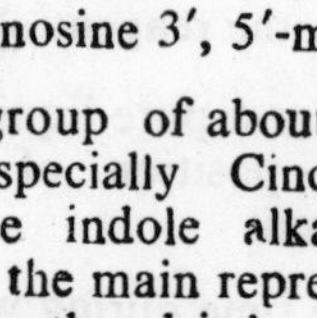

1, 8-Cineole

Cistron. A section of DNA which codes for the amino acid sequence of a polypeptide chain. It thus corresponds to a gene.

Citral. A doubly unsaturated monoterpene aldehyde. A mixture of the cis and trans isomers is a component of many essential oil. Fig. on next page

Citrate (si)-synthase, Citrate Condensing Enzyme, Citrogenase (EC 4.1.3.7). The tricarboxylic-acid-cycle enzyme which catalyses the synthesis of citrate from oxaloacetate and acetyl-coenzyme A in an aldol condensation.

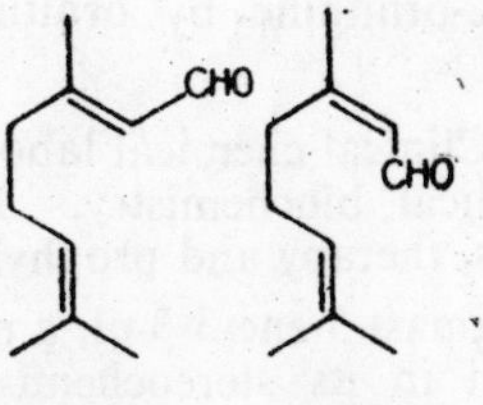

***trans*-Citral** ***cis*-Citral**

Citric Acid. A key metabolic intermediate, m.p. 153—155°C. Citrate is the starting point of the tricarboxylic acid cycle. Its concentration also coordinates several other metabolic pathways.

Citronellal. An unsaturated monoterpene aldehyde found in both optical isomers and the isopropylidene form.

(+)-Citronellal

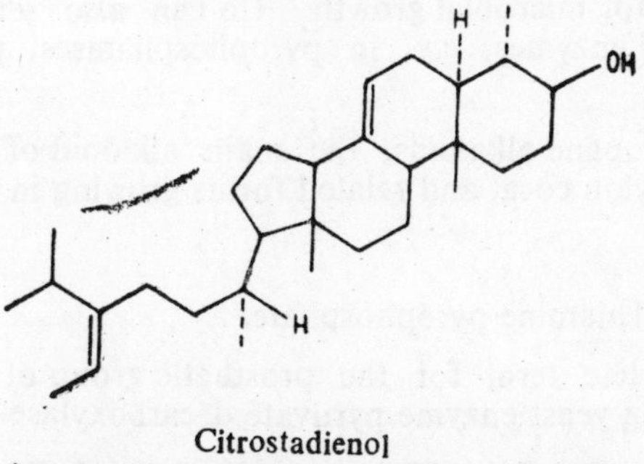

Citrostadienol

Citrostadienol. 4α-methyl-5α-stigmasta-7, 24(28)-diene-3B-ol, a phytosterol. It is found in the oils of graperfruit and orange peels. It is also considered one of the tetracyclic triterpenes and is ar intermediate in the synthesis of sterols.

l-Citrulline: N_5-(aminocarbonyl)-L-ornithine, α-amino-δ-ureidovaleric acid, H_2N-CO-NH-$(CH_2)_3$-CH(NH_2)-COOH. A nonprotein amino acid. It is synthesized in the liver from carbamoyl phosphate and L-ornithine by ornithine carbamoyltransferase (EC 2.1 3.3).

Clinical Chemistry. Clinical chemical laboratory diagnostics, part of clinical and medical biochemistry. Their results are important for the diagnosis, therapy and prophylaxis of diseases.

Clionasterol: (24S)-stigmast-5-ene-3-β-ol, a marine zoosterol. It differs from β-sitosterol in its stereochemistry at C atom 24. It is found in sponges, for example Cliona celata and Spongilla lacustris.

Clostripain (EC 3 4 22 8): SH-dependent, trypsin-like protease (M_r 50000) with endopeptidase and amidase-esterase activity isolated from culture filtrates of Clostridium histolyticum.

CMP. Abb. for cytidine 5′ monophosphate.

CoA, CoA-SH. Abb. for coenzyme A.

Coagulation Factors. See Blood coagulation.

Coagulation Vitamin. Absolute term for vitamin K.

Cobalt Co. An important bioelement which is present in traces in plants, animals and microorganisms. The ligands in cobalt complexes in living cells are often the corrin ring system of vitamin B_{12}, benzimidazole or sugar components. As a component of certain coenzymes, Co is also required for the symbiotic fixation of atmospheric nitrogen. Traces of Co are required for microbial growth. Co can also serve as cofactor of several enzymes, *e.g.* in pyrophosphatases, peptidases and arginase.

Cocaine. A tropane alkaloids, the main alkaloid of the coca plant, Erythroxylon coca, and related forms growing in the tropics.

Cocarboxylase

(1) See Thiamine pyrophosphate;

(2) obsolete term for the prosthetic group of the decarboxylating yeast enzyme pyruvate decarboxylase.

CO_2-compensation point. The concentration of CO_2 at which the rate of photosynthesis (CO_2 incorporation) and the rate of respiration (CO_2 production) are balanced. The value varies with illumination and must be quoted for a given light intensity.

Codeine. An opium alkaloid. Opium consists of about 40% codeine but it is found in other poppy species as well. M_r 299.37, m.p. 154 to 156°C $[\alpha]_D^{20}$ −137.7° (alcohol). Codeine is morphine 3-methyl ether, and is converted to morphine as the poppy ripens.

Codogenic Strand. The strand of a DNA double helix from which the genetic information is transcribed onto RNA.

Codon, Code Triplet. The linear sequence of three adjacent nucleotides in RNA which specify a particular amino acid.

Coenzyme. In the narrow sense, the dissociable, low-molecular-weight active group of an enzyme which transfers chemical groups (Group transfer) or hydrogen or electrons. Coenzymes in this sense couple two otherwise independent reactions, and

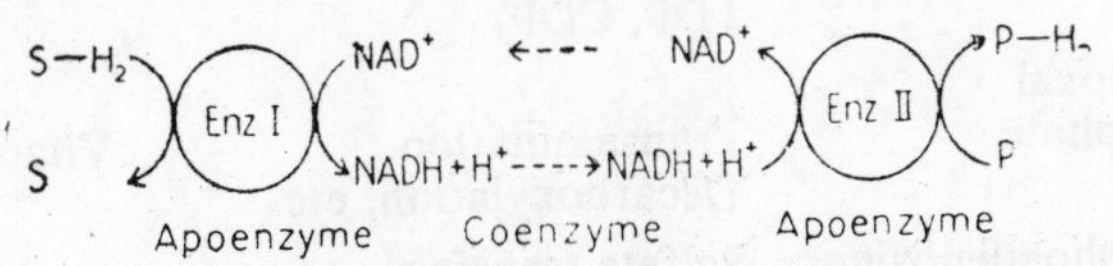

S—H₂ → S (Enz I) ; NAD⁺ ←--- NAD⁺ ; NADH + H⁺ ----→ NADH + H⁺ ; P → P—H₂ (Enz II)

Apoenzyme Coenzyme Apoenzyme

can thus be regarded as transport metabolites. In a wider sense, a coenzymes can be thought of as any catalytically active, low-molecular-weight component of an enzyme.

TABLE

Classification, metabolic function and source of the coenzymes

Coenzyme	Function	Vitamin source
(1) Oxidoreduction coenzymes		
NAD	Hydrogen and electron transport	Nicotinic acid
NADP	Hydrogen and electron transport	Nicotinic acid
FMN	Hydrogen and electron transport	Riboflavin
FAD	Hydrogen and electron transport	Riboflavin
Ubiquinone (Coenzyme(Q)	Hydrogen and electron transport	—
Lipoic acid	Hydrogen and acyl transfer	—

Heme coenzymes (Cytochromes)	Electron transport	—
Ferredoxins	Electron transport Hydrogen activation	— —
Thioredoxins	Hydrogen transport	
(2) Group transfer coenzymes		
Nucleoside diphosphates	Transfer of phosphorylcholine (CDP) and sugars (UDP, GDP, TDP, CDP)	
Pyridoxal phosphate	Transamination, Decarboxylation, etc.	Vitamin B_6
Phosphoadenosine phosphosulfate	Sulfate transfer	
Adenosine triphosphate	Phosphorylation, Pyrophosphorylation, Transfer of adenosyl group	
S-Adenosyl-L-methionine	Transmethylation	(Methionine)
Tetrahydrofolic acid and conjugates	Transfer of formyl, hydroxymethyl and methyl group	Folic acid
Biotin or CO_2-biotin enzymes	Carboxylation, transcarboxylation, decarboxylation	Biotin
Coenzyme A	Transacylation, etc.	Pantothenic acid
Thiamine pyrophosphate	C_2-group transfer, rarely, C_1-group transfer	Thiamine (Aneurin)
(3) Isomerization coenzymes		
Coenzyme form of vitamin B_{12}	Carboxyl shifts	Vitamin B_{12}
Uridine diphosphate	Sugar isomerization	

Coenzyme I. Nicotinamide adenine dinucleotide.

Coenzyme II. Nicotinamide adenine dinucleotic phosphate.

Coenzyme A. Abb. CoA, also CoA-SH: the coenzyme of acylation. CoA consists of adenosine 3′, 5′-diphosphate linked via the 5′-phosphate, to the phosphate of pantotheine 4′-phosphate. (Fig. 1). The Thiol group of the cysteamine is responsible for

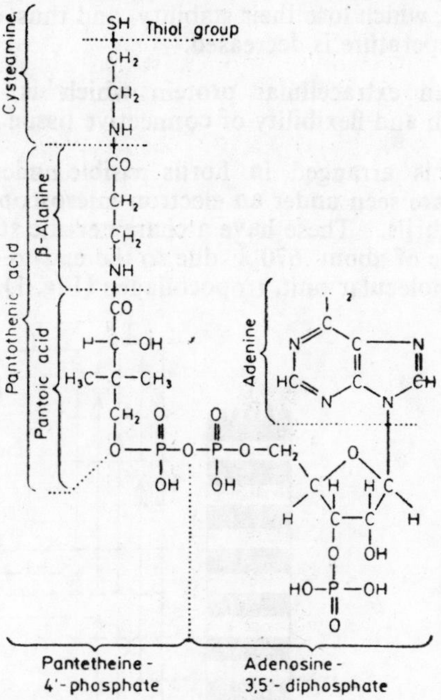

Coenzyme A

the biological activity of CoA. The metabolic significance of CoA rests on its ability to form high-energy thioester bonds.

Colchicine. The most important of the Colchicum alkaloids. It is extracted from Colchicum autumnale L. and is used mainly in plant genetics to inhibit mitosis and induce polyploidy. In small doses, it relieves pain and suppresses inflammation, but it is highly toxic, 20 mg being a lethal dose. It has been used against neoplastic growth.

Colchicum Alkaloid. A group of Isoquinoline alkaloids in which the nitrogen is present as a substituted amino group on a tricyclic skeleton.

These are biosynthesized via a 1-phenylethylisoquinoline alkaloid from which androcymbin is formed by hydroxylation, methoxylation, attack of phenol oxidases and oxidative couplings. Androcymbin is then converted to demecolcine and further to colchicine.

Cold-sensitive Enzymes. A group of oligomeri (consisting of several polypeptide chains) enzymes, of which about 25 are presently known, which lose their stability and thus enzymatic activity as the temperature is decreased.

Collagen. An extracellular protein which is responsible for the strength and flexibility of connective tissue.

It is arranged in fibrils visible under a light microscope, which are seen under an electron microscope to be composed of microfibrils. These have a characteristic striation with a repeat distance of about 670 Å, due to the end-to-end alignment of the basic molecular unit, tropocollagen (Fig. 1).

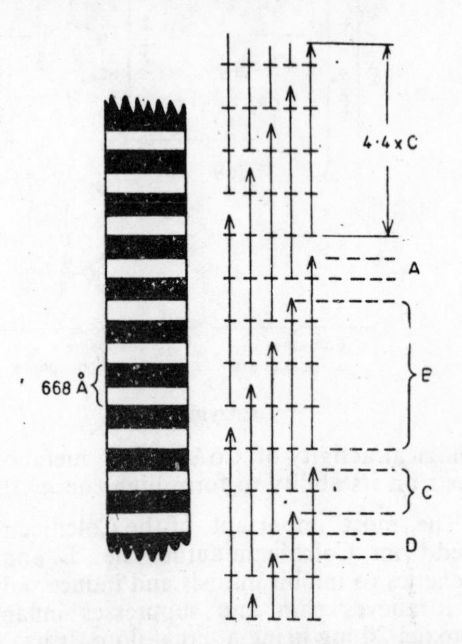

Schematic structure of the collagen microfibril

A : a region of short overlap. B : a long overlap region. C : an overlap region corresponding to one hole zone and one region of short overlap, and giving rise to the distance of 668 Å on the banded structure of the microfibril. D : a hole zone. Each single arrow (length 4.4 × C) represents a tropocollagen molecule.

Collagen is synthesized in fibroblasts as a precrusor, procollagen, which also consists of three chains.

Collagenase. A proteolytic enzyme, the only enzyme capable of degrading native collagen to soluble, low-molecular weight peptides.

Colophony. The residue from turpentine distillation.

Column Chromatography. A chromatographic separation method, in which the carrier material is packed in the form of a column inside a tube (usually of glass, and known as a chromatography column).

Commensalism The close spatial coexistence of two organisms of different species which is neither beneficial for the partners (commensals), as in symbiosis, nor detrimental to either partner, as in parasitism.

Compartment. A geometrically bounded portion of the cell which is structurally or biochemically separate from the rest of the cell space.

Competitive Inhibition. See effectors.

Complementary DNA, cDNA. DNA complementary to a mRNA. It is prepared in the laboratory as a probe for hybridization studies by incubating the mRNA with dATP, dGTP, dTTP and dCTP in the presence of a reverse transcriptase (RNA-dependent DNA-polymerase).

Complementary Structure. Two structures which define one another, for example, the two polynucleotide chains in the double helix of DNA. The base pairs adenine and thymine (or uracil, in RNA) and guanine and cytosine are complementary, which results in fact that the nucleotide sequence in one polynucleotide chain defines a unique sequence in the complementary strand through the formation of base pairs.

Complement Binding Reaction. The binding of the Cl component of complement to the Fc fragments of antibodies which are bound to the surface antigens of erythrocytes or bacteria. The complement system is activated by this step.

Complement System. A system in vertebrate serum consisting of at least nine protein components (C 1 to C 9) which, after activation by antibodies bound to the surface of erythrocytes, bacteria or protozoa, dissolves these cells by perforation of their membranes (hemolysis or cytolysis).

Concentration Variables, Fundamental Variables, Primary Variables. Those substances in an enzymatic system whose concentrations can be directly controlled by the experimenter, for example the substrates, products and effectors.

Coniine, 2-propylpiperidine. The most important of the Conium alkaloids. It is the toxic-principle of the poison hemlock, Conium maculatum. It is extremely poisonous and can cause death in humans at a dose of 0 5 to 1 g. The poison hemlock was used in ancient Athens to put Socrates to death.

Conium Alkaloids. A group of simple piperidine alkaloids found only in poison hemlock, Conium maculatum. The main alkaloids are Coniine and γ-coniceine (M_r 12 .22, b.p. 168 °C); the secondary alkaloids are N-methyl and hydroxy derivatives of coniine.

Constitutive Enzymes. The enzymes which, in contrast to the inducible enzymes, are constantly produced by the cell, irrespective of the growth conditions.

Cooperative Oligomeric Enzymes, Allosteric Enzymes. The enzymes composed of several subunits and displaying cooperativity.

Cooperativity. This is displayed by oligomeric or monomeric enzymes which possess more than one binding site for a certain ligand. The cooperative binding may be negative or positive, and it may occur for the same ligand (homotropic cooperativity) or for a different ligand (heterotropic cooperativity).

Cooperativity Model. A functional and structural model of cooperative oligomeric enzymes which is intended to describe and explain the Cooperativity in the turnover of substrates or binding of effectors. This can be used to derive binding potentials and equations which can be compared with the experimental data.

Copolymer. A polymeric molecule containing more than one kind of monomer unit. An example would be a synthetic polynucleotide obtained by incubation of two or more nucleoside di- or triphosphates with the appropriate polymerase. Using RNA polymerase, Khorana was able to synthesize C. according to the following scheme (the choice of starter molecule was arbitrary and could be varied according to need):

RNA polymerase AUGAUGAUG...

Copper Cu. An important bioelement in plants and animals, which is frequently involved in electron transport processes in membranes and particles (mitochondria). In plants, Cu is necessary for the synthesis of chlorophyll and is a component of a number of enzymes, such as cytochrome oxidase, ascorbic acid oxidase, tyrosinase and monophenol monoxygenase.

The blood contains a number of Copper proteins. The synthesis of hemoglobin is dependent on Cu, although it does not contain the metal.

Copper Proteins. Metalloproteins, often blue in color, which usually contain a mixture of mono- and mostly divalent copper in their molecules.

Coprogen. A Siderochrome synthesized by Penicillium and Neurospora spp.

Coprostane. Obsolete term for 5β-cholestane steroids.

Coprostanol. 5β-cholestan-3β-ol. A sterol alcohol. It is the main sterol in the feces, where it arises by reduction of cholesterol by intestinal bacteria.

Cordycepin, 3′-deoxyadenosine, Adenine 9-cordyceposide. A purine antibiotic synthesized by Cordyceps militaris and Aspergillus nidulans Nucleoside antibiotics). As an antimetabolite of adenosine, it inhibits purine biosynthesis, but is not very toxic. It can be phosphorylated to the monophosphate.

Core Particles. The particles released from Chromatin by partial enzymatic digestion. They contain about 140 base pairs of DNA and 2 molecules each of the "inner histones" H2A, H2B, H3 and H4.

Corrinoids. The chemical compounds based on the corrin ring, which is similar to the porphyrin ring. It consists of four pyrrole rings linked in a large ring. Three of the links between pyrroles are methenyl groups; the fourth is a direct bond between the two pyrroles. The pyrroles carry acetate, propionate and methyl substituents.

Cortexolone, Reichstein's Substance S. 17α-hydroxy-11-deoxy-corticosterone, 17α, 21-dihydroxypregn-4-ene-4, 20-dione. A mineralocorticoid hormone from the adrenal cortex. It differs structurally from Corticosterone in that the 11β-hydroxyl group is lacking, and there is an hydroxyl group at the 17α position. It is an intermediate in the biosynthesis of cortisol and cortisone from progesterone.

Cortexone, Reichstein's Substance Q, 11-deoxycorticosterone abb. DOC. 21-hydroxypregn-4-ene-3, 20-ditone. A mineralocorticoid hormone from the adrenal cortex. Its acetetate or glucoside is used in treatment of Addison's disease and of stock. It is biosynthesized from progesterone, which can also be convert into C. by partial synthesis, or microbiologically.

Corticosterone Reichstein's Substance H, Kendall's Substance B. 11β-21-dihydroxypregn-4-ene-3, 20-dione. An adrenal cortex hormone and a glucocorticoid. It is biosynthesized from progesterone, which is hydroxylated first to cortexone and then in the 11 position to C.

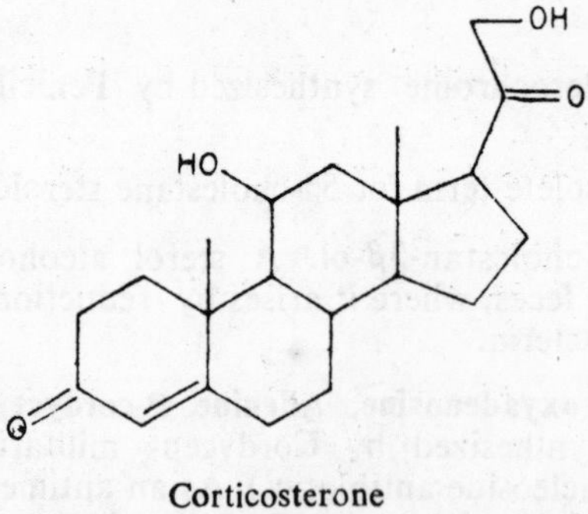

Corticosterone

Corticotropin, Adrenocorticotropin, Adrenocorticotropic Hormone, abb. ACTH. A polypeptide hormone secreted by the pituitary. The primary structure of the human hormone is Ser Tyr-Ser-Met-Glu-His-phe-Arg-Trp-Gly-Lys-Pro-Val-Gly-Lys-Lys-Arg-Arg-Pro-Val-Lys-Val-Tyr-Pro-Asn-Gly-Ala-Glu-Asp-Glu-Leu-Glu-Phe.

Cortisol, Reichstein's Substance M. Kendal's Substance F. 11£, 17α, 21-trihydroxypregn-4-ene-3, 20-dione. It is a glucocorticoid

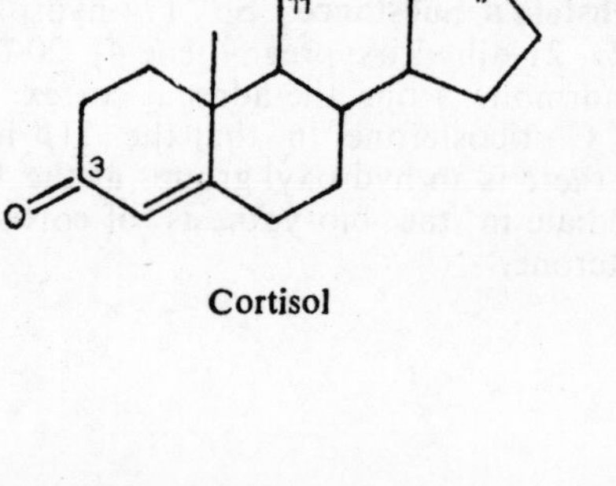

Cortisol

hormone from the adrenal cortex. It is found in the blood, bound to the protein transcortin, and in the urine of all mammals, including humans. Like cortisone, it stimulates the conversion of proteins to carbohydrates, increases the blood sugar level and promotes glycogen storage in the liver. It is biosynthesized from progesterone via 17 α-progesterone and cortexolone.

Cortisone, Reichstein's Substance F, Kendall's Substance E. 11-dehydro-17 α-hydroxycorticosterone, 17 α, 21-dihydroxypregn-4-ene-3, 11, 20-trione. A glucocorticoid hormone from the adrenal cortex. Like cortisol, it stimulates the formation of carbohydrate from proteins, promotes glycogen storage in the liver and raises the blood sugar level. It is biosynthesized from cortisol by enzymatic dehydrogenation of the 11 β-hydroxyl group. It can be obtained by partial synthesis from other pregnase compounds, bile acids and steroid sapogenins.

Cosubstrate. A Coenzyme which enters an enzymatic reaction as a second substrate.

Coumarin. o-hydroxycinnamic acid. A pleasant-smelling compound found in woodruff, sweet clover, tonka beans, lavender oil and many other plants. It acts as an inhibitor in many biotests.

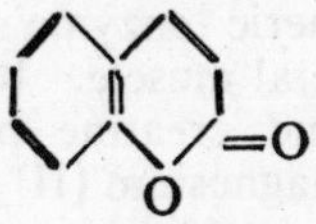

=O
O

Coumarin

Coupling Factors. See Respiratory chain.

Cozymase. See Nicotinamide adenine dinucleotide.

C_3 plants, Calvin Plants. The green plants which produce C_3 compounds as the first product of CO_2 fixation. The reaction is catalysed by the ribulose 1, 5-bisphosphate carboxylase (EC 4.1.1.39): ribulose 1, 5-bisphosphate+$CO_2 \rightarrow$2 (3-phosphoglycerate). The products of the light reaction, ATP and NADPH, are used to reduce the 3-phosphoglycerate to 3-phosphoglyceraldehyde which enters the Calvin cycle. The vast majority of green plants are C_3 plants.

C_4 Plants. Green plants which fix CO_2 into 4-carbon compounds, such as oxaloacetate, aspartate and malate. The carboxylase enzyme is phosphoenolpyruvate carboxylase (EC 4.1.1.31) which carboxylates phosphoenolpyruvate to oxaloacetic acid (Hatch-Slack-Kortschak cycle). These plants also have the Calvin cycle; the C_4 compounds are decarboxylated to supply CO_2 for the Calvin cycle in the cells of the vascular boundles. The C_4 reactions are found in many tropical plants and in unrelated dicotyledons which are adapted to a hot, dry climate.

Crabtree Effect. A reduction in the rate of respiration after addition of glucose. The cause is the increased intracellular glucose 6-phosphate level, which inhibits hexokinase, the first enzyme in glycolysis. Glucose also inhibits the enzymes of the tricarboxylic acid cycle and the respiratoy chain.

Creatine. β-methylguanidoacetic acid, methyl-glycocyamine, N-amidosarcosine,

$$H_2N-\underset{\underset{NH}{\|}}{C}-N(CH_3)-CH_2-COOH$$

a biochemically important amino acid derivative. It is very easily converted to creatinine. Over 90% of the creatine in an adult human is localized in muscle. The concentration is particularly high when a large amount of chemical energy is being converted to mechanical energy.

Creatine Kinase, Creatine Phosphokinase. A phosphotransferase, EC 2.7.3.2. C.k. is a dimeric enzyme of M_r 82000 found specifically in heart and skeletal muscle. It catalyses the formation of ATP from ADP and creatine phosphate in a reversible reaction dependent on magnesium (II) or manganese (II). When ATP is present in excess, creatine kinase makes possible to storage of the chemical energy of the ATP in the energy-rich creatine ATP⇌creatine phosphate+ADP. When there is damage to the skeletal or heart musculature, the level of creatine kinase in the serum is increased. It is therefore used in the early diagnosis of heart infraction, the detection of muscle wastage and to distinguish heart infraction from a lung embolism, in which there is no increase in the creatine kinase concentration in the serum.

CRH. Abb, for corticotropin releasing hormone.

Crocetin. A brick-red C_{20} dicarboxylic acid with seven conjugated trans double bonds, 4 methyl branches and two terminal carboxyl groups. It is an apocarotenoid, an oxidatian product of carotenoids.

Cross-link. A covalent bond formed between two polymers. Cross-links may be artificially introduced by addition of bifunctional alkylating agents to molecules, such as DNA or soluble proteins, which are not naturally cross-linked. They are formed enzymatically in insoluble structural proteins and in the cell walls of bacteria.

C-toxiferin I : calabash toxiferin l, Curare alkaloids.

CTP. Abb. for cytidine 5′-triphosphate.

Cryptoxanthin. (3R)-β, β-caroten-3-ol. A carotenoid belonging to the group of xanthophyils. It is a commonpigment in plants and is found especially in fruits and berries, *e.g.*, bell peppers, oranges, tangerines and papayas, as well as in maize and egg yolk. As a provitamin A, it is half as effective as β-carotene.

Crystallins. The soluble proteins that account for almost 90% of total vertebrate lens protein. Mammalian crystallins are classified into three major groups, α, β, and γ, originally on the basis of their precipitibility, but now more commonly according to their M_r range as determined by gel filtration.

Cucurbitacins. A group of tetracyclic triterpenes found as glycosides in Cucurbitaceae and Cruciferae. These toxic, bitter compounds are structurally related to the parent hydrocarbon cucurbitane, 19 (10-9β) abeo-5α-lanostane, which differs from lanostane (Lanosterol) in the formal shift of the 10 methyl group to the 9β-position. Cucurbitacin E was isolated in 1831 in crystalline form under the name elaterin.

Cultivation of Microorganisms, Culture Techniques. Deliberate propagation of an organism. Cultivation of an organism may be discontinuous (static) or continuous.

CUMP. Abb. for cyclic uridine 3′, 5′-monophosphate.

Curare Alkaloids. The toxic principles of arrow and bait poisons used by South America Indians.

Calabash curare is obtained from plants of the Strychnos genus and is packed in hollow gourds. It contains a large number of alkaloids and is extremely poisonous. The alkaloids are derived from indole, and can be divided into types, the yohimbine type (*e.g.*, mavacurine), the strychnine type (*e.g.* Wieland-Gumlich aldehyde and the bisindole type (*e.g.* calabash toxiferin-1).

These alkaloids interrupt nervous impulses at the end plates of motor nerves by displacing acetylcholine, which leads to paralysis of the striated muscles.

Curcumin, Tumberic Yellow, Diferuloymethane. A yellow pigment from the roots and pods of Curcuma longa L., which is cultivated in Southeast Asia. m.p. 183°C. The dried root is used as medicine for liver and bile ailments, and is a component of curry powder. It is used as a food color, as a dye for textiles and as an indicator (curcumin-boric acid paper). It has been shown that phenylalanine and acetate/malonate are specific biosynthetic precursors.

Curcumin

Cyanides. Salts of hydrogen cyanide (HCN) with the general formula Me′CN. The soluble cyanides are hydrolysed in water to the metal hydroxide and hydrogen cyanide, and are thus highly poisonous. HCN and alkali cyanides are often components of insecticides and other biocides.

Cyanidin. 3,5,7,3′,4′-pentahydroxyflavylium [illegible] cation, the aglycon of many Anthocyanins. Glycosides of cyanidin and a few acylated derivatives are found in many plants; the oxonium salts are responsible for the deep red color of many flowers and fruits, such as red roses, geraniums, tulips, poppies and zinnias. Chelates with iron (III) or aluminium ions are deep blue in color. When bound to a polysaccharide carrier, they form chromosaccharides, such as protocyanin, the blue pigment of cornflowers.

Cyanocobalamin. One of the B_{12} vitamins.

Cyanogenic Glycosides. A group of O-glycosides formed from decarboxylated amino acids. The cyano group arises from the α-C atom and the amino group. Hydrocyanic acid is generated from cyanogenic glycosides by β-glucosidases (such as emulsin) and oxinitrilases.

Cyasterone. A phytoecdysone. It has been isolated from various plants, including Cyathula capitata, family Amaranthaceae, and Ajuga decumbens, family Labiatae.

Cyclic. N^6,O^2-dibutyryladenosine 3′,5′-monophosphate, DBCAMP: A synthetic derivative of cAMP.

Cyclitols. The compounds derived from 1,2,3,4.5,6- hexahydroxycyclohexane. They have the same chemical formula, $C_6H_{12}O_6$, as the hexoses, and are biosynthetically related to them, but they have an isocyclic rather than a heterocyclic ring.

Cycloalkanes. The saturated hydrocarbon rings with the general formula C_nH_{2n}. The tendency to form and the stability of the rings depends on the number of carbon atoms, with 5 and 6-membered rings being most stable and most easily formed.

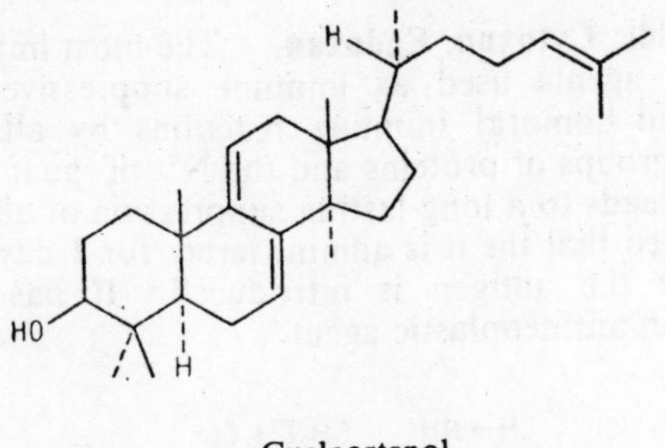

Cycloartenol

These are component of many natural products, especially the cyclic terpenes, in which derivatives of cyclobutane and cyclopropane are also found.

Cyclo-AMP. Abbreviation for cyclic adenoisine 3′,5′-monophosphate (Adenosine phosphates).

Cycloartenol. 9, 19-cyclo-5α, 9B-lanost-24-en-3β-ol. A tetracyclic triterpene alcohol. It is found in many plants, including the latex of Euphorbiacease, the potato (Solanum tuberosum) and nux vomica (Strychnos nuz vomica). It is biosynthesized from squalene via 2,3-epoxysqualene.

Cyclobuxamine H. A Buxus alkaloid.

Cyclo·GMP. Abbreviation for cyclic guanosine 3′5′-monophosphate.

Cycloheximide, Actidione. An antibiotic isolated from Streptomyces griseus. It is chemically a derivative of glutarimide and is water soluble. It inhibits protein biosynthesis on 80S ribosomes in eukaryotes by preventing the initiation and elongation reactions. The latter increases the number of monosomes

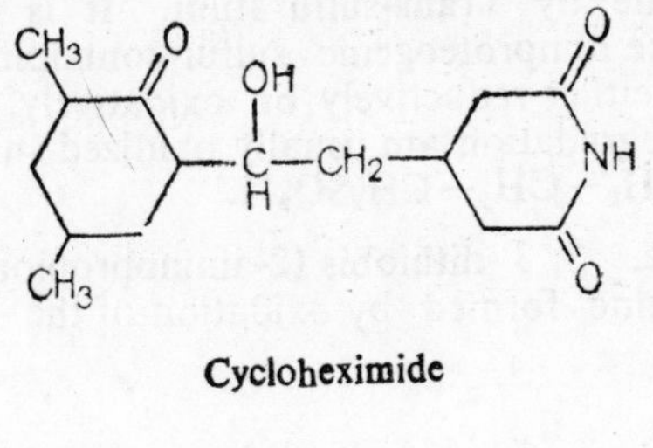

Cycloheximide

among the polyribosomes. It is used as a fungicide to control cherry leaf spot, turf diseases and rose powdery mildew.

Cyclo-IMP. Abbreviation for cyclic inosine 3′, 5′-monophasphte.

Cyclopentanoperhydrophenanthrene. See Steroids.

Cyclophosphamide, Cytoxan, Endoxan. The most important of the alkylating agents used as immune suppressives. It inhibits cellular and humoral immune reactions by alkylating the SH and NH_2 groups of proteins and the N^7 of guanine in nucleic acids. It leads to a long-lasting suppression of antibody synthesis, provided that the it is administered for 1 day before and 15 days after the antigen is introduced. It has also been proposed as an antineoplastic agent.

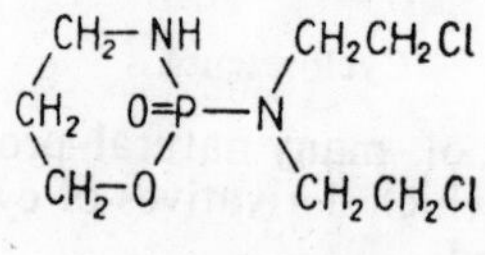

```
 CH2—NH       CH2CH2Cl
 /      \     /
CH2    O=P—N
 \      /     \
 CH2—O        CH2CH2Cl
```

Cyclophosphamide

Cyclo-UMP. Abbreviation for cyclic uridine 3′, 5′ -monophosphate.

Cyd. Abbreviation for Cytidine.

Cyproterone Acetate. 17α-acetoxy-6-chloro-1α, 2α-methylenepregna-4, 6-diene-3, 20-dione. An antiandrogen steroid derived from pregnane. It counteracts the effects of the male sex hormone testosterone, and is used in cases of hypersexuality ("hormonal castration").

Cys. Abbreviation for L-Cysteine.

L-Cysteine. Abbreviation **Cys.** L-2-amino-3-mercapto-propionic acid, HS-CH_2-CH(NH_2)-COOH. A sulfur-containing, proteogenic amino acid which is the central compound in Sulfur metabolism.

Cys is synthesized in the course of Sulfate assimilation or from Methionine by Trans-sulfuration. It is presumably the precursor of the nonproteogenic sulfur-containing amino acids. It is degraded either reductively or oxidatively. The end products of Cys degradation are usually oxidized sulfur compounds like taurine, $NH_2-CH_2-CH_2SO_3H$.

L-Cystine, Dicysteine. 3, 3'-dithiobis (2-aminopropionic acid). The dimer of cysteine formed by oxidation of the -SH group to a disulfied -S-S-.

Cyt. Abbreviation for cytosine.

Cytidine, Abb. Cyd, Cytosine Riboside, 3-D-ribofuranosyl-cytosine. A β-glycosidic Nucleside consisting of D-ribose and the pyrimidine base cytosine. The Cytidine phosphates have a very important role in the metabolism of all organisms.

Cytidine Diphosphoscholline. Abb. **CDP-choline, Active Choline.** The activated form of choline formed from phosphocholine and CTP.

Cytidine Diphosphoglyceride. Abb. **CDP-gly-ceride.** Phosphatide biosynthesis.

Cytidine Phosphates, Phosphoric Esters of Cytidine. Cytidine 5-monophosphate, abbreviation. CMP, cytidylic acid, cytidine 5' diphosphate, abbreviation CDP, and cytidine 5'-triphosphate, abbreviation CTP, are important in Phosphatide biosynthesis. CDP has been called the coenzyme of this synthesis. Activated choline is CDP-choline, and the sugar alcohol ribitol is also activated by bonding to CDP.

The reduction of ribose to deoxyribose in the synthesis of deoxyribonucleotides also occurs most often at the level of CDP.

Cytochalasins. A group of mold metabolites with cytostatic activity. Six chemically similar C, have been isolated from Helminthosporium dematioideum, Metarrhizium anisopliae and Rosellinia necatrix.

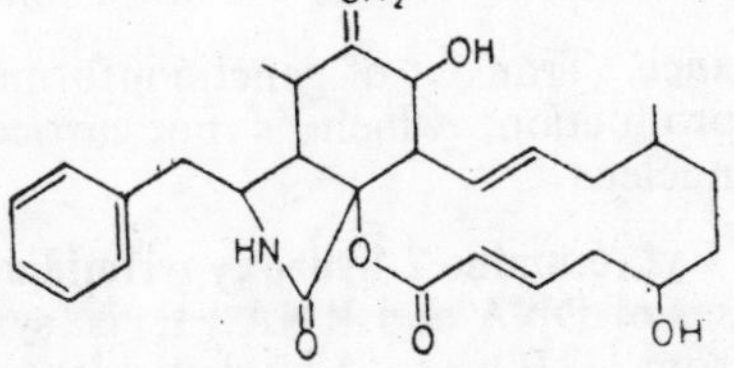

Structure of phomin, a typical cytochalasin

Cytochrome Oxidase. The last member of the electron transport chain. It normally reacts with oxygen, but can be inhibited by reaction with CN^- or CO. Its prosthetic group is hemin a, or cytohemin, which has a lipophilic C_{12} side chain, an aldehyde and a vinyl group on the porphyrin ring. The reaction with O_2 involves both the heme iron and copper :

$$2\,Fe^{2+} \xrightarrow{\frac{1}{2}O_2} 2\,Fe^{3+} \quad \text{and} \quad 2\,Cu^{2+} \xrightarrow{2e^-} 2\,Cu^+.$$

Electron microscopic studies have shown that the protein part of the molecule undergoes conformational changes during the redox process; the oxidized form is crystalline, but the reduced form is amorphous.

Cytochromes. A group of heme proteins which serve particle-bound redox catalysts in respiration, energy conservation, photosynthesis and some processes in anaerobic bacteria. They act as electron donors and acceptors by reversible valence changes in the iron atom at the center of their porphyrin complex:

$$Fe^{3+} \overset{+e^-}{\rightleftharpoons} Fe^{2+}$$

Cytokins, Phytokinis, Kinins. A group of plant hormones which promote cell division and generally stimulate plant metabolism, in particular RNA and protein synthesis. They are generally N-substituted derivatives of adenine. Together with other plant hormones (gibberellins and auxins), they mediate the plant's response to environmental factors, such as light.

They are synthesized mainly in the roots of higher plants, and they are not subject to much translocation.

The most important cytokinins are Kinetin Zeatin and Dihydrozeatin.

Cytolysosomes. See Intracellular digestion.

Cytoplasmon. The total extranuclear genetic information of a eukaryotic cell, excluding that in the mitochondria and plastids.

Cytoplasmic Inheritance. Transfer of genetic information in eukaryotic sexual reproduction, which is not carried by the chromosomes of the nucleus.

Cytosine, abb. C **or Cyt, 6-amine 2-hydroxypyrimidine.** One of the pyrimidine bases of DNA and RNA. It is synthesized as the triphosphate from UTP by an ATP-dependent amination.

Cytosine Arabinoside. See arabinosides.

Cytoxan. See Cyclophosphamide.

D

Deamination. Removal of the amino group, $-NH_2$ from a chemical compound :

(a) Oxidative D. is catalysed by Flavin enzymes and pyridine nucleotide enzymes. Amino acids may be oxidatively deaminated to keto acids.

(b) Transamination involves the transfer of an amino group from an amino to a keto compound, and is important in the synthesis of amino acids from tricarboxylic acid cycle intermediates.

Decarboxylases. Enzymes which catalyse the cleavage of CO_2 from α-ketoacids. One of the most important is Pyruvate decarboxylase which decarboxylates pyruvate to acetaldehyde. It requires thiamine pyrophosphate as a coenzyme.

Decarboxylation. The cleavage of the carboxyl group from a carboxylic acid to produce CO_2. The process accurs several times in the course of the Tricarboxylic acid cycle.

Defective Organism. A term sometimes applied to an Auxotrophic mutant.

Defficiency Mutants. See Auxotrophic mutants.

Dehydrobufotenin. A Toad poison found in Bufo marinus.

7-Dehydrocholesterol, Provitamin D_3, Cholesta-5, 7-dien-3β-ol. A zoosterol which occurs in relatively high concentrations in animal and human skin, where it can be converted to vitamin D_3 by ultravoilet radiation.

11-Dehydrocortiocsterone, Kendall's Substance A, 21-hydroxypregn-4-ene-3, 11, 20-trione. A glucocorticoid hormone from the adrenal cortex.

Dehydrogenases. Redox enzymes which extract hydrogen ($2H^+ + 2e^-$) from a substrate (hydrogen donor) and transfer it to a second substrate (hydrogen acceptor).

Dehydrogenation. Removal of hydrogen from a reduced substrate, which thereby becomes oxidized. Enzymatic dehydrogenation is catalysed by dehydrogenases or oxidases.

Deletion. Loss of one or more nucleotides of DNA, or of an entire segment of a chromosome; a form of mutation.

Delphinidin. 3, 3′, 4′, 5, 5′, 7-hexahydroxyflavylium cation. The aglycon of many Anthocyanins. Their glycosides are widespread among plants and are responsible for the mauve and blue colors of many flowers and fruits.

Demecolcine. See Colchicum alkaloids.

Demissine. One of the Solanum alkaloids, a steroid found in wild potatoes (Solanum demissum). It is a glycoalkaloid, consisting of the steroid base demissidine, 5α-solanidan-3β-ol.

Denaturation. Structural change in biopolymers which destroys the native, active configuration. It is brought about by heat, pH changes or chemical agents.

Dendrobium Alkaloids. A group of terpene alkaloids from various species in the orchid family Dendrobium.

Denitrification. See Nitrate reduction.

De Novo Purine Synthesis. See Purine biosynthesis.

De Novo Pyrimidine Synthesis. See Pyrimidone biosynthesis.

Density Gradient Centrifugation. A method of separating macromolecules on the basis of their density.

5′-Deoxyadenosine. A β-glycosidic deoxynucleoside which is important as a component of vitamin B_{12} (5′-Deoxyadenosylcobalamine).

5′-Deoxyadenosylcobalamine, B_{12} Coenzyme, DBC Coenzyme (DBC= dimethylbenzimidizole cobamide)- The coenzyme form of vitamin B_{12}. In this compound, the 6th coordination position of the cobalt atom in the center of the corrinoid ring is covalently bound to the 5′C atom of the deoxyadenosine. Other cobamide coenzymes contain an N-heterocyclic base other than dimethylbenzimidazole.

Deoxycholic Acid, 3α, 12α-dihydroxy-5β cholan-24-oic acid. A bile acid. It is found in the bile of most mammals, including man, dog. ox, sheep, and rabbit. It can be used as starting material for the partial synthesis of therapeutically important steroid hormones.

Deoxyribonuclease I (EC 3.1.21.1.). An enzyme which preferentially attacks double-stranded DNA and produces 5′-phosphodi- and -oligonucleotide by endonucleolytic cleavage. It is inhibited by a protein.

Deoxyribonuclease II (EC 3.1.22.1). A pancreatlic enzyme which requires calcium for activity and stability.

Deoxyribonucleic Acid, abb. DNA. Previously also called thymus DNA; a polymer of deoxyribonucleotides found in all living cells and some viruses. It is the carrier of genetic information,

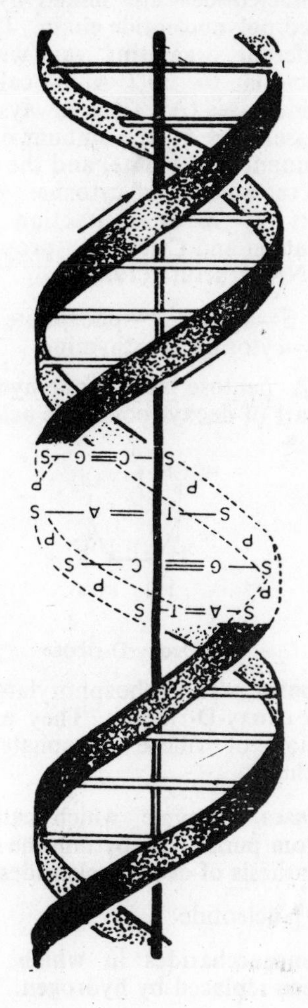

which is passed on from generation to generation by exact replication of the DNA molecule.

The mononucleotides of DNA consist of phosphorylated 2-deoxyribose which is N-glycosidically linked to one of four bases: adenine, guanine, cytosine or thymine. In the DNA of higher organisms, some of the cytosine is methylated in the 5 position, and in bacteriophage DNA, some is replaced by hydroxymethylcytosine.

The mononucleotides are linked by 3', 5'-diester bridges in an unbranched polynucleotide chain. The base contents of DNA from different organisms vary widely (from 8% A. in tuberculosis bacteria to 30% A. in calf thymus), but the number of purine bases (A+G) is always equal to the number of pyrimidine bases (C+T), the amount of adenine is always equal to the amount of thymine, and the amount of guanine is always equal to the amount of cytosine (Chargaff). This fact and the results of X-ray diffraction studies (Wilkins and Frankin) led Watson and Crick to propose the double-helix model of the DNA structure (1953).

DNA Double Helix. S=sugar, P=phosphate bridge, A=adenine, G=guanine, C=cytosine, T=thymine.

2-Deoxy-D-ribose. A pentose lacking one hydroxyl group. It is the carbohydrate part of deoxyribonucleic acids.

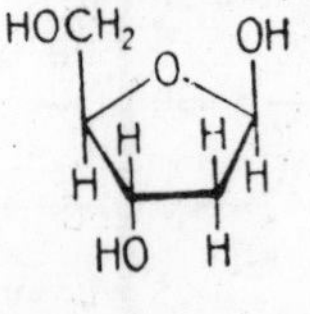

β-2-Deoxy-D-ribose

Deoxyribose Phosphates. The phosphorylated derivatives of the doxypentose 2-deoxy-D-ribose. They are synthesized biologically by reduction of ribose phosphates in the course of nucleotide synthesis.

Deoxyrlbosyltransferases. Enzyme which catalyse the transfer of deoxyribose from purine and pyrimidine deoxyribosides to free bases in the synthesis of deoxynucleosides.

Deoxyribotide. See Nucleotide.

Deoxy Sugars. Monosaccharides in which one or more hydroxyl groups have been replaced by hydrogen. There are two types, those with a methyle group in the terminal position, such as the 6-deoxyhexoses L-fucose and L-rhamnose, and those with the hydroxyl missing from the middle of the molecule, such as the DNA component 2-deoxy-D-ribose D. are often components of glycosides; D-digitoxose, for example, is the sugar component of many digitalis glycosides.

Deoxythymidine. See Thymidine.

Deoxythymidylic Acid. See Thymidine phosphate.

Deposiston. See Ovulation inhibitors.

Depsipeptide. The polypeptides whiich contain ester bonds as well as peptides. The naturally occurring D. are usually cyclic peptides, also called peptolides, which generally have α- or β-hydroxyacids as heterocomponents.

Derepression. The release of an operon from repression of transcription. In prokaryotic cells it occurs by inactivation of a repressor.

Dermatan Sulfate. Formerly called β-heparin, and chondroitin sulfate B. It is a mucopolysaccharide containing L-iduronic acid linked αl, 3 to N-acetyl-D-galactosamine 4-sulfate; the letter is linked αl, 4 to the next iduronic acid residue.

Desmosterol, 24-dehydrocholestrol, 5α-cholesta-5, 24-diene-3β-ol A zoosterol. It differs from Cholesterol in having an extra double bond between C24 and C 25. It has been isolated from the barancle Balanus glandula, chicken-embryos and rat skin.

Desulfuricants. Anaerobic bacteria of the genera Desulfovibrio and Desulfotomaculum, whose Sulfate respiration contributes to the process of Desul-furication.

Desulfurication. The anaerobic degradation of sulfur-containing organic compounds to inorganic sulfur.

Detergent Degradation. The process by which microorganisms digest synthetic detergents, thus removing them from the environment.

Dexamethasone, 9α-fluoro-17α-methylprednisolone, 9α-fluoro-16α-mythyl-11β; 17α, 21-trihydroxypregna-1, 4-diene-3, 20-dione. A synthetic pregnane derivative which is highly antiinflammatory but has little mineralocorticoid activity. It is used for arthritis. It is synthesized from cortisole.

Dextrans. High-molecular-weight polysaccarides synthesized by certain microorganisms. They consist of D-glucose linked α-glycosidically, primarily in 1, 6 bonds, but with some 1, 3 and 1, 4.

Dextrin 6-α-D-glucanohydrolase, Oligo-1, 6-glucosidase (EC 3.2.1.10). A glycosidase present in the digestive juices of the small intestine; formerly known as limit dextrinase, or isomaltase; It is specific for the hydrolysis of the α-1, 6 bonds in isomaltose and in the oligosaccharides produced by the action of α-amylase on starch and glycogen.

Dextrins. Water-soluble degradation products of starch. They are classified according to the color of their reaction product with iodine as amylodextrins (blue), erythrodextrins (red), and low-molecular-weight achroodextrins, which have no colored reaction product with iodine.

DHF. Abb. for dihydrofolic acid.

Diacetyl, Butane-2, 3-dione: CH_3-CO-CO-CH_3. A diketone produced as a byproduct of carbohydrate degradation. It is a component of the butter aroma, and has been found in many biological materials.

6-Diazo-5-oxo-L-norleucine. Abb. DON: $\overset{\ominus}{N}=\overset{\oplus}{N}=CH{\cdot}CO{\cdot}CH_2)$-CH (NH_2) COOH an antagonist of glutamine. It inhibits de novo purine synthesis in bacteria and mammals. It prevents the growth of experimental tumors, but is toxic for animals.

Dicarboxylic Acid Cycle. A cyclic pathway for the utilization of glyoxylic acid or one of its precursors (*e.g.* glycolic acid) as a carbohydrate source for the growth of microorganisms. The D. a.c. includes some of the reactions of the tricarboxylic acid cycle, and malate synthase (EC 4.1.3.2), of the glyoxylate cycle, acts here as a respiratory enzyme. It also provides starting materials for the synthesis of cell components.

Dichlorophenyl Dimethyl Urea, DCMU, Diuron. An herbicide which blocks electron transport from photosystem II to photosystem I.

Dichrostachinoic Acid. A sulfur amino acid which contains both reduced and oxidized sulfur.

Dictyosomes. Components of the Golgi apparatus, especially in plants. The terminology varies. They have also been called lipochondria and osmiophilic material.

Digestion. The totality of mechanical and chemical processes occurring in the digestive tract, that result in the degradation of foodstuffs to low M_r absorbable, nonantigenic substances. The digestive tract, especially in mammals. shows considerable structural and biochemical adaptation to the nutritional physiology of the organism, *e.g.* in carnivores, herbivores and omnivores.

Digestive Enzyme. Hydrolases present in the digestive tract of all animals, which catalyse hydrolysis of mechanically disrupted foodstuffs (proteins, carbohydrates, fats, nucleic acids) to their absorbable components.

Diginin. A digitanol composed of the pregnane derivative (Steroids) diginigenin, and the deoxysugar D-diginose. D. occurs together with cardial glycosides in Digitalis, *e.g.* Digitalis purpurea, from which it was isolated in 1936 by Karrer.

Digifolcin, which also occurs in Digitalis species, has the aglycon digifologenin, M_r 360 45, m.p. 176°C, $[\alpha]_D$—269°, which differs from diginigenin in having an extra 2β-hydroxyl group.

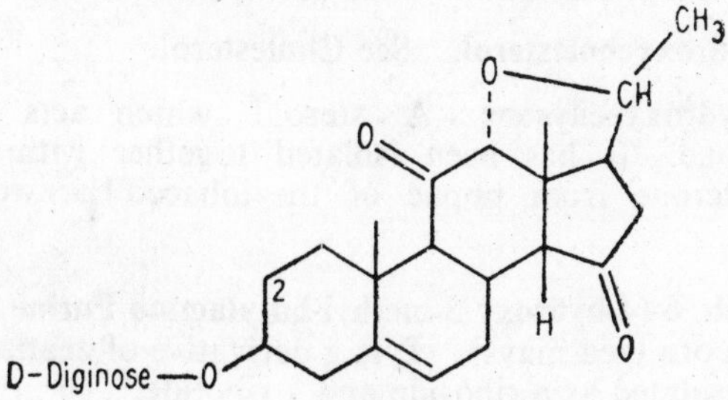

CH_3 O CH O 2 H O D-Diginose—O

Diginin

Digitalis Glycosides. The cardiac glycosides of the cardenolide group found in the leaves of foxgloves, Digitalis purpurea and Digitalis lanata.

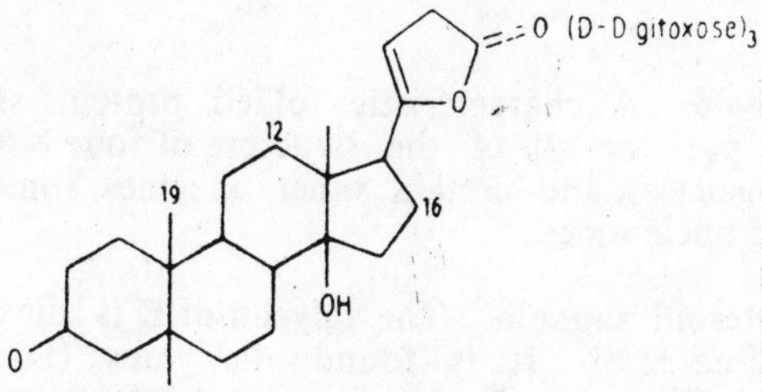

O (D-Digitoxose)$_3$ O 12 19 16 OH O H

Digitoxin

They are indispensible cardiotonic agents, used for long-term treatment of chromic heart weakeness are replacing the leaf powders or extracts formerly used.

Digitanols, Digitanol Glycosides. A group of plant glycosides with pregnane type Steroids as aglyca, for example, Diginin and digifolein. They occur together with cardiac glycosides, but have no cardiotomic activity themselves. They are biosynthesized from pregnenolone.

Digitonin. A mixture of four different steroid saponins from the seeds of the purple foxglove, Digitalis purpurea. It is a strong hemolytic poison, due to its affinity to blood cholesterol. It is used as a precipitating agent for cholesterol and other sterols, in their isolation and quantitative determination.

Dihydroorotate. An intermediate in Pyrimidine biosynthesis.

Dihydrouracil. An intermediate in Pyrimidine degradation.

Dihydroxyacetone Phosphate. See Triose phosphates.

20, 22-Dihydroxycholesterol. See Cholesterol.

20, 26 Dihydroxyecdysone. A steroid which acts as a molting hormone. It has been isolated together with Ecdysone and ecdysterone from pupae of the tobacco hornworm, Manduca sexta.

Dihydrozeath, 6-(4-hydroxy-3-methyl-butylamino Purine. A cytokinin from corn (zea mays). It is a derivative of zeatin and has also been isolated as a riboside and a ribotide.

N_6(γ, γ-dimethylallyl) Adenosine, N_6-isopentenyladenosine. One of the Rare bases in nucletic acids found in certain transfer RNAs, for example, serine tRNA. It also acts as a Cytokinin.

Dimethylallylpyrophosphate. An intermediate in Trepene biosynthesis.

Dinucleotide Fold. A characteristic folded protein structure constituting part or all of the structure of four NAD-dependent dehydrogenases, and certain other enzymes, some of which do not bind nucleotides.

Dioscin. A steroid saponin. The aglycon of D is diosgenin, (25-R) spirost-5-en-3β-ol. It is found in yams (Dioscorea) and triliums. Diosgenin is an important starting material for partial synthesis of steroid hormones.

Diphosphopyridine Nucleotide. See Nicotinamide adenine dinucleotide.

Disaccharidases. A group of enzymes which hydrolyse disaccharides. They are most abundant in ripe fruits, microorganisms (yeast) and the intestinal mucosa.

Disaccharides. See Carbohydrates.

Disulfide Bridges, Cystine Bridges. A term referring to disulfide bonds,-S-S-, in proteins and peptides formed by oxidation of two sulfhydryl groups : 2-SH→-S-S-. These are the major factors responsible for formation and maintenance of secondary structure in proteins. Proteins which contain a large number of these bridges are very resistant to denaturation by heat, acid or alkali, detergents, etc. and to hydrolysis by proteolytic enzymes.

Diterpene Alkaloids. A group of Terpene alkaloids.

Diterpenes. Terpenes built from four isoprene units ($C_{20}H_{32}$), Phytol, an aliphatic diterpene is important as the ester component of chlorophyl and as a part of vitamins K and E.

TABLE

Diterpenes and their significance

Class	Representatives
Hormones	Gibberellins, trisporic acids, antheridiogens
Vitamins	Vitamin A
Chromophore of visual purple	Retinol
Resenic acids	Abietic acid
Alkaloids	Cassaine, aconitine
Sweet substance	Stevioside

Diurnal Acid Rhythm. The fluctuation in the content of organic acids in the chlorophyll-containing parts of succulent plants, such as cacti. At night, CO_2 is fixed by carboxylation of phosphoenolpyruvate to oxaloacetate, which is then reduced to malate, leading to a considerable acidification of the tissue. By day, the malate is oxidatively decarboxylated to pyruvate, and the CO_2 is fixed photosynthetically. The mechanism allows the plants to absorb CO_2 from the air at night, when they can open their stomata without suffering the high water loss which would occur during the day. It is thus an adaptation to a hot, arid climate.

DNA. Abb. for deoxyribonucleic acid.

DNA Nucleotidytransferase. See DNA polymerase.

DNA Polymerase, DNA Nucleotidyltransferase (EC 2.7.7.7). An enzyme which catalyses the synthesis of DNA polynucleotide chains on a pre-existing DNA matrix (DNA replication). The precursors are the four 3′ deoxyribonucleotide triphosphates. In vitro, the enzyme can also synthesize homo and copolymers from triphosphates. The DNA polymerases from different organisms have different specificities, some for single and others for double strands of DNA as primers.

DNA-RNA Hybrids. Double-stranded molecules, of which one strand is DNA, and the other the complementary RNA. They are presumably the intermediate form in RNA transcription and in the multiplication of on cogenic RNA viruses but they can also be produced in vitro by Hybridization. They are resistant to ribonuclease.

Dolichol Phosphate. A coenzyme involved in the glycosylation of proteins and lipids. It is a membrane bound polyprenol phosphate which accepts glycosyl units from soluble donors (uridine-or guanosine-diphosphate glycosides) and donates them in turn to membrane proteins or lipids.

DON. Abb. for 6-diazo-5-oxo-L-norleucine.

Donor Position. The binding site for the peptidyl-tRNA on the ribosome during Protein synthesis.

Dopa. Abb. for 3, 4-dihydroxyphenylalanine.

Dopamine, Hydroxytramine, 3, 4-dihydroxy-phenethylamine. One of the catecholamines. It is formed by decarboxylation of 3, 4-dihydroxylphenylalanine (dopa), which in turn is formed by hydroxylation of tyrosine. It is the precursor of the hormones noradrenalin and adrenalin. In the liver, lungs and intestines, it is the end product of tyrosine metabolism. It is a neurol transmitter in the central nervous system.

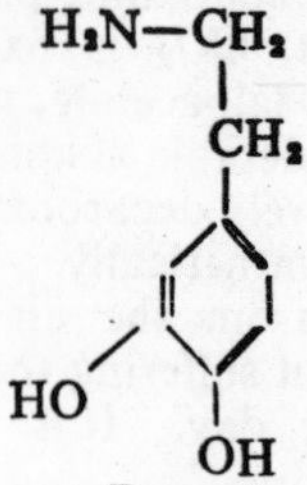

H_2N-CH_2

CH_2

HO

OH

Dopamine

Doping. The administration of foreign substances to increase the performance of athletes.

Double Strand Break. A break in a double-stranded DNA molecuie in which both strands are broken without separating from each other.

Dose. The amount of a pharmaceutical agent administered at a time. In animal experiments, the dose is given as the amount having a specified effect on a stated fraction of the individuals tested.

TABLE

Dose-effect Relationships

Term	Abb.	Definition
Median effective dose	ED_{50}	D. at which 50% of the experimental subjects show an effect
Effective, therapeutic or curative D.	CD_{50}	Same as above; used with therapeutically applied compounds
Minimal lethal D.	LD_{05}	D. causing death of 5% of the subjects
Median lethal D.	LD_{50}	D. which causes death in 50% of the experimental animals
Lethal D. Absolute lethal D.	LD_{100}	D. which kills 100% of the experimental animals.

DPN. Abb. for diphosphopyridine nucleotide.

DPX 1840: 3, 3a-dihydro-2-(p-methoxyphenyl)-8H-pyrazolo-[5, 1-a]-isoindol-8-one. A synthetic growth regulator affecting auxin transport, the formation of fruit-tripening hormone and root growth, etc. in cotton and soybeans.

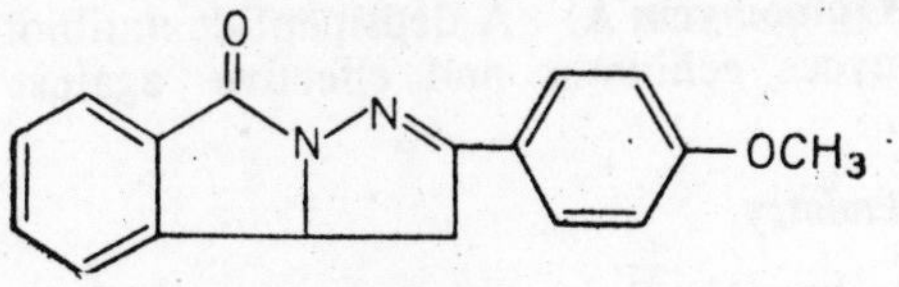

DPX 1840

Dry Weight. The weight of material, in g or mg, remaining after drying tissues, organisms, etc. at temperatures somewhat above 100C°.

DTDP-sugars. Sugars or sugar derivatives activated by bonding to deoxythymidine diphosphate sugars).

DThd. Abb. for deoxythymidine.

Dulcitol, Galactitol. An optically inactive sugar alcohol derived from galactose. It has 6 C atoms. It is found in algae, fungi and the sap and bark of various higher plants. It is produced synthetically by reduction of galactose or by isolation from dulcite or Madagascar mannu (Melampyrum nemorosum L.).

E

EAG. Abb. for electroantennogram.

Ecdysone, α-ecdysone, Molting Hormone, (22 R)-2β, 3β, 14, 22, 25-pentahydroxy-5-βcholest-7-en-6-one. A steriod hormone which stimulates molting of caterpillars, pupa formation and emergence from the pupa. Its first recognizable effect in the activation of certain genes.

Ecdysterone, β-ecdysone, Crustecdysone, 20-hydroxycedysone. A molting hormone, found together with Ecdysone in pupae of silk-worms (Bombyx mori) in amounts of 2.5 mg/500 kg.

Ecgonine. The principal part of the cocaine molecule and the basis of many Coca alkaloids.

Echinoderm Toxins, Echinoderm Saponins. Low M_r steriod toxins, produced in the glands of echinoderms.

Echinomycin, Quinomycin A. A depsipeptide antibiotic isolated from Streptomyces echinatus and effective against Gram-positive bacteria.

Ecological Chemistry

1. The investigation and optimization of the effects of man-made chemicals (*e.g.* pesticides, fertilizers) on the environment.

2. The study of the interactions of living organisms with each, other, and with their environment, of a chemical level.

Ectocarpene, all-cis-(1-cyclohepta-2′, 5′-dienyl)-but-1-ene. A sexual attractant excreted by the female gametes of the brown algae Ectocarpus siliculosus.

Edestin. A haxameric, globular protein from hemp seeds, Cannabis sativa. Each of the 6 subunits consists of two nonidentical polypeptide chains joined by disulfide bridges.

EDTA. Ethylenediaminetetraacetic acid.

Effectors. Chemical compounds, such as metabolites, hormones or cyclic AMP, which regulate the activity of a gene or enzyme, usually by allosteric interaction with a regulator protein or the enzyme protein.

EL-531, α-cyclopropyl-α-(4-methoxyphenyl)-5-pyrimidine Methanol. A synthetic growth retardant. It is a gibberellin antagonists which delays the growth of lettuce hypocotyls.

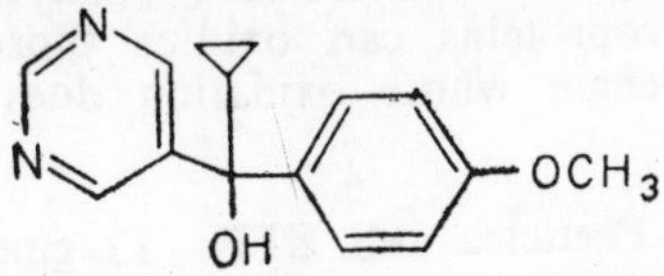

N N OH OCH_3

EL-531

Elastase (EC 3.4.21.11). An endopeptidase specific for the Elastin in animal elastic fibres. Its inactive precursor, proelastase, is formed in the vertebrate pancreas and transformed in the duodenum to elastate by the action of trypsin.

Elastin. A structural protein which is the main component of the elastic fibres of the tendons, ligaments, bronchi and arterial walls. It owes its elasticity to a high content of glycine, alanine and proline, and of valine (17%), leucine and isoleucine (12% together). The sequences-Gly-Val-Pro-Gly and-Gly-Val-Pro- are frequent in the protein.

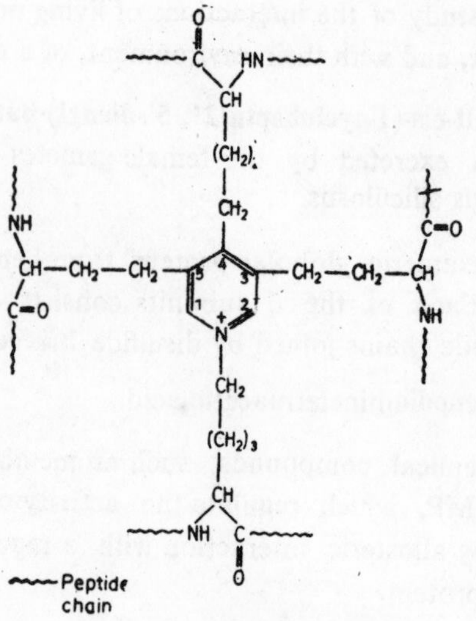

Desmosine

Electron Transfer Flavins, abb. **ETF.** The flavoproteins which mediate electron transport from reduced FADH to the cytochrome system. Flavoproteins can oxidize those substrates of the Respiratory chain whose oxidation does not involve pyridine nucleotides.

Electron Transport Particles. abb. **ETP.** Fragments of mitochondria, obtained by ultrasonic or detergent treatment, which are capable and transporting electrons. ETP contain the complete electron transport system of the Respiratory chain. In the electron microscope, ETP appear as membrane-enclosed vesicles, which are thought by some authors to be a giant molecule of defined composition.

Electrophoresis. A method of separating charged particles or macromolecules by allowing them to migrate in an electric field. The method is used most frequently in biochemistry to separate delicate macromolecules, usually proteins or nucleic acids.

Elongation. The phase of Protein biosynthesis in which the amino acid chain is extended by addition of new residues.

Elongation Factors, Transfer Factors. The proteins catalysing the elongation of peptide chains in Protein biosynthesis.

Embden Ester. A mixture of D-glucose-6-phosphate and D-fructose-6-phosphate, both of which are intermediate in glycolysis.

Embden-meyerhof-parnas Pathway. See Glycolysis.

Embelin. See Benzoquinones.

Embryonal Inducers. Compounds which induce the differentation of organs in the course of embryonic development.

Emerson Effect, Enhancement Effect. An increase in the photosynthetic quantum yield from long-wave red light (700 nm) obtained by simultaneous irradiation with shorter wavelengths (< 670 nm). There is a sharp drop-off in photosynthetic efficiency around 700 nm, but light of this wavelength can be utilized synergistically with light of shorter wavelengths.

Emetine. A dimeric isoquinoline alkaloid which is the principal alkaloid of ipecac, the ground roots of Uragoga ipecacuanha. It is very poisonous, exerting a strong simulus on mucuous membranes.

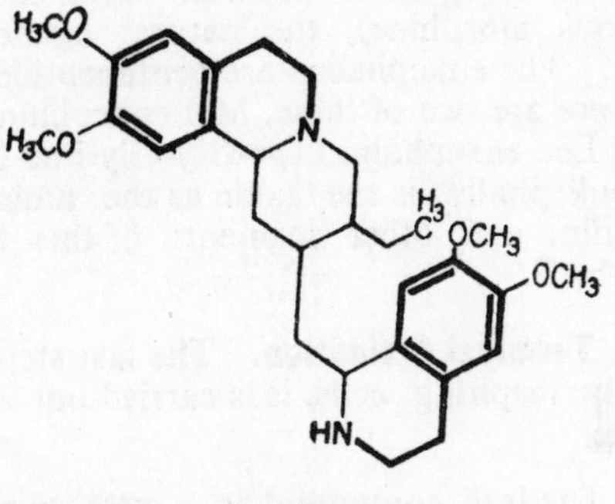

H_3CO, H_3CO, N, H_3C, OCH_3, OCH_3, HN

Emetine

Emodin. An orange anthraquinone pigment, m.p. 255°C. It often occurs as a glycoside, or as the dimer, skyrin. It is used as a cathartic.

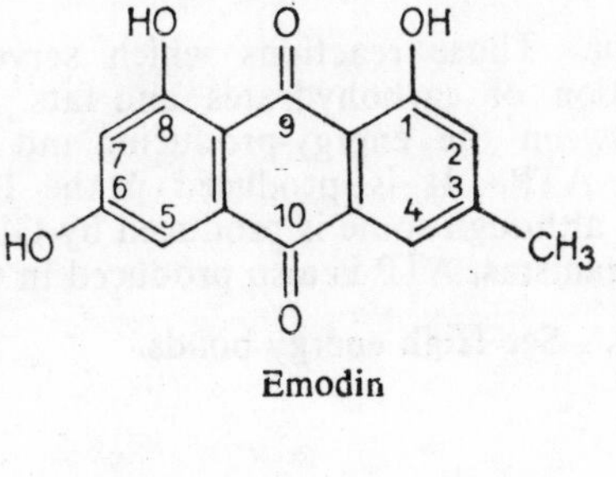

HO, O, OH, 8, 9, 1, 7, 2, 6, 3, 5, 10, 4, HO, CH_3, O

Emodin

Encephalitogenic Protein, Myelin Protein A 1' The most important myelin protein of the mammalian central nervous system. On injection into guinea pigs, rabbits or rats it induces allergic autoimmune encephalomyelitis, abb. EAE, an inflammation of the brain and spinal column.

Endocrine Hormones. Hormones produced by specialized cells in endocrine glands, and released into the blood stream.

Endorinlogy. The study of hormones. It includes the chemistry, metabolism, effects on cells, organs, individuals and populations of hormones. Ecoendocrinology is concerned with the interactions of endocrine systems with the environment.

Endoplasmic Reticulum. Abb. **ER**; a net-like system of double membranes located in the cytoplasm of eukaryotic cells. The ER appears to be continuous with the outer nuclear membrane and with the secretory system of the cell (the Golgi apparatus. If the ER is associated with ribosomes it is called rough ER; otherwise it is called smooth ER.

Endorphins. The endogenous peptides with morphine-like effects (endogenous morphine), the natural ligands for the opiate receptors. The enkephalins are pentapeptides isolated from pig brain. There are two of these, Met enkephain (Tyr-Gly-Gly-Phe-Met) and Leu enkephalin (Tyr-Gly-Gly-Phe-Leu). The sequence of Met eukephalin is the same as the amino acids 61 to 65 of β-lipotropin, and other segments of this hormone have been found to be E.

End Oxidation, Terminal Oxidation. The last step in catabolism. In aerobically respiring cells, it is carried out via the tricarboxylic acid cycle.

End Prodoct. The last compound in a metabolic pathway, which is irreversible as written.

End Product Repression. Inhibition of the synthesis of the enzymes of a reaction sequence (Enzyme repression), or inhibition of the activity of the first enzyme in the sequence.

Energy Metabolism. Those reactions which serve to release energy by degradation of carbohydrates and fats. The most important link between the energy-producing and energy-consuming reactions is ATP. It is produced in the largest amounts by Respiration although some is produced by Glycosis. In photosynthetic organisms, ATP is also produced in the light reactions.

Energy-rich Bonds. See High energy bonds.

Energy-rich Phosphates, High Energy Phosphates. The phosphorylated compounds which high energies of hydrolysis of the phosphate ester. Chemical energy in biological systems is stored in energy-rich phosphates. These may be acid anhydrates of phosphoric acid (*e.g.* ATP), enol phoshate (*e.g.* phosphoenolpyruvate) or amidine phosphates (*e.g.* Phosphagens). A high energy phosphate bond is represented as R ~ p, instead of the usual hyphen between groups.

Enniatins. The ring-shaped depsipeptide antibiotics produced by the fungus Fusarium orthoceras var. enniatum and other Fusaria.

Enolase (EC 4.2.1.11). An enzyme of Glycolysis which catalyses the reversible dehydration of 2-phosphoglycerate to phospho enol pyruvate.

Enterobacteriaceae. A family of bacteria capable of formic acid fermentation. They are Gram-negative rods which do not form spores. Motility is provided by peritrichal flagella.

Enteropeptidase, Enterokinase (EC 3.4.21.9). A highly specific duodenal protease which acts only on trypsinogen. It consists of two covalently bound glycopeptides. It removes the N-terminal peptide from trypsinogen (Val-[Asp_4-Lys) to activate it to trypsin.

Enzyme Graph, Enzyme Network. A way of representing the stoichiometry of an enzyme reaction as a network. The nodes of the network are the Enzyme species which are connected by arrows showing the direction of the reaction. These are labelled with the corresponding rate constant.

Enzyme Induction. Stimulation of the synthesis of enzymes by inducers. In bacteria, many catabolic enzymes are only synthesized when the appropriate substrate is available in the medium.

Enzyme Isomerization. The reversible changes in enzyme conformation in the course of a catalytic cycle.

Enzyme Kinetic Parameters, Enzyme Parameters. The parameters of the rate equations which remain constant, so long as temperature, pressure, pH value and buffer composition are constant. They are derived from the rate constants of the rate equations, and are frequently used to characterize the enzyme functionally.

Enzyme Kinetics. The mathematically treatment of enzyme-catalysed reactions. A great deal of information about reaction mechanisms can be obtained from kinetic experiments and evaluation of the data so obtained.

Entner-douderoff Pathway. A degradation pathway for carbohydrates in microorganisms, especially Pseudomonas species, which lack the enzymes hexokinase, phosphofructokinase and glyceraldehyde-3-phosphate dehydrogenase Balance : Glucose

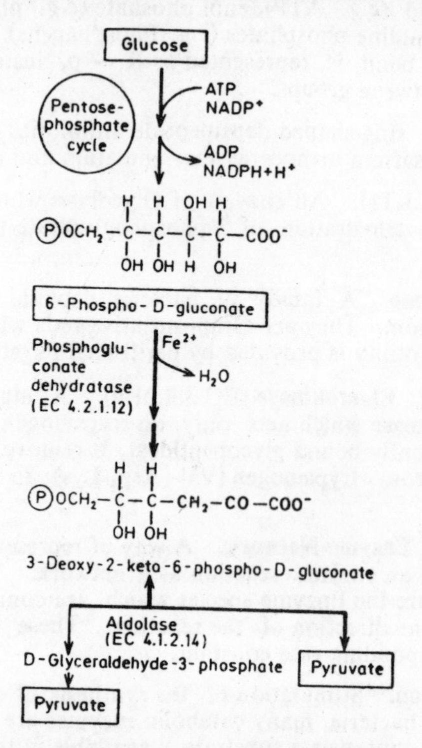

$(C_6H_{12}O_6) + ADP + P_i + 2\,NAD(P)^+ \rightarrow$
2 pyruvate $+ 2H^+ + H_2O$.

Enzyme Repression. The blockage of the synthesis of anabolic enzymes by the end product of the biosynthetic pathway to which they contribute. This type of regulation is found in prokaryotes for operons which synthesize various amino acids which may be present in the medium. If the amino acid present, the synthesis of all the enzymes in the operon is turned off, but if it is in short supply, the operon is derepressed.

Enzymes. The protein catalysts. About 90% of cellular proteins are enzymes but some structural proteins (for example, actin and myosin) also catalyse reactions. A protein is an enzymes, if it is known to catalyse a reaction, but it is always possible that a given protein catalyses an unrecognized reaction.

Enzymes differ from chemical catalysts most strikingly in their specificity. They catalyse specific reactions of one or a

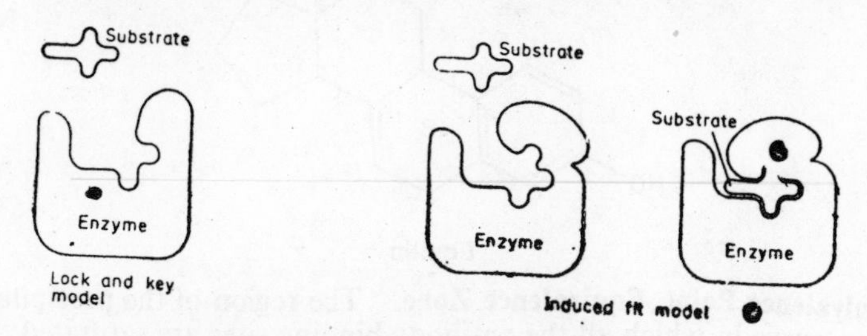

Models of Enzyme-substrate Binding

few closely related compounds distinguishing absolutely between stereoisomers. The lock and key model suggests that an enzyme, like a complicared lock, can only be fit by a substrat to (key) of precisely the correct shape. This view is modified in the induced fit model, in which the E, changes its conformation when it binds the substrate in such a way as to make the fit more complete. The substrate adheres to the enzyme because the arrangement of charged groups on the enzymes exactly complements the charged groups on the substrate, or the hydrophobic parts of the substrate exactly fit hydrophobic grooves or pockets in the enzymes, or the E may carry groups in the right positions to form hydrogen bonds with hydroxyl or amino groups of the substrate.

Enzyme Species Intermediates. All covalent and noncovalent complexes between an enzyme and a substrate and or product or effector, and all the various enzyme isomers (Enzyme isomerization). The concentrations of the E.s. cannot be measured directly by means of steady state Enzyme kinetics but their steady-state concentrations can be calculated from the kinetic equations for definite values of the concentration variables. In principle, the concentrations of the E.s. could be determined by the methods of presteady-state kinetics.

Equilenin. 3-hydroxyestra-1,3,5 (10), 6,8-pentaen-17-one, an estrogen. It found together with Equilin in the urine of pregnant mares. It has $^1/_{25}$ of the biological activity of estrone.

Equilin. 3-hydroxyestra-1,3,5 (10), 7-tetraen-17- one. An estrogen. It is isolated from the urine of pregnent mares and has $^1/_{20}$ the biological activity of estrone.

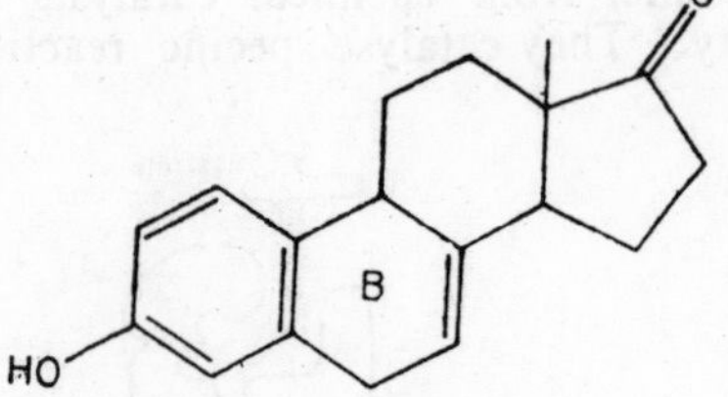

HO B O

Equilin

Equivalence Point, Equivalence Zone. The region of the precipitation curve in which all the antibody binding sites are saturated with the antigenic determinants, and the antibodies have been quantitatively precipitated.

Erepsin. An outdated term for the amino and dipeptidases secreted by the mucous membranes of the small intestine.

Ergocalciferol. See Vitamins (vitamin D_2).

Ergochromes, Secalonic Acids. A group of weakly acid, bright yellow natural pigments based on a dimeric 5-hydroxychromanone skeleton. They have been isolated from a number of molds and lichens and are poisonous.

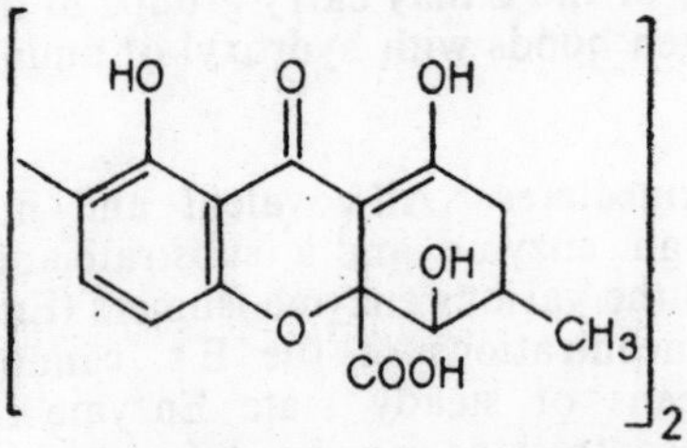

HO O OH OH O CH_3 COOH]2

Structural system of the ergochromes

Ergot Alkaloids, Claviceps Alkaloids, Ergoline Alkaloids. A group of over 30 indole alkaloids, possessing the ergoline ring system. They are produced by various species of the fungal genus Claviceps (family Ascomycetes) which parasitize rye and wild grasses. Following infection by spores of Claviceps purpurea, the ears of rye develop 1—3 cm long dark violet (almost black), highly poisonous sclerotia, containing up to 1% E.a. More recentyl, E. a. have been found in higher plants.

Ergosterol, Provitamin D_2, Ergosta-5,7, 22-triene-3β-ol. The most important of the mycosterols. It is converted by irradiation with UV to vitamin D_2 and is used for the production of this vitamin.

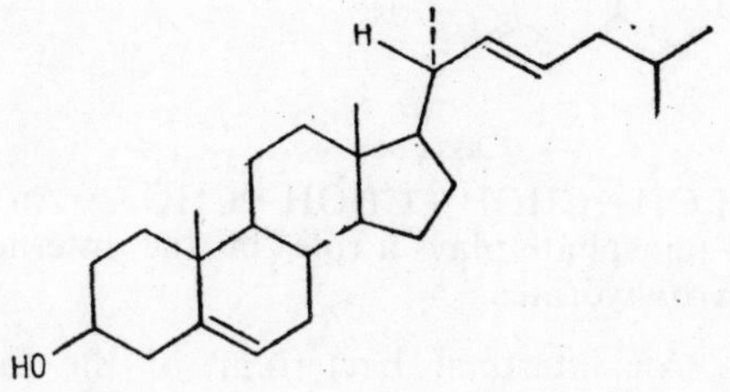

Ergosterol

Erucic Acid, Z-13-docosenoic acid, Δ^{13-14} docosenoic acid :

$$CH_3-(CH_2)_7-CH=CH-(CH_2)_{11}-COOH$$

A fatty acid. M_r 338.56, cis-form, m.p. 34°C, $b.p._{10}$ 254°C. It is a component of the glycerides in many seed oils of Cruciferae and Tropaeolaceae. It constitutes 40 to 50% of the total fatty acids of rapeseed, mustard and wall-flower seeds, and 80% of fatty acids of nasturtium seeds.

Erythrina Alkaloids. A group of isoquinoline alkaloids, usually tetracyclic, found exclusively in the legume genus Erythrina.

d-Erythritol. $CH_2OH-CHOH-CHOH-CH_2OH$. An optically inactive sugar alcohol derived from D-erythrose. It is found in some algae, fungi, lichens and grasses.

Erythrocruorin. A hemoglobin-like protein found in many invertebrates. In some snails and worms it is a high-molecular-weight extracellular respiratory pigment.

Erythrophleum Alkaloids. A group of terpene alkaloids from Erythrophleum guineense and Erythrophleum ivorense. They are esters or amides of the diterpene cassainic acid with substituted ethanolamines.

Erythropoietin. A hormone which stimulates production and release of red blood corpuscles. It is a glycoprotein (M_r 24000 to 46000) containing sialic acid. It is released from the α-globulin fraction of the blood by an enzyme produced from the juxtaglomerular cells of the kidneys when there is a lack of oxygen.

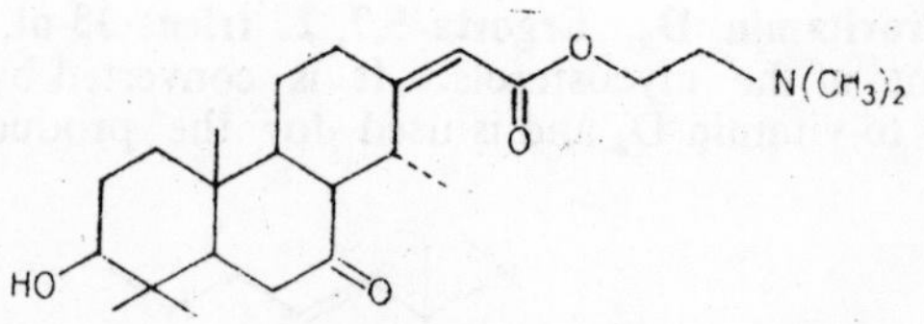

Cassaine

d-Erythrose. $CH_2OH-CHOH-CHOH-CHO$. An aldotetrose. Erythrose 4-phosphate plays a role in the intermediary metabolism of carbohydrates.

Escherichia Coli. An intestinal bacterium of the family Enterobacteriaceae. It is the most common experimental organism in molecular biology. A single cell has dimensions of about $1 \times 1 \times 3$ μm, a volume of about 2.25 μm^3, and a wet weight of about 10^{-12}g, which corresponds to a dry weight of about $2.5 10^{-13}$g. Under optimal conditions, the organism doubles about every 20 minutes. The wild type can grow on a completely synthetic medium containing salts, glucose and ammonia. The chemical composition, biosynthetic capacity and energy expenditures are listed in the table. The majority (70%) of the dry mass is protein, but the number of (smaller) lipid molecules is much larger. When the cell is growing logarithmically, it must synthesize 12500 lipid molecules and 1400 protein molecules per second Assuming an average of 400 peptide bonds per protein, this amounts to 560000 peptide bonds per second, for which the cell expends 88% of its biosynthetic energy.

Esperin. A circular antibiotic depsipeptide from Bacillus mesentericus. There is a higher α-hydroxy fatty acid on the N-terminus of the sequence Glu-Leu-Leu-Val Asp-Leu-Leu which simultaneously forms a lactone structure with the γ-carboxyl group of the Asp residue.

Essential Oils. An extremely heterogeneous mixture of volatile, lipophilic plant products with characteristic odors. The International Standard Organization (ISO) has defined the E o. in the strict sense as the steam distillates of plants or oils obtained by pressing out the rinds of a few citrus fruits. However, in practice the products of extraction with organic solvents, enfleurage or maceration of blossoms (flower oils) and the resinoids which can be extracted from other plant parts, resins and balsams are included by the term E.o.

Esterases A large group of hydrolases acting on ester bonds :

$$R^1COOR^2 + H_2O \longrightarrow R^1COOH + R^2OH$$

(R^1 is the acid residue and R^2 the alcohol residue). The acid may be a carboxylic, a phosphoric or a sulfuric acid, and the ester may be an alcoholic or a thiol ester.

Estradiol Oestradiol. 17β-estradiol, estra-1, 3, 5 (10)-triene-3, 17-β-diol. The most potent natural estrogen. M_r 272.39, m.p. 178°C, $[\alpha]_D$+81°. It occurs in high concentration in pregnancy urine, Graafin follicides and the placenta. In the organism E. and Estrone are interconvertible. Stereoisomeric 17α-estradial, m.p. 222°C, $[\alpha]_D$+54°, has only 1/350 of the activity of E. Therapeutically it is derived totally from laboratory synthesis It is used for the treatment of menstrual disorders, menopasual problems, etc.

Estradiol

Estriol, Oestriol. Estra-1, 3, 5 (10)-trien-3, 16α, 17β-triol; an estrogen. It has been isolated from human female urine (especially during pregnancy) and placenta, from mare's urine, and from willow catkins. It is formed in the organism from estrone and estradiol

Estrogens, Oestrogens. A group of female sex hormones. The ring system of E. is estrane (Steroids). They have an aromatic A ring with a phenolic 3-hydroxy group, and an oxygen function on C-17. The chief estrogens are Estrone, Estradiol and Estriol.

Estrone Oestrone. 3-hydroxyestra-1, 3, 5 (10) trien-17-one. An estrogen, which occurs in human urine, ovaries and placenta, and it has also been found in pomegrante seeds (17 mg/kg) and palm oil.

Etamycin, Viridogrisein A cyclic peptide lactone antibiotic from Streptomyces griseus and related species. It is effective against Gram-positive bacteria and Mycobacterium tuberculosis.

Ethanol, Ethyl Alcohol. CH_3-CH_2-OH. The end product of alcoholic fermentation of carbohydrates. It is produced by decarboxylation of pyruvate, and is found in small amounts in many organisms. In humans the normal level is 0.002 to 0.005% in the blood. It is degraded in the liver to acetaldehyde and acetate.

Ethanolamine, Aminoethanol. $H_2N-CH_2-CH_2-OH$. A biogenic amine produced by decarboxylation of L-serine. It is a common constituent of phospholipids in which it is esterfied to an acyl glycerol phosphate moiety to form phosphatidyly ethanolamine.

Ethylenediaminetetraacetic Acid. abb. EDTA, Ethylenedinitrolotetraacetic Acid. A chelating agent, used in biochemical systems in vitro for the chelation of divalent metal ions. Oral administration of EDTA may be used for the chelation of lead in cases of lead poisoning.

Ethynylestradiol, 19-nor-17α-pregna-1, 3, 5 (10)-trien-20-yne-3, 17β-diol. A synthetic estrogen. Subcutaneously it has the same biological activity as the natural hormones estradiol, but when administered orally, it is much more active and is therefore used in oral contraceptives.

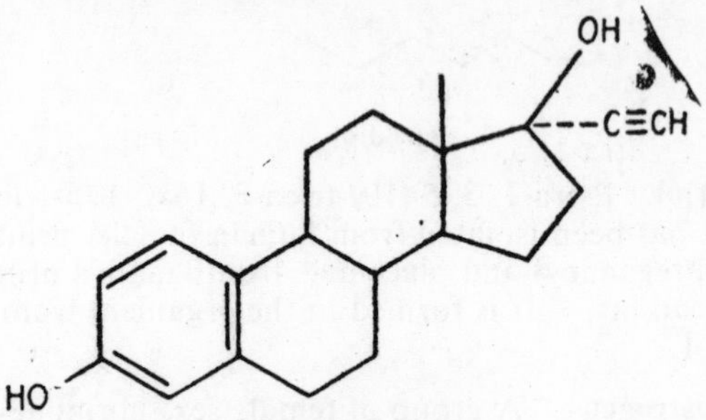

Ethynylestradiol

ETP. Abb. for Electron transport particle.

Eukaryote. An organism with eukaryotic cell.

Eumelanins. Brown or black melanin pigments found in animals. These are heterogeneous cross-linked polymers built of 5, 6-dihydroxyindole units, and usually linked to proteins. They are very widespread in skin, hair, feathers, insect cuticula, etc.

Evolution. The process by which organisms have come into existence.

Extrachromosomal Genes : DNA molecules located outside the nucleus. The most important are in the Plastids and Mitochondria. The corresponding extra chromosomal genes in bacteria are the Plastids.

F

Facilitated Diffusion. The passive transport, the movement of specific compounds across a bio-membrane from higher to lower concentration, but at a rate greater than simple diffusion.

Factor II. Prothrombin.

Factor VIII. Antihemophilic factor.

Factor XIII, Fibrin-stabilizing Factor. Refers to last clotting factor to act. It is an α_2-plasma globulin and contains 2 α and 2 β-chains. It is activated by thrombin in the presence of calcium ions to factor XIIIa, which catalyses the formation of γ-glutamyl-E-lysine peptide bonds in a calcium-dependent transmidation reaction. These bonds serve to cross-link the fibrin chains into a three-dimensional network, the clot.

FAD. Abb. for flavin adenine dinucleotide.

Farnesol. 3, 7, 11-trimethyl-2, 6, 10-dodecatrienl-ol. An acyclic sesquiterpene alcohol. It occurs widely in nature in essential oils, and also takes the place of phytol in the bacterial chlorophyll of the genus Chlorobium.

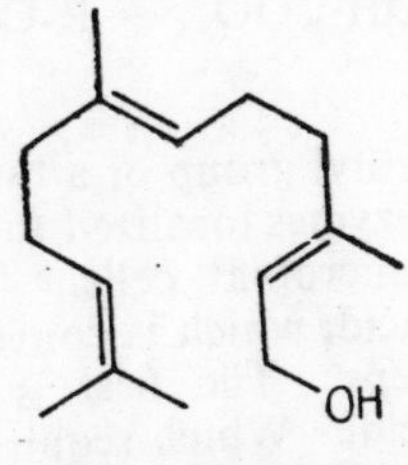

OH

trans-trans-Farnesol

Fat Biosynthesis, Triacylglyceride Synthesis. Refers to biosynthesis of neutral fats from fatty acids and glycerol. Fatty acids and glycerol are synthesized by separate pathways and then combined in a series of reactions in which the acyl moieties of fatty acyl-CoA molecules get transferred to glycerol phosphate.

Fat Degradation. Refers to the hydrolysis of the ester bonds in neutral fats to form fatty acids and diacylglycerol, which is further degraded to fatty acids and glycerol. The enzymes responsible are lipases.

Fats, Triacyglycerides. Esters of glycerol with saturated or unsaturated fatty acids. Fats are neutral compounds.

Almost without exception, the fatty acids of neutral fats are unbranched and have an even number of carbon atoms, usually between 4 and 26.

Fatty Acid Biosynthesis. The stepwise construction of fatty acids from acetyl units. The reactions are catalysed by a classical example of a multienzyme complex.

Fatty Acid Degradation. The catabolic pathways for fatty acids. The most common to β-oxidation; minor pathways are α-and γ-oxidation.

1. β-Oxidation, the stepwise degradation of fatty acids from the carboxyl end. In each turn of the cycle, two carbon atoms are removed as acetyl-CoA, and the β-carbon atom is oxidized; $R\text{-}CH_2CH_2\text{-}CO\cdot SCoA$ HS-CoA R-Co·CoA + $CH_3\text{-}CO\text{-}CoA$.

2. α-Oxidation of fatty acids occurs in germinating plant seeds. A fatty acid peroxidase (EC 1.11.1.3) catalyses the decarboxylation and simultaneous formation of an aldehyde, in which H_2O_2 acts as hydrogen acceptor. The aldehyde can either be oxidized to a fatty acid or reduced to a fatty alcohol $R\cdot CHOH\cdot COO^- \rightarrow R\cdot CO\cdot COO^- \rightarrow R\cdot COO^- + CO_2$.

3. The terminal methyl group of a fatty acid may be oxidized (ω·oxidation) enzymes localized in the microsomal fraction of animal and microbial cells. The substrate is usually a C_8 to C_{12} fatty acid, which is converted to the decarboxylic acid in two steps. The first is hydroxylation to an ω-hydroxy fatty acid. Which requires oxygen and NADPH. The second step is catalysed by a soluble, non-microsomal enzyme, which is usually NAD^+-dependent.

Patty Acids. The carboxylic acids found in fats and oils with the general formula C_nH_{2n+1} COOH for saturated fatty acids. The unsaturated fatty acids have double bonds between two or more carbons, and correspondingly fewer H atoms.

Fatty Alcohols. Unbranched, aliphatic mono-alcohols with 10 to 20 C atoms. They are found in nature as components of waxes. They are produced by reduction of fatty acids, and are used in the production of surface-active agents and emulsifiers.

FCCP. Abb. for carbonyl cyanide-p-trifluoromethoxyphenylhydrazone.

Fd. Abb. for ferredoxin.

Febrifugine. A quinolazine alkaloid isolated from the shrub Dichroa febrifuga, which has been used in Chinese folk medicine since ancient times. It is a strong antimaterial and fever-sinking agent, but is too toxic for human use.

Fermentation. A form of metabolism in which the end products could be further oxidized. Per unit of substrate. It yields far less energy than respiration. For example, a yeast cell obtains 2 molecules ATP per molecule of glucose when it ferments it to ethanol, while complete respiration would yield 38 molecules of ATP.

Fermentation Products. The end products of anaerobic microbial metabolism. The products listed in the table are (or were) industrially significant.

TABLE

Fermentation Products

Product	Organism	Use
Ethanol	Saccharomyces cerevisiae	Industrial solvent, beverages
Glycerol	Saccharomyces cerevisiae	Production of explosives
Lactic acid	Lactobacillus delbrueckii Lactobacillus bulgaricus	Food and pharmaceuticals
Acetone and Butanol	Clostridium acetobutylicum Clostridium butylicum	Solvents

(*Contd. on page* 117)

2, 3-Butylene glycol (2, 3-butane diol)	Bacillus polymyxa Bacillus subtilis	After conversion to butadiene production of synthetic rubber;
	Aerobacter aerogenes	Antifreeze

Fermentation Techniques, Microbial Production Techniques. Techniques for large-scale production of microbial products. These must both provide in optimum environment for the microbial synthesis of the desired product and be economically feasible on a large scale. They can be divided into surface (emersion) and submersion techniques. The latter may be run in batches or continuously.

Fermenters. Large-scale vessels in which micro-organisms are cultured to obtain some product of their metabolism. The main vessel is usually cylindrical; the requirements of the particular microorganism and the desired product dictate the accessories required for stirring, cooling, aeration, input of medium and harvesting.

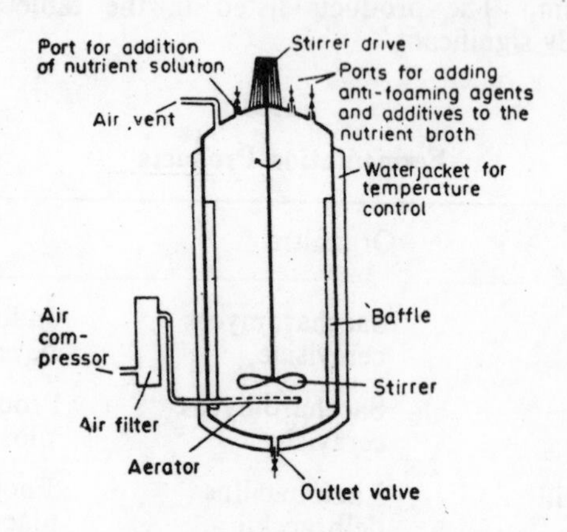

Cross section of a fermentor

Ferredoxins, abb. **Fd.** Low-molecular-weight iron-sulfur proteins which transfer electrons from one enzyme system to another, without possessing any enzyme activity themselves.

Ferritin. The most important iron-storage protein in mammals. Together with the related Memosiderin, it contains 25% of the iron in the body. It is found in the spleen, liver, bone, marrow and reticulocytes, where excess iron is stored intracellularly and can be mobilized at need. It has also been found in molluscs, plants and fungi.

Fervenulin. 6, 8-dimethylpyrimido 5. 4-e-1, 2, 4-triazine-5, 7-(6H, 8H) dione, M*r* 193.17, m.p. 177, a pyrimidine antibiotic synthesized by Streptomyces fervens. Its structure is analogous to that of toxoflavin, but it is less toxic. It has a broad spectrum of action, especially against cocci, Gram-negative and phytopathogenic bacteria and trichomonads.

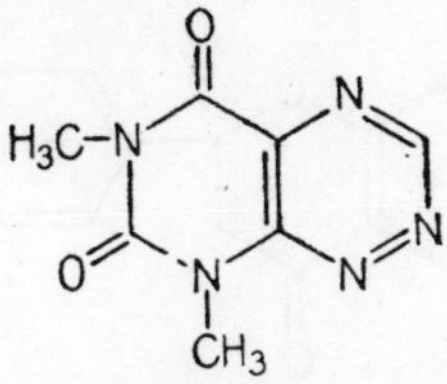

Fervenulin

Fibrin. The protein endproduct of blood coagulation. It is generated from the plasma protein Fibrinogen by thombin in the presence of Ca(II) ions. Thrombin removes two fibrinopeptide pairs A_2 and B_2 from fibrinogen. After the A peptide has been removed, the protein polymerizes in a pH-dependent, linear fashion. Thereafter, the B peptide is removed. At this point the fibrin bundle is still soluble in urea and has a cross-banding pattern visible in the electron microscope. It is establized by factor XIII, which must first be activated by thrombin and Ca^{2+}, Factor, XIIIa removes the carbohydrate and generates two intermolecular E-(γ-glutamyl) lysinebonds per molecule between the γ and α chains of the fibrin bundle. A cross-linked fibrin clot is insoluble both in water and in ȣ M urea, and provides a stable would closure (Blood coagulation). It can be degraded to soluble eleavage products by Plasmin.

Fibrinogen. The precursor of Fibrin. It is the only coagulable protein in the blood plasma of vertebrates and some arthropods. Its concentration in human plasma is 200 to 300 mg/100 ml. On electrophoresis of the plasma proteins, it migrates between the β and γ fractions.

Fibrinopeptides. The two pairs of peptides (A and B) cleaved off the amino ends of the 2α and 2β chains of fibrinongen by thrombin.

Flagellin. The main protein component of bacterial flagella. The filaments of flagella have an a-keratin structure, which is reversibly converted to the β-keratin structure on stretching of the aggregate. It has 304 amino acid residues, but no cysteine or tryptophan and only traces of proline and histidine.

Flevanones. Colorless flavonoids with a ring system derived from flavonone. Hydroxylated flavanones are found free or as glycosides in higher plants, especially in the rose, rue and composite families. Flavanones with a hydroxyl group in position 3 are called flavanonols.

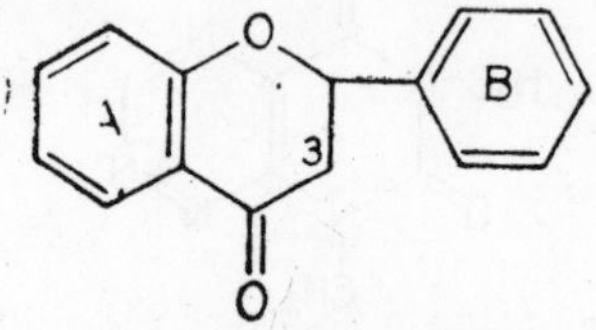

Flavanone Ring System

Flavin Nucleotides, Flavocoenzymes. The coenzymes of the flavin enzymes. Strictly speaking, they are Prosthetic groups, but some flavin enzymes can be easily separated from them. The two flavin nuclotides are Flavin mononucleotide and Flavin adenine dinucleotide.

Flavin Adenine Dinucleotide. Abb. FAD, Riboflavin Adenosine Diphosphate. A flavin nucleotide which is the active group of many Flavin enzymes.

Flavin Enzymes, Flavoproteins, Yellow Enzymes. A diverse group of more than 70 oxidoreductases found in animals, plants and microorganisms which have Flavin adenine dinucleotide or Flavin mononucleotide as prosthetic groups. These coenzymes can be reversibly reduced by hydrogen atoms, which they can accept either directly from a substrate (for example, succinate dehydrogenase) or from NAD(P)H. The group name refers to the yellow color of the oxidized riboflavin portion. The properties of the flavin adenine dinucleotide (FAD) or flavin mononucleotide (FMN) are strongly dependent on the protein molecule around them. This group of proteins is structurally and functionally very heterogeneous, and there is no single generalized type of flavin enzyme. The metalloflavin enzymes (metalloflavoproteins) also contain metals, such as Fe, Mg, Cu or Mo, which are involved in the binding of the Fe. to the mitochondrion and in the redox reaction.

Flavin Mononucleotide, abb. **FMN. Riboflavin 5-phosphate.** The prosthetic group of various flavin enzymes. $E'_0 = -0.219$ V (pH 7.3, 30°C), fluorescence maximum at 530 nm. FMN is composed of 6,7-dimethylisoalloxazine (flavin) and a ribitol residue linked to N, FMN occurs as the free acid or the sodium salt and usually contains 2 to 3 molecules H_2O. The phosphate ester bond is hydrolysed in acid solution, and the isoalloxazine-ribitol bond is unstable in alkaline solution. Thel compound is photolabile over the entire pH range, but is particularly so in alkaline solutions.

FMN is formed from riboflavin and ATP by a flavokinase. It is hydrolysed by acid and alkaline phosphatases.

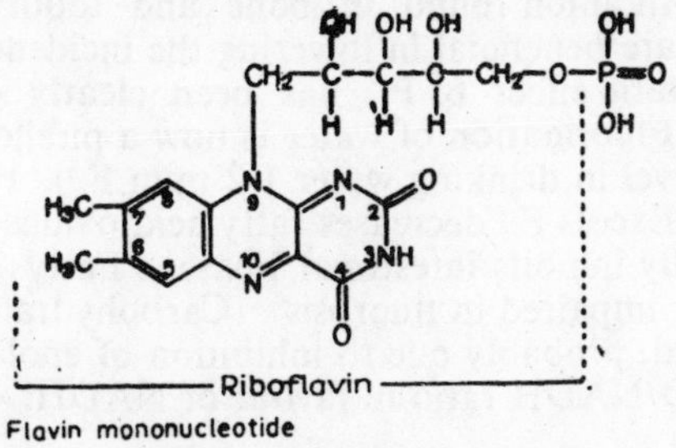

Riboflavin

Flavin mononucleotide

Flavin Mononucleotide

Flavones, Flavone Pigments. A group of plant pigments containing the flavone ring system. This consists of two substituted phenyl rings (A and B) and the pyrone ring C, fused to ring A, which is responsible for the typical reactions of the flavones.

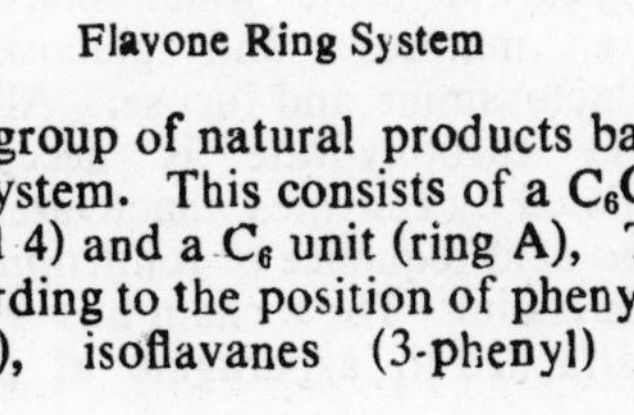

Flavone Ring System

Flavonoids. A large group of natural products based on the phenylchromane ring system. This consists of a C_6C_3 unit (ring B and carbons 2, 3 and 4) and a C_6 unit (ring A), The flavonoids are subdivided according to the position of phenyl ring into flavanes (2-phenyl, Fig.), isoflavanes (3-phenyl) and neoflavanes (4-phenyl).

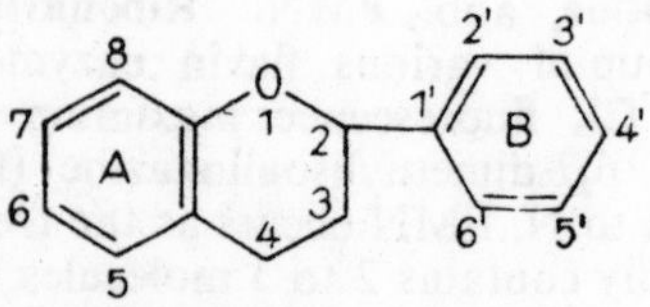

Flaving ring system

Flowering Hormone, Anthesin, Florigen, Vernaline. A principle which has been demonstrated physiologically but never isolated or chemically characterized.

Fluoride, F^-. An anion found in bone and tooth apatite. Small quantities are beneficial in lowering the incidence of caries, and this cariostatic effect of F^- has been clearly demonstrated in humans. Fluoridation of water is now a public health measure (optimal level in drinking water 1-2 ppm F^-). F^- affects several enzymes. Excess F^- decreases fatty acid oxidase in rat kidney, and partially inhibits intestimal lipase. Fatty acid utilization is generally impaired in fluorosis. Carbohydrate metabolism is also affected, probably due to inhibition of enolase and a shift of the NAD/NADH ratio in favour of NADH.

Fluoroacetic Acid. $CH_2F—COOH$. A very poisonous carboxylic acid which is converted in the tricarboxylic acid cycle to fluorocitrate. This is a strong inhibitor of acconitase, thereby stopping the Tricarboxylic acid cycle with lethal results.

Fly Agaric Toxins. The toxins of Amanita muscaria. Poisonings by this mushroom are rarely fatal. These toxins include Muscarin and other quaternary ammonium bases, such as muscaridin; indole compounds; Ibotenic acid and its easily derivatives, Muscimol and Muscazone. Muscimol and ibotenic acid inhibit motor functions, and muscimol is psychotropic.

Follicle-stimulating Hormone, Follitropin. An acidic glycoprotein containing many glutamate, threonine and cysteine residues. It contains 27% carbohydrate which consists chiefly of sialic acid, galactose, mannose and glucosamine, with smallar amounts of galactosamine and fucose. All of the sialic acid is acetylated. The carbohydrate is necessary for biological activity, as removal by enzymes inactivates the hormone The complete amino acid sequence of follitropin is known. There are two peptide chain. The α chain has 92 amino acids, with carbohydrate attached to asparagine at positions 52 and 78.

This chain is almost identical to the α chain of Human chorionic gonadotropin, but it differs from the α chain of human luteinizing hormone.

Formycins. A group of pyridine antibiotics (Nucleoside antibiotics) synthesized by Nocardia interforma.

n-Formylglycinamide Ribotide. abb. **FGAR.** An intermediate in Purine biosynthesis.

n-Formylgycinamidine Ribotide. An intermediate in Purine biosynthesis.

Fragment Reaction. A reaction used to assay the activity of peptidyl transferase. In a cell-free system containing 70 or 80S ribosomes, the growing peptide chain is transferred to Puromycin and released as peptidyl-puromycin.

FRH. Abb. for follicle-stimulating hormone releasing hormone.

Friedelin. A pentacyclic triterpene ketone. It is abundant in the bark of the cork oak (1%), in grapefruit rinds and some lichens.

Fructans. High-molecular-weight polysaccharides of D-fructose linked 1, 2- or 1, 6-glycosidically. They are common in plants. Examples are inulin and phlein, and the branched triticin, hordecin and graminin.

d-Fructose, Fruit Sugar, Levulose. A ketohexose, it tastes sweeter than any other carbohydrate and is fermentable by yeast. It crystallizes as β-pyranose, but forms compounds as furanose.

Fructose 1,6-bisphosphate, Fructose 1,6-diphosphate, Harden-Young Ester. A derivative of fructose in which the OH groups of C atoms 1 and 6 are esterified to phosphoric acid. It is an important metabolic intermediate.

Fructose 2,6-bisphosphate. The low Mr stimulator of phosphofructokinase. It has been purified from rat liver.

Fructose-bisphosphate Aldolase, Aldolase (EC 4.1 2.13) A tetrameric lyase which reversibly cleaves fructose 1,6-bisphosphate into the two triose phosphates dihydroxyacetone phosphate and D-glyceraldehyde phosphate. The reaction is analogous to the aldol condensation

$$(CH_3CHO + CH_3CHO \rightarrow CH_3-CHOH-CH_2-CHO),$$

hence the name of the enzyme.

Fructose 6-phosphate, Neuberg Ester. A phosphoric acid ester of fructose. It is an intermediate in Glycolysis produced by isomerization of glucose 6-phosphate. It can also be produced by transketolalion from erythrose 4-phosphate.

β-h-Fructosidase. See Invertage.

Fruit-ripening Hormone Ethylene, $CH_2=CH_2$. gaseous plant hormone. It occurs widely in plant tissues, and accelerates the ripening of of fruit. dropping of leaves and fruit and aging of the plant.

Fruit Sugar. See D Fructose.

FSH. See Follicle-stimulating hormone.

l-Fucose. 6-deoxy-L-galactose. It is a component of the blood-group substances A, B and O and of various of oligosaccharides in human milk, sea weed, plant mucilages. It is also found in assorted glycosides and antibiotics.

```
        CHO
         |
    HC—C—H
         |
     H—C—OH
         |
     H—C—OH
         |
    HO—C—H
         |
        CH3
```
L-Fucose

Fucosterol. (24E)-stigmasta-5,24 (28)-dien-3β-ol, a phytosterol. It is characteristic of marine brown algae and has also been isolated from fresh-water algae.

Fucoxanthin. A carotenoid pigment with an allene, an epoxy and a carbonyl group, and three hydroxyl groups, one of them acetylated. It is found in many marine algae, especially brown algae (Phaeophyta), and it is the most abundant of the naturally occurring carotenoids.

Fumarase, Fumarate Hydratase (EC 4.2.1.2). The tricarboxylic acid cycle enzyme which converts fumarate to malate by adding water to the double bond. The reaction is reversible.

Fumaric Acid. Trans-ethylene dicarboxylic acid. It occurs widely in the free form in plants. It is an intermediate in the Tricarboxylic acid cycle, and is the form in which the carbon skeletons of aspartate phenylalanine and tyrosine (via fumarylacetoacetate) are fed into the cyele. The cis-isomer is maleic acid.

```
H          COOH
  \        /
    C = C
  /        \
HOOC        H
```

Fumaric acid

Fumigatin. See Benzoquinones.

Fundamental Variable. See Concentration variable.

Funtumia Alkaloids. A group of steroid alkaloids characteristic of the genus Funtumia of the dogbane Apocynaceae) family. The F.a. are derived from pregnane and have an amino or methylamino group on carbon 3 and/or 20. These are biosynthesized from cholesterol and pregnenolone.

Funtumidine. See Funtumia alkaloids.

Funtumine. See Funtumia alkaloids.

Furanoses. See Carbohydrates.

Fusel Oil. Unpleasant-tasting side product of alcoholic fermentation. It consists mainly of amyl, isoamyl, isobutyl and propyl alcohols. The compounds are formed from amino acids, especially leucine, isoleucine and tyrosine, by deamination and decarboxylation. Tyrosol, which is formed from tyrosine, is a component of beer.

Fusidic Acid. A tetracyclic triterpene antibiotic. It is isolated from culture filtrates of Fusidium coccineum and, like structurally related antibiotics Cephalosporin P_1 and helvolic acid, it is effective against Gram-positive organism. It is biosyntheszied from squalene via 2,3-epoxysqualene. It inhibits protein synthesis by preventing the reaction of the elongation factor EFG with the small ribosomal subunit.

G

Gadoleinic Acid, $^9\Delta$-eicosenoic Acid. $CH_3-(CH_2)_9-CH=CH-(CH_2)_7$ COOH a fatty acid. It is found as a component of glycerides in plant and fish oils, and of phosphatides.

Galactans. Refer to polysaccharides of D-galactose found in plants. They are usually unbranched and have high molecular weights. Examples are agar-agar and carrageenan.

D-Galactosamine, Chondrosamine, 2-amino-2-deoxy-D-galactose. An amino sugar. $[\alpha]^{20}/D + 125° \rightarrow +98°$ (water) It is derived from D-galactose by replacement of the OH group on C2 by an amino group. It is usually found in nature as the N-acetyl derivative and is a component of a few mucopolysaccharides such as chondroitin sulfate, blood group substance A, etc. It is also found in mucoproteins.

Galactose. An aldohexose occurring naturally in D and L-forms. It is especially widespread in animals and is a component of oligosaccharides, such as lactose, and of the cerebrosides and

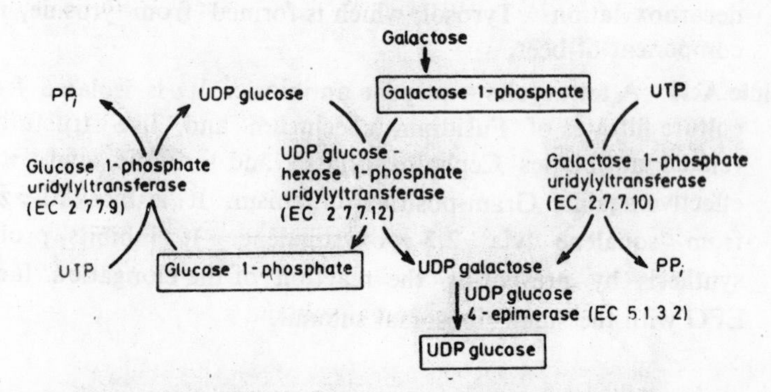

Relationship between galactose and glucose metabolism

gangliosides of nervous tissues. In plants, it is a component of melibiose, raffinose and stachyose, and of the Galactans, it is also the sugar component of some glycosides.

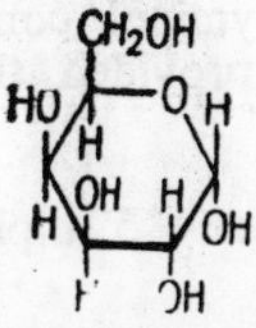

α-D-Galactose

It is synthesized as uridine diphosphate (UDP)—G. from UDP-glucose. The epimerization on C4 is catalysed by UDP glucose 4-epimerase (EC 5.1.3.2). This reaction is reversible, and the degradation of UDP galactose occurs via UDP glucose. G. is fed into general glucose metabolism by the pathway shown in the figure. Galactose 1-phosphate can also react directly with UTP to form UDP galactose.

D-Galacturonic Acid. M_r 194.14, m p. 159°C, α-form $[\alpha]^{20}/D+98° \rightarrow 50.8°$ (water), β form $[\alpha]^{20}/D+27° \rightarrow 50.8$ (water). Pectins contain 40 to 60% G. a.; the compound is a component of a few other plant polysaccharides as well.

Galegine, (3-methyl-2-butenyl)guanidine. A guanidine derivative found together with 4-hydroxygalegine in the seeds of goat's rue, Galega officinalis. It is synthesized in the shoot and accumulated in the seeds.

Gas Chromatography. Refers to a separation technique which is based on the distribution of gaseous compounds between a mobile gaseous phase and a stationary adsorbant phase. A prerequisite for separation is that the compounds in question can be vaporized or reproducibly converted to volatile derivatives.

Gastrin Inhibitory Peptide, GIP. A polypeptide hormone (for structure, Secretin), purified from crude preparations of cholecystokinn-pancreozymin. It shows potent enterogastrone activity, *i.e.* it inhibits secretion of acid and pepsin by the stomach, and inhibits gastric motility.

Gastrin. A hexadecapeptide hormone from the gastric antrum. Human gastrin I is Pyr Gly-Pro-Trp-Leu-Glu-Glu-Glu-Glu-Glu-Ala-Gly-Tyr-Trp-Met-Asp-Phe-NH_2, M_r 2116. Human gastrin II has an additional sulfate group on Tyr 12. Leu 5 is replaced by Met in porcine gastrin and by Val in the sheep and bovine hormones. In bovine gastrin, there is also an Ala instead of Glu in position 10.

GC Content. The amount of guanine + cytosine in nucleic acids, expressed in mol % of the total bases. The GC content plus the AT (adenine + thymine) content of a nucleic acid is thus equal to 100 mol %, provided that the molecule is double-stranded.

GDP. Abb. for Guanosine 5′-diphosphate Guanosine phosphates.

Gelatins. Proteins obtained by extraction from tissues rich in Collagen.

Gene. A section of DNA coding for a single polypeptide chain (structural gene), a particular species of transfer or ribosomal RNA, or a sequence which is recognized by and interacts with regulator proteins (regulatory gene).

Gene Activation, Gene Expression. Gene activation deals with the mechanisms which determine which genes will be expressed at a given time. In prokaryotes, the genes for the enzymes of a single metabolic pathway are often regulated as a unit, a the Operon.

In multicellular organisms, it is crucial to growth and differentiation.

It can be initiated by various effectors, which probably combine with specific nuclear proteins rather than acting directly on the genes. Such effectors may control the rate of transcription of particular genes, in addition to initiating or terminating transcription.

Gene Amplification. The production of extrachromosomal coupies of the genes for ribosomal RNA.

Gene Synthesis, Cell-free Gene Synthesis. The term includes cell-free replication of bacteriophage genome, which was first achieved in 1967 by Goulian, Kornberg and Sinsheimer with ΦX-174. Khorana and coworkers later synthesized the gene for alanine tRNA by a combination of chemical and enzymatic methods.

Gene Expression. The use of the information in a particular gene through transcription and translation. It is subject to regulation at several levels : RNA synthesis, processing, and transport and during protein biosynthesis. In another sense, it is the appearance in the phenotype of a particular trait determined by the gene in question.

Genetic Code, Code. The relationship between the sequence of bases in a nucleic acid and the sequence of amino acids in the polypeptide synthesized from it.

Genetic Code in Mitochondrial mRNA. In mitochondrial mRNA some of the codons are different from those in the G.c. described above, which was established for cytoplasmic and bacterial mRNA. For example, in human mitochondrial mRNA, AGA add AGG are used as termination codons. Other differences are as follows :

Some codons in mitochondrial mRNA

mRNA	5′ CAA	GUC	AUA	CUA	UGA3′
Yeast mitochondria	Gly	Val	Ile	Thr	Trp
Human mitochondria	Gly	Val	Met	Leu	Trp
Neurospora mitochondria	Gly	Val	Ile	Leu	Trp

Genetic Material. The carrier of hereditary information. In higher orgnisms, bacteria and some virus, it is double-stranded DNA, in other virus it is single-stranded DNA and in still other virus, it is RNA.

Genome. The sum of all chromosomal genes in a haploid cell (including prokaryotes) or the heploid set of chromosomes in an eukaryotic cell.

Genotype. The sum of an individual's genes, both dominant and recessive.

Gentiana Alkaloids. Terpene alkaloids with a pyridine skeleton (they therefore are also pyridine alkaloids) found primarily in gentian (Gentiana) species. The biogenetic precursors of these alkaloids are probably bicyclic monoterpenes.

Gentianine

1. A terpene and pyridine alkaloid, which occurs in many plants of the gentian family (Gentianaceae).

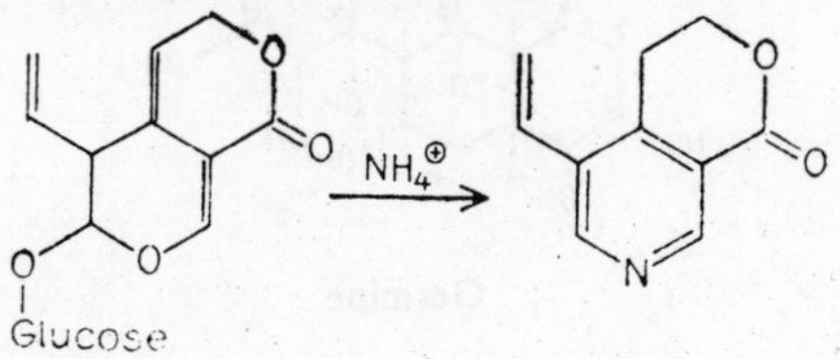

$$\text{Gentiopicroside} \xrightarrow{NH_4^{\oplus}} \text{Gentianine}$$

Glucose

Gentiopicroside Gentianine

2. Gentisin, gentianic acid, 1, 7-dihydroxy-3-methoxy-9H-xanthene-9-one. A yellow pigment from the yellow gentian (Gentiana lutea).
3. An anthocyan pigment with a delphinidin structure from Gentiana acaulis.

Gentianose. A nonreducing triaccharide composed of two D-glucose and D-fructose units. One glucose and the fructose are joined as in sucrose; the second glucose is linked β-1, 6 to the first. G. is a storage compound found in the roots of gentians.

Gentiobiose. A reducing disaccharide consisting of two molecules of D-glucopyranose linked β-1, 6.

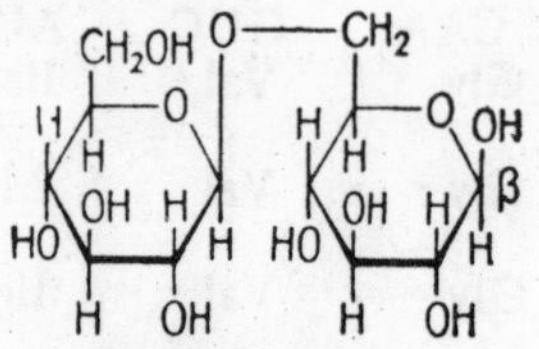

Gentiobiose

Geranial. The trans isomer of Citral.

Geraniol, 2, 6 dimethylocta-2, 6-dien-8-ol. The most important of the doubly unsaturated monoterpene alcohols. Nerol is the double bond isomer with the cis configuration at position 2. The structural isomer linaool has its OH group on C3 instead of C 1. It is a component of many essential oils, and makes up to 60% of oil of roses. If has the fragrance of roses. It is used, primarily as the acetate (b.p. 242 °C), in the perfume industry.

Germine. A Veratrum alkaloid with a C-nor-D-homo ceveratrum type structure. It has been found in Veratrum album, Veratrum

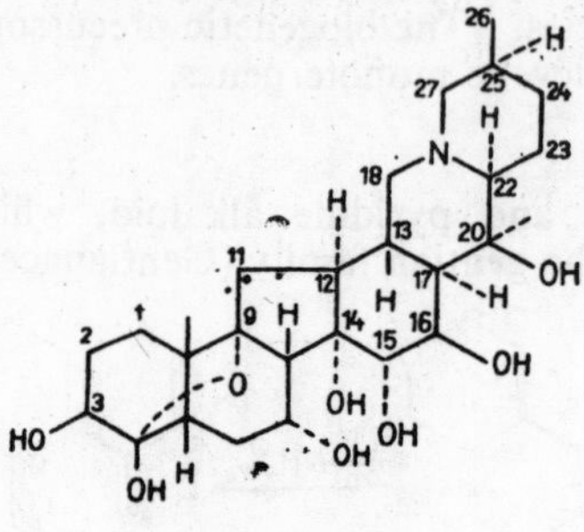

Germine

viride and Veratrum nigrum and Zygadenus venosus in the form of ester alkaloids. The most common acid components are acetate, angelate and tiglate. These ester alkaloids reduce blood pressure.

Gerontology. The study of aging, or those processes which occur progressively in the life of an organism, especially those which begin after it has reached maturity. Geriatrics is the branch of medicine concerned with old people. Aging consists of a gradual decline of physiological functions, eventually causing death. For humans, the probability of a neutral death increases logarithmically with increasing rate.

Gestagens, Progestins, Gastins. A group of female gonadal hormones, including Progesterone and other natural and synthetic steroids with progesterone-like effects (*e.g.* Norgestrel and Chlormadione acetate). Oral gestagens are used to correct irregularities in the menstrual cycle, repeated abortion and as components of Ovulation inhibitors. They are also being used increasingly to regulate animal reproduction.

GH. Abb. for growth hormone.

Giant Chromosome, Polytene Chromosome. An especially large type of chromosome present in the Diptera. Giant chromosomes in the salivary glands of Drosophila have been extensively studied. There have a cable-like appearance, due to multiple endomitotic duplication of the chromosomes, without separation of the chromatids. They may consist of many thousand single chromosomes, and may reach a length of 0.5 mm and a thickness of 25 μm. The supercoiled DNA molecules are much more extended than in ordinary chromosomes, and the duplicated molecules remain in register as they lie side by side. This gives rise to a banded structure, consisting of DNA-rich bands with DNA-poor regions in between.

Gibbs Effect. After the brief photosynthetic assimilation of $^{14}CO_2$ by Chlorella, the resulting fructose bisphosphate is labelled symmetrically, as expected from the operation of the Calvin cycle. In contrast, glucose phosphates and the glucose moiety of starch are asymmetrically labelled, C—4 containing significantly more ^{14}C than C—4, and C—1 and 2 containing significantly more than C—5 and 6. This apparently anomalous labelling of glucose is known as the Gibbs effect.

Gibberellin Antagonists. Inhibitors of the effects of gibberellins on plants. Their effects can be at least partially overcome by gibberellins The term is used independently of the mechanism of inhibition. Competitive inhibitors of gibberellins, compounds which bind to the same active sites as the hormones, are called antigibbrellins.

Gibberellins. A class of widely occurring plant hormones which stimulate extension growth. Chemically, they are diterpenoid acids. The first G. to be discovered was isolated from Gibberella fujikuroi (Fusarium moniliforme), the pathogen of the rice disease Bakanae, in 1938. The first pure G. was G A_3 (gibberellic acid), crystallized in 1954. Further research revealed that G are widespread natural plant growth regulators.

All of the known 42 gibberellins can be derived from the tetracyclic gibbane skeleton. These are referred to order of their discovery, as G. A_1 through A_{42}. There are two main groups :

1. G. with 20 C atoms (ent-gibberellanes). These include G. A_{12} through A_{15}, A_{17} through A_{19}, A_{23} through A_{25}, A_{27}, A_{28} A_{36} through A_{30}, A_{41} and A_{42}.
2. G with 19 C atoms (ent-20.nor-gibberellanes) A_{20}-A_{22}-A_{26}-A_{29}-A_{35} and A_{40}.

Gibbane ent-Gibberellane

GA_{14} GA_3 (Gibberellic acid

Gitonin. A steroid saponin. The aglycon is gitogenin, (25 R)-spirostan-2α, 3β-diol, the sugar chain consists of 2 galactose, 1 glucose and 1 xylose units. Gitogenin differs from digitogenin in lacking the 15β-hydroxyl group. It has been isolated from Digitalis purpurea and Digitalis germanicum. Free gitogenin has also been isolated from agave und yucca species.

Gla. Abb. for L-4-carboxyglutamic acid.

Gln. Abb. for L-glutamine.

Globulins. A group of simple proteins which are insoluble in pure water but soluble in dilute salt solutions (salting in effect). They are found in all animal and plant cells and body fluids, including serum and milk. The group includes many enzymes and most glycoproteins.

Glomerine. A quinazoline alkaloid in the defense secretion of the insect Glomeris marginata. m p. 204 °C. It is one of the few animal alkaloids. The secretion is exuded from 8 pores arranged in a row on the 1.5 cm animal. It contains about 50 μg G. and another alkaloid, homoglomerine (m.p. 149°C).

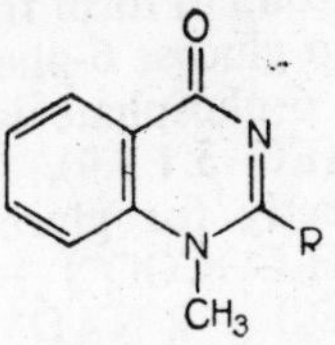

Glomerine: R = C_2H_5
Homoglomerine: R = CH_3

Glu. Abb. for L-glutamic acid.

Glucagon A pancreatic polypeptide hormone consisting of a single chain of 29 amino acids. The primary structure is given in the Fig. It is produced in the A cells of the islets of Langerhans in the pancreas in response to a drop in the blood sugar concentration. It promotes the hydrolysis of glycogen and lipids, and raises the blood sugar level. It is degraded in the liver. It can be detected in blood serum by radioimmunological techniques with a sensitivity in the pg/ml range.

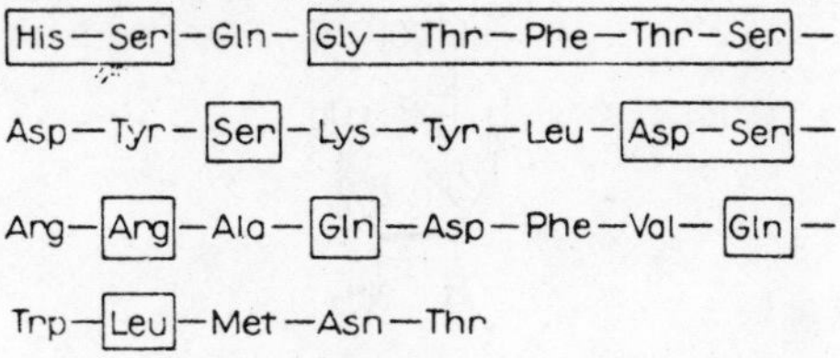

His—Ser—Gln—Gly—Thr—Phe—Thr—Ser—

Asp—Tyr—Ser—Lys—Tyr—Leu—Asp—Ser—

Arg—Arg—Ala—Gln—Asp—Phe—Val—Gln—

Trp—Leu—Met—Asn—Thr

Glucagon

The boxed amino acids are identical with the secretin sequence.

Glucans. The polysaccarides which are composed of D-glucose. They may be either straight or branched chains. The glucosidic linkages may be α-1, 4, as in amylose and bacterial dextran, β-1, 4, as in cellulose, β-1, 3, as in leucosin and callose, or 1, 6 as in pustulan The branched glucans include amylopectin (α-1, 4 and α-1, 6 bonds), dextran, laminarin and lichenin.

Gluconeogenesis. The synthesis of glucose from pyruvate or Amino acids. It cannot occur by a simple reversal of Glycolysis because the equilibria of those reactions are too unfavourable under physiological conditions. Instead, the pyruvate is carboxylated to oxaloacetate (either directly or indirectly, via malate).

The oxaloacetate is then decarboxylated and phosphorylated simultaneously by the phosphoenolpyruvate carboxykinase (ATP) (EC 4.1.1.49) to form phosphoenolpyruvate. Reversal of the glycolytic reactions then yields fructose 1, 6-bisphosphate from the phosphoenolpyruvate. The phosphofructokinase reaction is not reversible; instead fructose-bisphosphatase (EC 3.1.3.11) removes one phosphate group to form fructose 6-phosphate. This is converted readily to glucose 6-phosphate. If the blood sugar level is low, glucose 6-phosphate is hydrolysed to glucose by glucose-6-phosphate (EC 3.1.3.9). Otherwise, the glucose 6-phosphate is used directly for glycogen synthesis. Overall reaction : 2 pyruvate (2 CH_3COCOO^-) + 2 NADH + 4 H^+ + 6 ATP → glucose ($C_6H_{12}O_6$) + 2 NAD^+ + 6 ADP + 6 P_i. The energy required for G. can be obtained by oxidizing 20 to 30% of the lactate to CO_2 and H_2O.

Glucoplastic Amino Acids. Amino acids whose degradation products can contribute to Gluconeogenesis.

Glucosamine, 2-amino-2-deoxyglucose, Chitosamine. A widely occurring aminosugar. It is a component of chitin, mucopolysaccharides like heparin, chondroitin and mucoitin sulfate, and of blood group substances and other complex polysaccharides. It is usually present as N-acetylglucosamine.

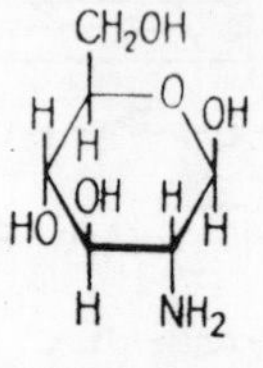

β-Glucosamine

D-Glucose, Dextrose, Grape Sugar, Blood Sugar. A hexose. M_r 180.16, β-form m.p. 146°C, $[\alpha]_D$ + 112.2 → + 52.7° (c = 10 in water); β-form, m.p. 148 to 155°C, $[\alpha]_D$ + 18.7 → + 52.7° (c = 10 in water). The most stable configuration for the pyranose form is

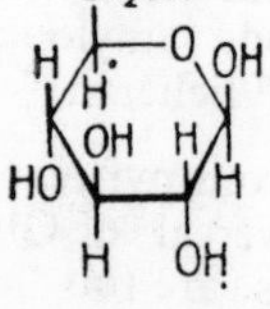

α-D-Glucose

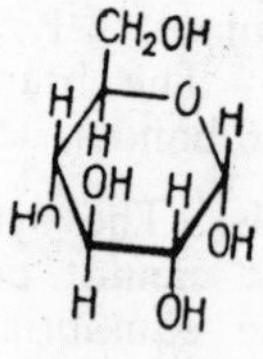

β-D-Glucose

the chair, in which all the hydroxy groups of the β-form are equatorial. In the α-form, the two hydroxyl groups at positions 1 and 2 are cis. It is susceptible to various forms of anaerobic and aerobic fermentation to alcohol, lactate, acetate, or citrate. It is the most important animal monosaccharide, and is also the most abundant natural organic compound.

Glucose 1, 6 bisphosphate. A glucose derivative which is an important intermediate in glycolysis. It is synthesized in yeast, plants and muscles by the reaction glucose 1-phosphate + ATP Mg^{2+} glucose 1, 6-bisphosphate + ADP; and in Escherichia coli and muscles by the reaction 2 glucose 1-phosphate ⇌ glucose 1, 6-bisphosphate + glucose. It is the cosubstrate of phosphoglucomutase (EC 2.7.5.1), which catalyses the interconversion of glucose 1- and 5-phosphates.

Glucose 1,6-bisphosphate (Ⓟ donor) + Glucose 1-phosphate (Substrate) $\xrightleftharpoons[\text{Phosphoglucomutase}]{Mg^{2+}}$ Glucose 6-phosphate (Product) + Glucose 1,6-bisphosphate (regenerated Ⓟ donor)

Phosphoglucomutase reaction

D (+)-Glucose Oxidase (EC 1.1.3.4). A plant and microbial flavin enzyme which oxidizes β-D-glucose in the presence of O_2 to glucuronic acid and H_2O_2.

Glucose 1-phosphate, Cori Ester. The product of phosphorolysis of Glycogen and Starch. It is converted to glucose 6-phosphate by phosphoglucomutase (EC 2.7.5.1).

Glucose 6-phosphate, Robinson Ester. The key intermediate in Carbohydrate metabolism.

Glucose-6-phosphate Dehydrogenase, GPDH (EC 1.1.1.49). The key enzyme of the Pentose phosphate cycle. It occurs widely in plants and animals and has been shown to be formed from inactive precursor subunits. It is a tetrameric enzyme with Mr ranging from 206 000 in Neurospora to 240 000 in erythrocytes. In humans, 50 hereditary variants of the erythrocyte GPDH are known.

Glucuronate Pathway, Glucuronate-xylose Cycle, D-glucuronate-L-gulonate Pathway. A pathway in carbohydrate metabolism by which myo-inositol and ascorbate are synthesized and degraded. (Fig. 1). Glucose is oxidized at position 6 to D-glucuronate probably via UDP glucose (Nucleoside diphosphate sugars). Glucuronate, which is also the product of myo-inositol oxygenase, is the starting material for the synthesis of glucuronides. It is degraded by reduction to L-gulonate. Since the C-6 of glucuronate becomes the C-1 of gulonate, the latter belongs to the L series of carbohydrates. L-Gulonate is diverted into the

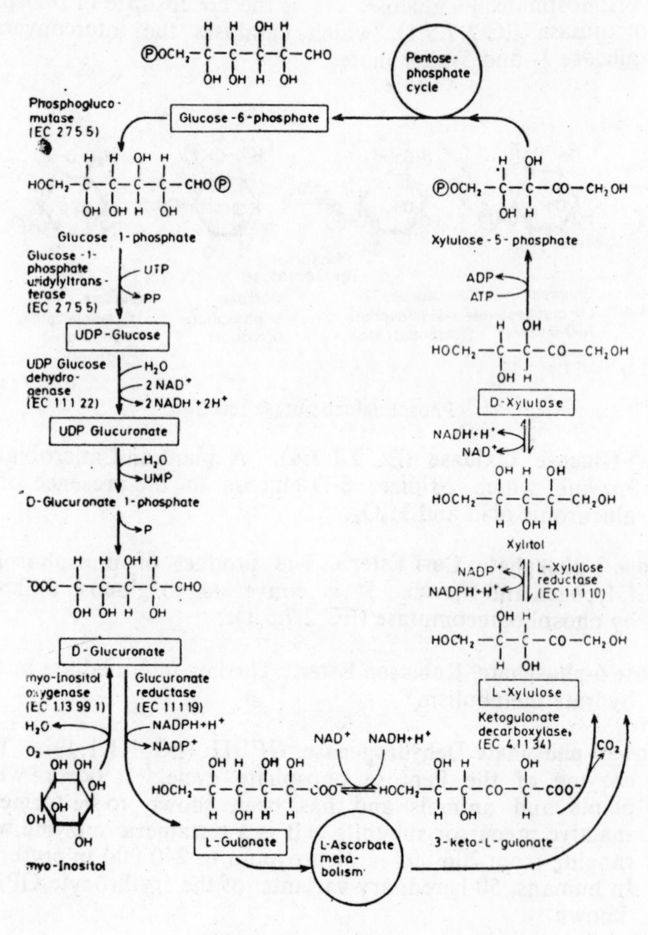

Glucuronate pathway

L-ascorbate pathway, or it is oxidized to 3-keto-L-gulonate, which is decarboxylated to form L-xylulose. Xylulose is reduced to the sugar alcohol xylitol, which is reoxidized, to D-xylulose. The change in configuration is again accomplished by an end-for-end shift in which the C 5 of xylulose is formed from the C-1 of xylitol. D-Xylulose is phosphorylated to xylulose-5-phosphate, which is a member of the Pentose phosphate cycle. Glucose 6- phosphate, the precursor of UDP glucose, is regenerated from xylulose. 5-phosphate in the pentose phosphate cycle.

Bacteria follow an alternative pathway for the degradation of glucuronate which goes via several intermediates to glyceraldehyde 3-phosphate and pyruvate (Fig. 2)

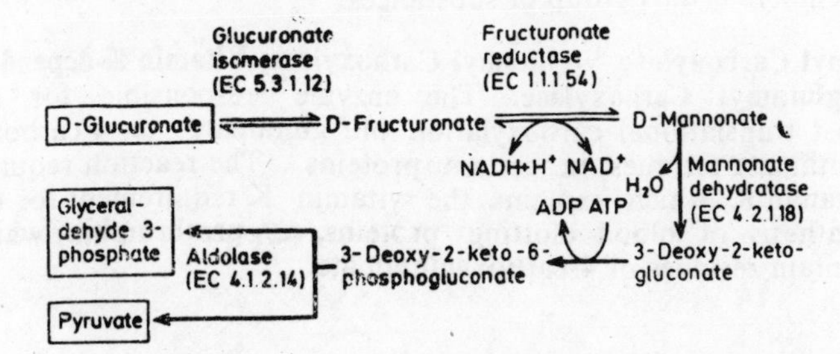

Alternative Catabolic Pathway for Glucuronate in Bacteria

l-Glucuronic Acid. A derivative of glucose. It is a component of mucopolysaccharides like hyaluronic acid and chondroitin sulfate. It is synthesized from D-glucose.

d-Glutamic Acid, Abb. **Lα aminoglutaric Acid.** A proteognic amino acid with two carboxyl groups. Since only the α-carboxyl group forms peptide bonds in proteins, the remaining free carboxyl group gives the polypeptide an acid character.

Cylization of L-glutamic Acid

Glu is present in nearly all proteins, especially seed proteins It is easily cyclized to L-pyrrolidone carboxylic acid on heating; this compound is also formed as a post translational modification of Glu residues in proteins.

l-glutamine, Abb. **Gln or Glu-NH$_2$** A proteogenic amino acid, the 5-amide of L-glutamic acid. Gln is synthesized from glutamic acid and ammonium in an endergonic reaction :

$$Glu + NH_4 + ATP \xrightarrow[Mg^2]{} Gln + ADP + P_i$$

catalysed by glutamine synthetase (EC 6.3.1.2). The synthesis is made thermodynamically favourable by coupling with hydrolysis of ATP.

Glutamine Antagonists, Glutamine Analogs. The structural analogs of L-glutamine which competitively inhibit glutamine-dependent enzyme reactions. 6-Diazo-5-oxo-L-norleucine and Albizziin are members of this group of substances.

4-glutamyl Carboxylase, γ-glutamyl Carboxylase, Vitamin K-dependent γ-glutamyl Carboxylase. The enzyme responsible for the post translational carboxylation of glutamate to 4-carboxyglutamate residues in certain proteins. The reaction requires vitamin K, which explains the vitamin K requirement for the synthesis of blood clotting proteins, *e.g.* prothrombin, which contain residues of 4-carboxyglutamate.

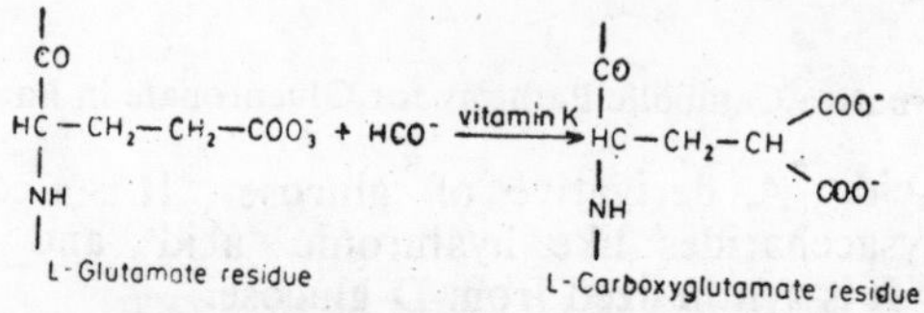

$$\underset{\text{L-Glutamate residue}}{\overset{|}{\underset{|}{\overset{CO}{\underset{NH}{HC}}}}-CH_2-CH_2-COO^-} + HCO^- \xrightarrow{\text{vitamin K}} \underset{\text{L-Carboxyglutamate residue}}{\overset{|}{\underset{|}{\overset{CO}{\underset{NH}{HC}}}}-CH_2-CH(COO^-)_2}$$

Glutathione: HOOC-CH(NH$_2$)-CH$_2$-CH$_2$-CO-NH-CH(CH$_2$SH)-CO-NH-CH$_2$-COOH. A naturally occurring peptide with a γ-peptide bond. It serves as a biological redox agent, as a coenzyme and cofactor, and as a substrate in certain coupling reactions catalysed by Glutathione S-transferase.

Glutathione S-transferase, Ligandin, (EC 2.5.1.18). A group of enzymes of the liver cytosol, which catalyse reaction of glutathione (acting as a nucleophile) with a wide range of electrophilic substrates :

$$RX + GSH \rightarrow HX + RSG$$

where GSH is reduced glutathione, **R** may be an aliphatic, aromatic or heterocyclic radical, and X may be a sulfate, nitrite, halide, epoxy, or ethene group, or the cyanide group of a thiocyanate.

Glutelins. A group of simple proteins from grain. They may contain up to 45% glutamic acid. The best known are glutenin in wheat, orycenin in rice and hordenin in barley.

Gluten. A mixture of about equal parts of the simple proteins glutelins and prolamins. It makes flour capable of rising when made into bread.

Gly. Abb. for glycine.

Glycerate Pathway. An anaplerotic pathway for utilization of glyoxylate in plants and microorganisms. Two molecules of glyoxylate are converted to tartronate semialdehyde by tartronate semialdehyde synthase (EC 4.1.1.47). The semialdehyde then gets reduced to D-glycerate and phosphorylated by glycerate kinase (EC 2.7.1.31) to 3-phosphoglycerate. This is converted by phosphoglyceromutase (EC 2.7.5.3) and enolase (EC 4.2.1.11) to phosphoenolpyruvate, which enters the general metabolism. Balance :

$$2\ \text{glyoxylate} + ATP + NAD(P)H + H^{+} \rightarrow$$
$$\text{phosphoenolpyruvate} + ADP + CO_2 + NAD(P)$$

Glyceric Acid: $HOCH_2$-CHOH-COOH. An hydroxyacid which is metabolically important, especially in its phosphorylated from, Glycerate is formed from glyoxylate via the Glycerate pathway or from serine via hydroxypyruvate.

Glycerides, Acylglycerides. Esters of fatty acids with glycerol. Mono-and diacylglycerides usually occur only as metabolic intermediates. Mixtures of triglycerides are Fats.

Glycerol, Propan-1, 2,-3-triol: CH_2OH-CHOH-CH_2OH. A syrupy, sweet-tasting fluid. It occurs most often in nature as its esters in fats, fatty oils and Phospholipids. About 3% G is formed as a byproduct of alcoholic fermentation by reduction of dihydroxyacetone phosphate or glyceraldehyde 3-phosphate and hydrolysis of the phosphate group.

Glycinamide Ribonucleotide. Abb. Gar: an intermediate of Purine biosynthesis.

Glycine, Abb. Glu, Aminoacetic Acid: H_2N-CH_2-COOH. The simplest proteogenic amino acid. Gly is not essential in the diet. Its amino nitrogen can easily be exchanged, so it is added to amino acid diets in large amounts to provide nitrogen for the synthesis of other amino acids. Gly is converted by transamination or oxidative deamination to glyoxylic acid, which is further metabolized to formic acid.

Glycine Allantoin Cycle, Purine Cycle. A series of reactions leading to the synthesis of urea in the lungfish (Dipnoi) and certain urea-accumulating plants. The glyoxylate and urea produced by purine degradation are reassimilated at different rates.

Glycinin. The chief protein component of soybeans. It is stored in subcellular particles, the "protein bodies". The dimer (M_r 350000) consists of 12 subunits (M_r 28500 each). 6 of these chains are acidic (IP 3 0 to 3 4) and 6 are basic (IP 8.0 to 8.5). It is structurally related to the protein Arachin.

Glycogen. An animal polysaccharide which, like amylopectin, consists of D-glucose units. Most of the glycosidic bonds are α 1, 4. but at branching points there are also α-1, 6 links. The side chains consist of 6 to 12 glucose units. It is most abundant in the liver (up to 10%) and the muscles (up to 1%) and serves as a short-term storage substance.

Glycogen Metabolism. The formation and breakdown of glycogen The process has been well studied in mammalian muscle. Glycogen is synthesized from uridine diphosphate glucose (UDPG) and a starter molecule of the structure (α-1, 4-glucosyl)$_n$ by glycogen synthase (EC 2.4.1.11). The branching of the molecule is accomplished by 1, 4-α-glucan branching enzyme (EC 2 4.1.18), which transfers a segment of chain from a 4-OH to a 6-OH. (The enzyme is sometimes called Q enzyme).

Glycogen Storage Disease. A condition characterized by excessive storage of glycogen in the liver, muscles, or other organs due to the hereditary lack of one of the enzymes of glycogen degradation. Several different types are known; they are listed in the Table. The prognosis for Types I, III, V and VI is relatively favourable, while Types II and IV are fatal in early childhood.

TABLE

Types of glycogen storage disease

Type	Missing enzyme	Affected organs
I (von Gierke)	Glucose-6-P phosphatase	Liver, kidneys, small intestine
II (Pompe)	Amylo-1, 4-α-glucosidase	Skeletal muscles, liver, CNS, leukocytes

III (Cori)	Amylo-1, 6-α-glucosidase	Skeletal muscles, liver leukocytes
IV (Andersen)	Amylo-1, 4→1, 6-transglucosidase	Skeletal muscles, liver
V McArdle)	Muscle phosphorylase	Skeletal muscles
VI (Hers)	Liver phosphorylase	Liver leukocytes
VII	Phosphofructokinase	
VII	Phosphorylase kinase	

Glycolic Acid, Hydroxyacetic Acid: $HOCH_2$-COOH. An hydroxy-carboxylic acid found in young plant tissue and green fruits, such as gooseberries, grapes and apples. It is an intermediate in photosynthesis. Its precursor is active glycolaldehyde, which is formed in Transketolation reactions.

Glycolipids. Compounds in which one or more monosaccharide residues are glycosidically linked to a lipid part, either a mono- or diacylglycerol, a long-chain base (sphingoid) like sphingosine, or a ceramide.

Glycolysis, Embden-meyerhof-parnas Pathway. The main pathway for anaerobic degradation of carbohydrates found in all groups of organisms. For each mole of glucose consumed, 150.7 kJ (36 kcal) energy is released. The organism obtains a net yield of 2 moles ATP per mole glucose. The starting material is glycogen or starch, which is hydrolysed to glucose 1-phosphate or glucose monomers. It can be divided into four phases :

1. the formation of two molecules of triose phosphate (glyceraldehyde 3-phosphate and dihydroxyacetone phosphate) from one molecule of hexone. Two molecules of ATP are consumed in this step.

2. Dehydration of the triose phosphates to 2-phosphoglycerate; NAD^+ is oxidized in the process to NADH. One molecule ATP is generated per triose phosphate, which makes up for the ATP invested in step 1.

3. Conversion of 2-phosphoglycerate to Pyruvate via phosphoenolpyruvate. Another ATP is generated here for each molecule of pyruvate.

4. The reduction of pyruvate to regenerate NAD^+. In muscle, the pyruvate is converted to lactate, and in yeast, it is reductively decarboxylated to ethanol. In an aerobically

respiring cell, the NADH is oxidized ultimately by the respiratory chain (Hydrogen metabolism), and the pyruvate is further oxidized in the Tricarboxylic acid cycle. Under anaeraboic condition, G. can only be maintained continuously by the redox reactions.

Balance : Glucose ($C_6H_{12}O_6$) + 2 P_i + 2 ADP → 2 lactate $C_3H_6O_3$) + 2 ATP.

If the starting material is glycogen, which is degraded to glucose 1-phosphate, the yield is 3 ATP per glucose 1-phosphate consumed.

The key enzyme of G. is 6-phosphofructokinase (EC 2.7.1.11), which is inhibited by high concentrations of ATP and is activated by ADP or AMP. Its product, fructose bisphosphate, activates pyruvate kinase. The Pasteur effect is another form of regulation of glycolysis.

Glycophorins. A group of sialoglycopeptides found in human erythrocytes.

Glycoproteins. The proteins with covalently bonded sugar units. They are typical of plants and animals, but not bacteria. Typical glyco-proteins are either secreted into body fluids or are membrane proteins. They include many enzymes, most protein hormones, plasma proteins, all antibodies, complement factors, blood group and mucus components and many membrane proteins.

Glycosidases. A group of hydrolases which attack glycosidic bonds in carbohydrates, glycoproteins and glycolipids. They are not highly specific.

Glycosides. A group of compounds in which mono- or oligosaccharide units are linked by acetal bonds with hydroxyl groups of alcohols or phenols (O-glycosides) or with amino groups (N-glycosides). Acid hydrolysis cleaves these compounds into a sugar and a noncarbohydrate portion, the aglycon or genin.

Glycyrrhetic Acid, Glycyrrhetin. A pentacyclic triterpene with carboxylic acid and ketone functions. It is the aglycon of glycyrrhizic acid, an extremely sweet principle from the root of licorice, Glycyrrhiza glabra L. It is the aglycon of Saponins from other plants, *e.g.* the bark of Pradosia latescens and rhizomes of the fern Polypondium vulgare.

Glyxaolase, Aldoketomutase. A system of enzymes found in many organisms. It consists of lactoyl-glutathione lyase (G. I.) (EC

4.4.1.5), which condenses methylglyoxal and glutathione to S-lactoyl-glutathione, and hydroxyacylglutathione hydrolase (G. II) (EC 3.1.2.6), which hydrolyses the condensation product to lactate and glutathione.

Glyoxylate Cycle, Kerbs-kornberg Cycle. An alternative to the tricarboxylic acid cycle found in microorganisms and plants. In the

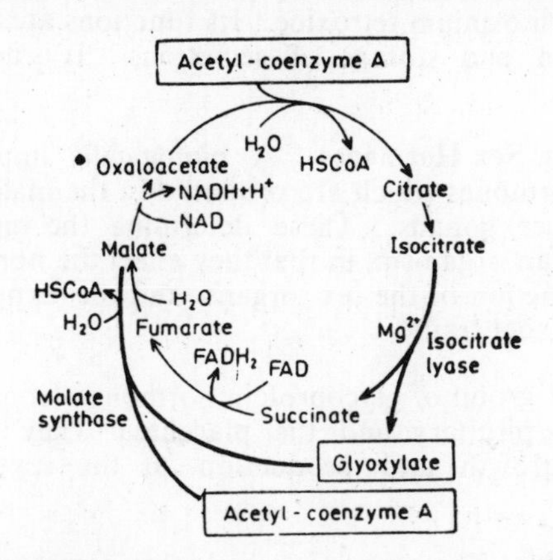

This oxaloacetate is generated from acetyl-coenzyme A (Fig.) The key enzymes are isocitrate lyase (isocitratase) (EC 4.1.3.1) and malate synthase (EC 4.1.3.2). Balance : 2 acetyl-CoA + NAD^+ + 2 H_2O → succinate + 2 CoA + NADH + H^+.

Glyoxylic Acid, Oxoacetic Acid, Glyoxalic Acid. CHO-COOH. A carboxylic acid found in green fruits, seedlings and young leaves. In some plant families (maple, borage, horse chestnut). It is present in the form of allantion and allantioc acid, which arise in the course of purine carbolism. It is synthesized by transamination or oxidative determination of glpcine, or from sarcosine. Decarboxylation of this acid yields active formate. It is the starting point of the Glyoxylate cycle.

GMP. Abb. for guanosine 5′-monophosphate.

GMP Reductase (EC 1.6.6.8). A flavin enzyme which catalyses the conversion of guanosine 5′-monophosphate to inosinic acid in a single step.

Goitrogens, Antithyroid Componnds. Substances that inhibit iodine peroxidase (conversion of iodide to "active iodine" : H_2O_2 + $2I^-$ + 2 H^- → 2"I" + 2 H_2O). the iodination of tyrosine residues, and the coupling of monoiodotyrosine and diiodotyrosine residues to form T_3 and T_4. The resulting low plasma levels of T_3 and T_4 cause enhanced release of thyrotropin from the anterior pituitary. This, in turn, results in a compensatory hypertrophy of the thyroid gland. Such an enlargement, without inflammation or malignancy, is called a goiter.

Golgi Apparatus. A structural element of eukaryotic cells which is stainable with osmium tetroxide. Its functions are the synthesis, concentration and storage of secretions. It is not present in prokaryote.

Gonadal Hormones, Sex Hormones A biologically important group of steriod hormones which are produced in the male (testes) and female (ovaries) gonads. These determine the male or female character of an organism, in that they effect the normal development and function of the sex organs and the expression of the secondary sexual traits.

Gonadotropins. A group of glycoprotein hormones from the anterior lobe of the pituitary and the placenta. They stimulate the gonads to growth and production of the sex-specific hormones.

Gonads, Sex Glands. The organs in which the gametes and sex hormones are formed. The male gonads are the testes, the female organs, the ovaries. The formation and secretion of the sex hormones are controlled by the pituitary gonado-tropins.

Gossypol. An aromatic triterpene from cotton seed (Gossypum hirsutum). Which is synthesized from mevalonic acid via neryl pyrophosphate and cis, cis farnesyl pyrophosphate.

Gougerotin, Aspiculamycin, Asteromycin. 1-cytosinyl-4-sacrosyl-D-serylamino-1. 4-dideoxy-β-D-glucopyranuramide, an important pyrimidine antibiotic synthesized by Streptomyces gougeroti. It inhibits protein biosynthesis on both eukaryotic and prokaryotic ribosomes.

Gramicidins. The cyclic peptide antibiotics produced by Bacillus brevis. Gramicidin S. cyclo-(-D-Phe-L-Pro-L-Val-L-Orn-L-Leu)$_2$ The primary structure was determined by Synge, and the first total synthesis was reported in 1965 by R. Schwyzer et. al. It is effective against Gram-positive, but not Gram-negative bacteria. Its structure is thought to be an antiparallel pleated sheet.

L-Leu
L-Orn D-Phe
L-Val L-Pro
L-Pro L-Val
D-Phe L-Orn
L-Leu

Gramicidin S

Grisein. An iron-containing antibiotic synthesized by Streptomyces griseus. It is a cyclic polypeptide containing cytosine. The iron ions are strongly bound as hydroxamate-iron (III) complexes. It is especially effective against Gram-negative bacteria, but is not effective against fungi.

Griseofulvin. An antifungal agent synthesized by Panicillium griseofulvi. It is a polyketide synthesized from one molecule of acetyl-CoA and 6 molecules of malonyl-Coa.

Group Transfer. The enzymatic transfer of a functional group from one molecule to another. In both anabolism, and catabolism, small groups of atoms are handled as units. Coenzymes often serve as corriers of the groups, hence the term "transport metabolites". Hydrolysis and phosphorolysis can be regarded as special cases of G.t. in which the group is transferred to water or a phosphate ion.

Growth. An irreversible increase in the mass of living material, usually accompanied by an increase in the size of a cell or organism, as well as an increase in the number of cells.

Growth Factor, Growth Supplement. A supplementary substance required in the Nutrient medium for growth of a specific organism. Growth factors are required, *e.g.* by Auxotrophic mutants; these have lost the ability to synthesize an essential metabolite, which then becomes a growth factor.

Growth Regulators. The organic compounds, which in small quantities inhibit, accelerate or in some way influence physiological processes in plants. G.r. include natural (endogenous) substances, *e.g.* Phytohormones and native inhibitors; and many synthetic compounds, especially herbicides and Growth retardants. The Auxins include natural and synthetic compounds.

Growth Retardants. Synthetic plant growth inhibitors, often used in agriculture for the control of plant growth. In particular, they cause stalk shortening in grasses, and some growth retardes are used for controlling stalk length in cereal crops.

Growth Vitamin. Vitamin A (Vitamins).

GTF. Abb. for glucose tolerance factor. Chromium.

GTP. Abb. for guanosine 5′-triphosphate, Guanosine phosphates.

Guanidine Derivatives. Compounds containing the strongly basic guanidine group (Fig. 1). Guaniodine itself, $H_2N\text{-}C(=NH)\text{-}NH_2$, is found in freeform only in a few plants. The guanidino group is synthesized de novo in the course of arginine biosynthesis; other G. are formed by Transamidination from arginine are

$HN{=}C(NH_2)(NH{-}R)$ $HN{=}C(NH_2)(N(R)(R_1))$ $HN{=}C(NH{-}R_1)(NH{-}R)$ $HN{=}C(N(R_1)(R_2))(NH{-}R)$

Types or guanidine compounds

limited to plants, *e.g.* L-Canavanine and Galegine; other G.d., such as Phosphagens, only occur in animals. The G.d., streptidin, is a component of the antibiotic, sreptomycin. G.d. are degraded by various enzymes :

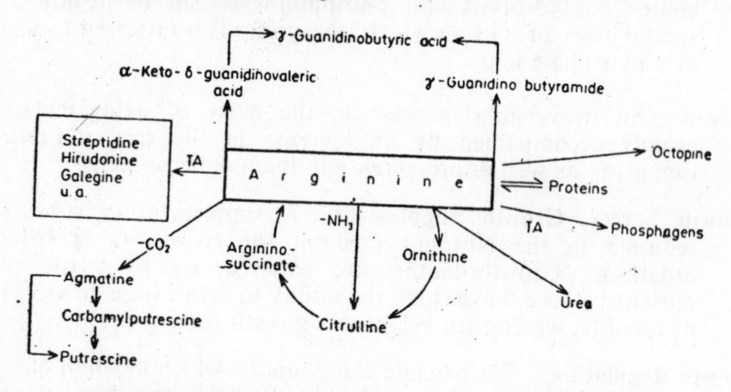

Metabolic fates of Arginine. TA=Transamidination

1. Transaminase can act as a catabolic enzyme, if the G d. is further degraded by other enzymes and the resulting amino compound is catabolized (Fig. 3);

2. Arginase and heteroarginase catalyse the hydrolysis of G.d. to produce urea. Arginase (L-arginine-ureohydrolase) cleaves L-arginine, L-canavanine and γ-hydroxyarginine, and it probably occurs in isoenzyme forms that possess heteroarginase activity. Heteroarginases differ markedly

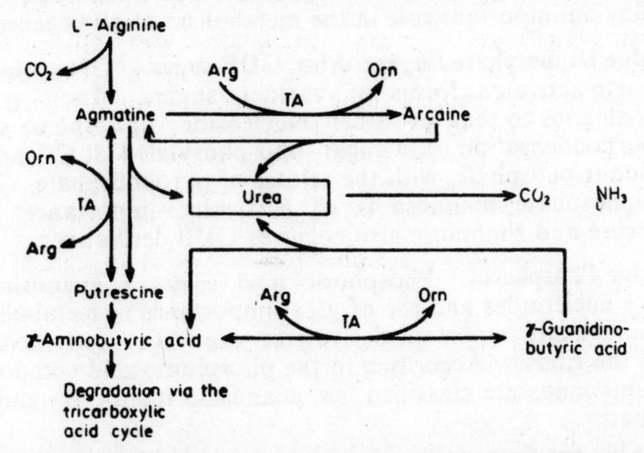

Arginine Degradation in a mushroom. TA—Transamidination, Arg = L-arginine, Orn — L-ornithine

from classical arginase with respect to substrate specificity; they cleave G.d. with chain lengths less than 6 C-atoms, *e.g.* γ-guanidobutyric acid, and they show wider specificity in that they cleave certain special G.d., *e.g.* γ-guanidobutyric acid (also cleaved by γ-guanidobutyrase from Streptomyces griseus), areain, ngmatine and streptomycin.

Arginine decarbozy-oxidase, hitherto described only from Streptomyces griseus and the pond snail (Limnaea stagnalis), degrades arginine to γ-guanidobutyramide, canavanine to β-guanido-cxypropionamide, and homoarginine to δ-guanidovaleramide. Other reactions for the degradation of L-eanavanine are found in bacteria.

Guanine, 2-amino-6-hydroxypurine, abb. **G** or **Gua.** One of the four nucleic acid bases. M_r 151.13, m.p. 360°C (d). It is also a

OH, N, N, H_2N, N, N, H

Cuanine

component of nucleotide coenzymes and the starting material for the biosynthesis of many natural products, including proteins and the vitamins folic acid and riboflavin.

Guanosine. Abb. **Guo.** 9-βD-ribofuranosylguanine, a β-glycosidic nucloside containing D-ribose and guanine. The Guanosine phosphates play an important role in the metabolism of all organisms.

Guanosine Diphosphate Sugars. Abb. **GDP-sugar.** Guanosine-diphosphate-activated forms of various sugars. Their synthesis is analogous to that of other Nucleoside diphosphate sugars *i.e.* the condensation of a sugar phosphorylated at C-1 and guanosine triphosphate, with the release of pyrophosphate. Guanosine diphosphate mannose is of particular importance; glucose, fucose and rhamnose also occur as GDP derivatives.

Guanosine Phosphates. Phosphoric acid esters of guanosine. They are nucleotides and are of great importance in metabolism. The biologically important derivatives are those esterified on C-5′ of the ribose. According to the phosphoric acid component, the compounds are classified as guanosine mono-, di- and triphosphates.

Guo. Abb. for Guanosine.

Gutta. A rubber-like polyterpene of about 100 isoprene units in which the double bonds are in the trans configuration. Gutta is less elastic than rubber, but it is more resistant to chemicals and environmental influences (insulating material). Depending on it source, it occurs in mixtures with other terpenes. The mixture with resins is called guttapercha, and that with triterpene alcohols is chicle (starting material for manufacture of chewing gum).

H

Haem. Same as heme.

Hallucinogens. Refers to a group of drugs which cause changes in mood perception, thinking and behavior. The group does not include addictive drugs, but some users become dependent on hallucinogens. Those hallucinogens with the strongest action are called psychedelics.

D-Hamamelose. A monosaccharide with a branched carbon chain. It occurs in higher plants for example hamamelis (which hazel) bark. Its biosynthesis, involves an intramolecular rearrangement of an unbranched hexose.

Haptens. Refer to the partial or incomplete antigens. These are either chemically defined molecules, *e g.*, dinitrophenol, or part of an antigen. They can bind specifically to the corresponding antibodies (Immunoglobulins), but they do not act by themselves as antigens. Only after coupling to a carrier protein do the haptanes become full antigens which are capable of eliciting an immune response.

Haptoglobin, Abb. Hp. An acid α_2-plasma glycoprotein, which binds specifically to free plasma oxyhemoglobin to form a high molecular weight complex, which cannot be filtered by the kidneys. The associated conformational change in the hemoglobin allows the heme α-methenyl oxygenase of the liver to remove the heme porphyrin ring. Thereafter the globin is degraded by the trypsin-like protease action of the β-chain of the haptoglobin. Haptoglobin is a tetramer consisting of two nonequivalent chain pairs, 2α and 2β held together by disulfide bridges. Human Hp. exists in three genetic variants Hp 1-1, 2-2 and 2-1 which differ in their electrophoretic patterns.

Har. Abbreviation for Homoarginine.

Harman Alkaloids. A group of indole alkaloids with a β carboline skeleton. These are formed biosynthetically from tryptophan and a carbonyl component. In medicine, these alkaloides are occasionally used for encephalitis and Parkinson's disease. Some of these alkaloids lead to hallucinutions and in toxication.

Hashish. The dried resin from the glandular hairs of the female hemp plant (Cannabis sativa L) Due to the psychoactive Δ^3-tetrahydrocannabinol (2 to 8%) which it contains, it is one of the most common Narcotics. Marihuana (often used synonymously with hashish) is the dried and chopped tips of the shoots of the female hemp with a content of 0 5 to 2% Δ^3-tetrahydrocannabinol. Both hemp drugs have been used for millenia in folk medicine, due to their intoxicating action. Today they are, along with alcohol, the most widely used drugs; the number of consumers is estimated as 300 million. The drugs are usually smoked alone or mixed with tabacco in cigarettes ("joints") or pipes.

Hatch-slack-kortschak Cycle, Abb. **HSK Cycle, C-acid Cycle.** The photosynthetic reaction cycle in C_4 plants. The enzyme which carries out photosynthetic carboxylation is phosphoenolpyruvate carboxylase :

1. Which carboxylates phosphoenolpyruvate to oxaloacetate. Oxaloacetate is reduced to L-malate by the NADP-dependent malate dehydrogenase.

2. L-Malate leaves the mesophyll cells of the C_4 plant and is oxidatively decarboxylated.

3. To pyruvate in the vascular cells. The carbon dioxide, which was originally fixed in the mesophyll cells as the β-carboxyl group of oxaloacetate, is thus released again and can be assimilated by the reactions of the Calvin cycle.

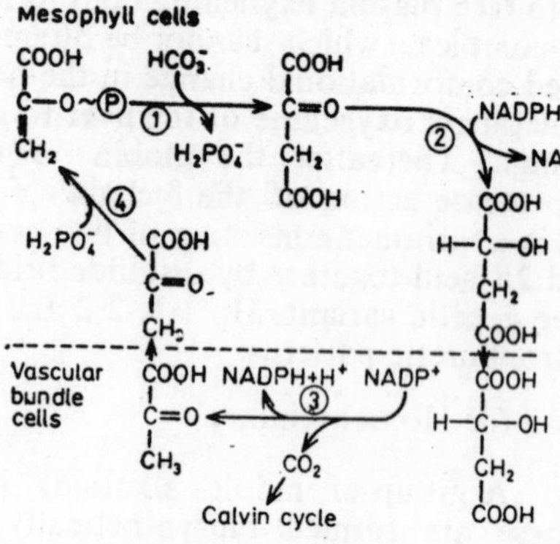

Mesophyll cells

COOH — C—O~P — CH_2 ; HCO_3^- ; ① ; $H_2PO_4^-$; COOH — C=O — CH_2 — COOH ; ② ; $NADPH+H^+$; $NADP^+$; COOH — H—C—OH — CH_2 — COOH ; ④ ; $H_2PO_4^-$; COOH — C=O — CH_2

Vascular bundle cells

COOH — C=O — CH_3 ; $NADPH+H^+$; $NADP^+$; ③ ; CO_2 ; Calvin cycle ; COOH — H—C—OH — CH_2 — COOH

The second reaction product, pyruvate, recenters the mesophyll cell, where it is converted back into phosphoenolpyruvate (PEP) by;

4 pyruvate, orthophosphate dikinase (ATP:pyrophosphate biphosphotransferase) in the following reaction :

$$\text{pyruvate}+\text{ATP}+P_i \rightarrow \text{PEP}+\text{AMP}+PP_i$$

Hb : Abbreviation for Hemoglobin.

HbS. Abbreviotion for Sickle-cell hemoglobin.

HCG. Abbreviation for Human chorionic gonadotropin, same as Chorionic gonadotropin.

Hcy. Abbreviation for Homocysteine.

Heavy Metals. All metals with a density greater than 5. In living organisms, they are usually present in stable organic complexes.

Biological functions of H.m. are listed in Table 1. Chemical ligands of H.m. are listed in Table 2.

TABLE 1.

Heavy Metals in biomolecules

Heavy metal	Type of biomolecule	Biological function
Iron	Hemoglobins; cytochromes	O_2-Transport; electron transport
	Flavin enzymes (metalloflavo-enzymes), *e.g.*, xanthine oxidase	Oxidation, dehydrogenation, and/or reduction
	Iron-sulfur protein *e.g.*, ferredoxin, complexes of the respiratory chain	Electron transfer
	Nitrogenase	Reduction of N_2 to ammonia
	Ferretin; conalbumin, transferrin (siderophilin)	Fe-storage; Fe-transport
	Siderochromes, mycobactin; enterobactin, etc.	Fe-transport in microorganisms
Cobalt	Vitamin B_{12} and its coenzyme forms	Reduction and methyl group transfer (methionine synthesis); isomerization
Copper	Laccase, cytochrome oxidase Cupreine, ceruloplasmin	Oxidation Storage and transport of copper
Manganese	Arginase	Arginine hydrolysis, urea cycle
	Decarboxylases and other enzymes	Release of CO_2, etc.
Molybdenum	Nitrogenase	Binding and activation of molecular nitrogen (subsequent reduction also requires iron)

	Nitrate reductase	Nitrate reduction
	Xanthine oxidase	Purine oxidation, etc.
Zinc	Carbonic anhydrase, peptidases, phosphatases, pyridine nucleotide enzymes and other proteins.	Zn functions in substrate binding (*e.g.*, for hydride transfer in ternary complexes with pyridine nucleotide enzymes), and in protein structure (*e.g.* alcohol dehydrogenase)
Insulin		Aggregation of the polypeptide

TABLE 2

Ligands of Heavy Metals in Biomolecules

Heavy metal	Ligand	Example
Iron	Porphyrin, imidazole Sulfur Phenolate	Myoglobin Ferredoxin Transferrin
Cobalt	Corrin, benzimidazole	B_{12} and derivatives
Copper	>N bases	Cupreine
Manganese	Carboxylate, phosphate, imidazole	Glycolysis and proteolysis-enzymes
Vanadium	No ligand yet identified	Vanadium proteins of tunicates
Zinc	Imidazole (His), carboxyl groups (Glu)	Carboxypeptidases
	R·S$^-$	Dehydrogenases
	Imidazole (His)	Carbonic anhydrase

Heliangin. A sesquilactone and a growth inhibitor, first isolated from the leaves of the Jerusalem artichoke (Helianthus tuberosus). It is a gibberelin antagonist and inhibits the growth of oat coleoptiles, and simulates root growth in beans (Phaseolus spp.).

Helix. The spiral arrangement of a biopolymeric compound, *e.g.*, starch, some proteins and DNA.

Helminthosporal. A natural product from the phytopathogenic fungus Helminthosporium sativum (syn. Bipolaris sarokiniana, Shoemaker), which has a physiological effect on plants similar that of gibberellins.

Helvolic Acid. Refers to one of the tetracyclic triterpene antibiotics. It differes structurally from Fusidic acid in having a Δ^1-3.6-diketo and a 7α-acetoxy function, and in lacking the 11α-hydroxyl group. It is produced by Aspergillus fumigatus and related fungal species. Like the structurally related triterpene antibiotics, fusidic acid and cephalosporin P_1, it is effective against Gram-positive organisms.

Heme a. The heme prosthetic group found in cytochromes a/a_3.

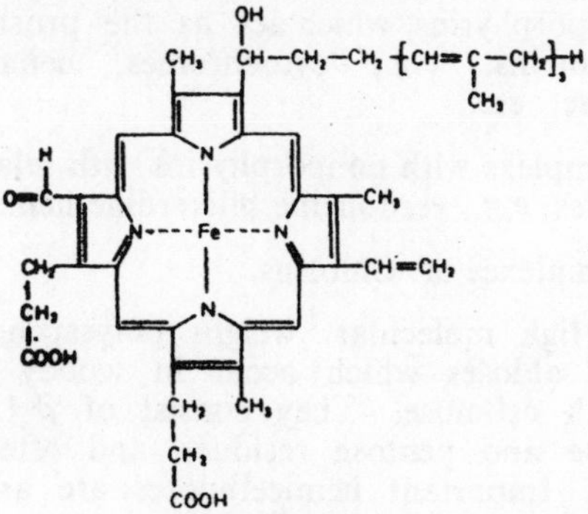

Heme a

Heme c. The heme prosthetic group of cytochromes c, b_4 and f.

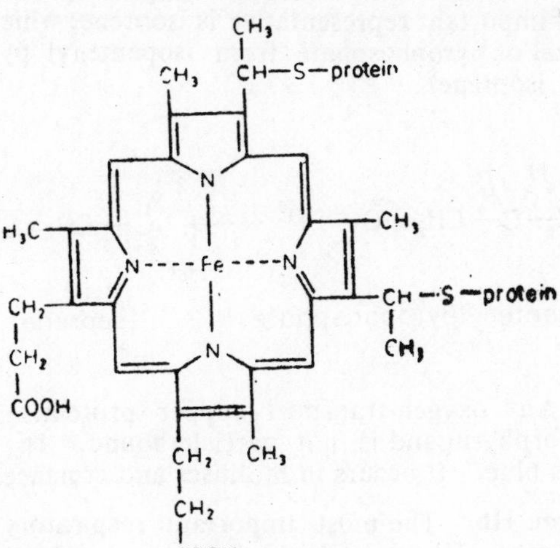

Heme c

Heme d. It was formerly heme a_1 : a heme prosthetic group found in the terminal oxidase system of many bacteria.

Heme Iron. Iron which is coordinately bound in porphyrins. It is present, for example, in the Hemoproteins.

Hemerythrin. A red-brown chromoprotein containing iron but not porphyrin. It transports oxygen and occurs as respiratory pigment in the blood cells of certain marine invertebrates, such as sipunculoid worms, polychete worms and lamp shells (Brachiopoda).

Hemes

1. Metalloporphyrins, which act as the prosthetic groups of hemoproteins, *e.g.*, cytochromes, hemoglobin, nitrite reductase, etc.
2. Iron complexs with nonporphyrins with related tetrapyrrole structures, *e.g.*, verdoheme biliverdine heme.
3. Iron complexes of Chlorins.

Hemicelluloses. High molecular weight polysaccharde complexes composed of aldoses which occur in woody parts of plants together with cellulose. They consist of β-1,4-glycosidically joined hexose and pentose residues and often also contain uronic acid. Important hemicelluloses are arabans, xylans, glucans, galactans, fructans and mannans.

Hemiterpenes. Terpenes composed of a single isoprene unit (C_5H_8). The most important representative is isoprene, which is formed by removal of pyrophosphate from isopentenyl pyrophosphate ("active" isoprene)

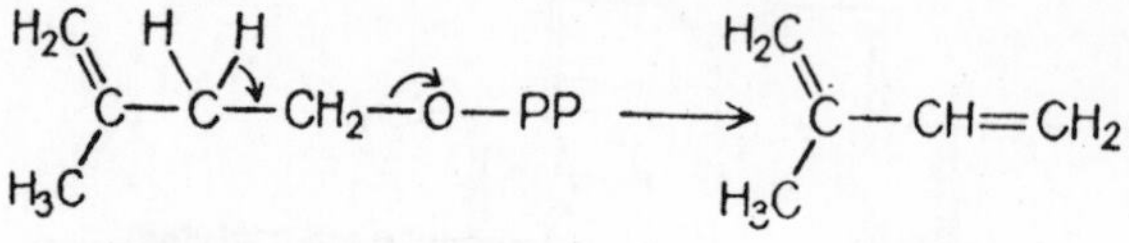

$$H_2C{=}C(H_3C){-}CH_2{-}CH_2{-}O{-}PP \longrightarrow H_2C{=}C(H_3C){-}CH{=}CH_2$$

Isopentenylpyrophosphate Isoprene

Hemocyanin. An oxygen-transport copper protein. It does not contain porphyrin and is not particle-bound. In its oxidized state, it is blue. It occurs in molluscs and crustacea.

Hemoglobin, abb. **Hb.** The most important respiratory protein of the vertebrates. Hb, the colored component of blood, is present as a 34% solution in the erythrocytes (red blood cells) and carries oxygen from the lungs to the tissues of the organism. It is a tetramer composed of 2 polypeptide chain pairs and 4 heme groups with M_r 64500. With a Fe^{2+} content of 0.334%, the total of 950 g Hb in a human represents 3.5 g or 80% of the total body iron. The Hb of the adult human being consists of 96.5 to 98.5% HbA_1 ($\alpha_2\beta_2$) and 1.5 to 3.5% HbA_2 ($\alpha_2\delta_2$).

Hemopexin. A single-chain, heme-binding β_1^- plasma glycoprotein. It contains 22% carbohydrate. In contrast to Haptoglobin. It binds neither hemoglobin nor cytochrome c, but only their prosthetic group, heme.

Hemoproteins. Refers to the ubiquitous chromoproteins, which, as respiratory pigments, are involved in oxygen transport and in oxygen storage. As catalase and peroxidase, they effect the reduction of peroxides, and as cytochromes, they are involved in electron transport between dehydrogenases and terminal acceptors.

Hemosiderin. An iron storage protein of the mammalian organism, functionally related to Ferritin. It is deposited in the liver and spleen (hemosiderosis), particularly in diseases associated with increased blood destruction, such as pernicious anemia, or with increased iron resorption (hemochromatosis), or even in hemorrhages.

Heparin. An acid mucopolysaccharide from animal tissues which prevents blood clotting. It consists of equal amounts of D-glucosamine and D-glucuronic acid, α-1, 4-glycosidically linked, and contains O-and N-sulfate residues. It prevents the clotting of blood by preventing the conversion of prothrombin to thrombin and of fibrinogen to fibrin. Clinically, it is applied parenterally for the treatment of thrombosis, phlebitis and embolism.

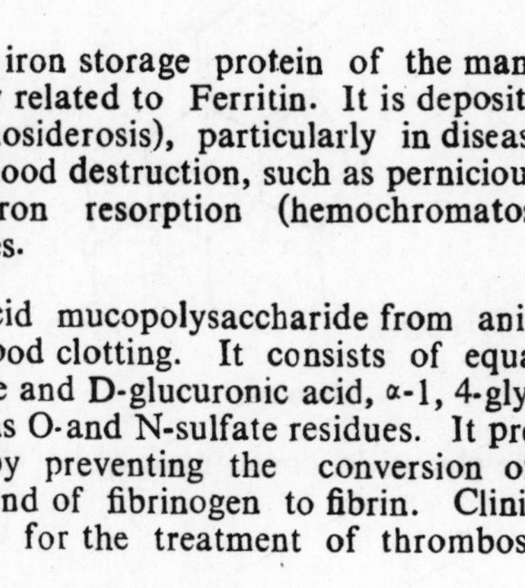

Heparin

Heparitin Sulfate, Heparan Sulfate. A monosulfate ester of an acetylated (N-acetyl) heparin As isolated from animal tissues (liver), it is probably a mixture of mucopolysaccharides with varying degrees of sulfatation or amino group acetylation.

Heptoses. Monosaccharides containing 7 atoms. The 7-phosphates of D-mannoheptulose and D-sedoheptulose are important in carbohydrate metabolism.

Heroin, Diacetylmorphine, Diamorphine. Refers to one of the most dangerous narcotics. It is not very stable and decomposes in boiling water. It is synthesized by acetylation of the two hydroxyl groups of morphine with acetyl chloride. This increases the analgesic effect by a factor of six. Due to theextreme danger of addiction, the the rapeutic use of H is for bidden in most countries.

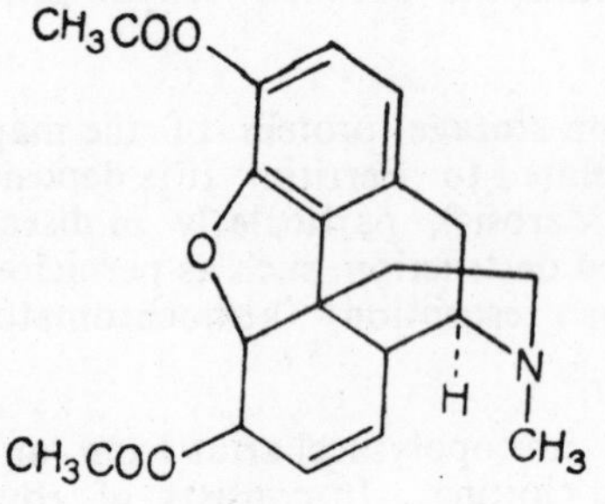

Heroin

Hesperidin. A flavone glycoside, constituting 8% of the dry weight of orange peel.

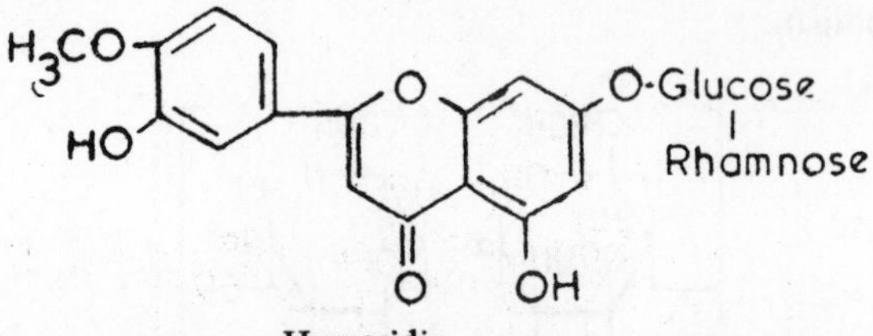

Hesperidin

Heterogeneity. In proteins, a term for differences in the structure of a species of protein which cause no change in the biological activity. It can be either genetically controlled or can arise through partial chemical or enzymatic modification.

Heteroglycans. The polysaccharides composed of two or more different carbohydrate residues, for example, pectins, plant mucilages, plant gums and mucopolysaccharides.

Heteroside. A compound of one or several carbohydrate residues and a component belonging to different class of substance, the aglycon or genin.

Herotrophy, Heterotrophic Nutrition. A nutritional dependence on organic compounds. In carbon heterotrophy, organic carbon compounds serve as sources of carbon and energy for the synthesis of body substituents and ATP.

Hexestrol. Meso-hexestrol, a synthetic compound with estrogenic activity. It is not a steroid, but is used therapeutically in the same way as natural estrogens.

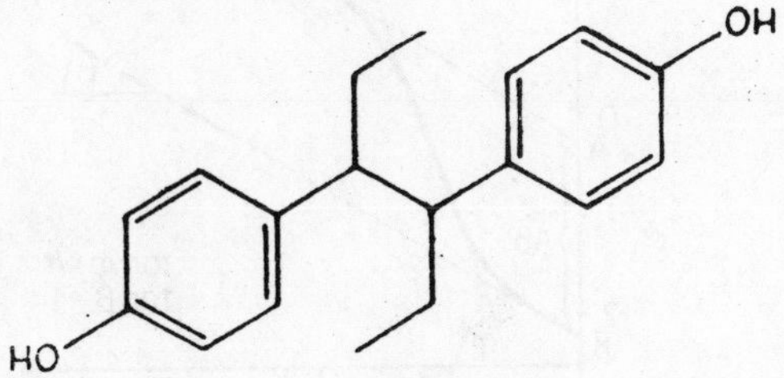

OH

HO

Hexestrol

Hexitols. Sugar alcohols with 6 C-atoms Of the 10 possible isomers. D-sorbital, dulcital, D-mannitol, iditol and allitol are found in nature.

Hexosans High-molecular-weight plant polysaccharides, belonging to the group of homoglycans, and composed of hexoses. Examples are glucans, fructans, mannans and galactans.

Hexoses. Aldoses containing 6 C-atoms; one of the important groups of monosaccharides. All possible stereoisomeric aldohexoses (there are four asymmetric C atoms) have been isolated or synthesized. D-Glucose, D-mannose, D-galactose and L- and D-talose are widespread in nature, both as free sugars and in bound form.

High Energy Bonds, Energy-rich Bonds. The chemical bonds which release more than 25 kJ/mol on hydrolysis. They are usually esters (enol, thio and phosphate esters), acid anhydrides, or amidine phosphates.

In biological systems, the energy released is used to transfer the hydrolysed residue to other metabolic compounds (group transfer).

Hill Plot. A graphic method for the determination of the degree of cooperativity of an enzyme. The plot of log $Y_s/(1-Y_s)$ versus log α is a curve with a slop of 1 for large or small α and a finite energy of interaction between the substrate-binding sites. Y_s is the saturation function, that is, the fraction of the enzyme in the enzyme-substrate complex, and $\alpha = S/K_m$. For the values of α usually obtained experimentally, an approximate straight line is obtained with the maximal stope h (Fig.), the Hill coefficient.

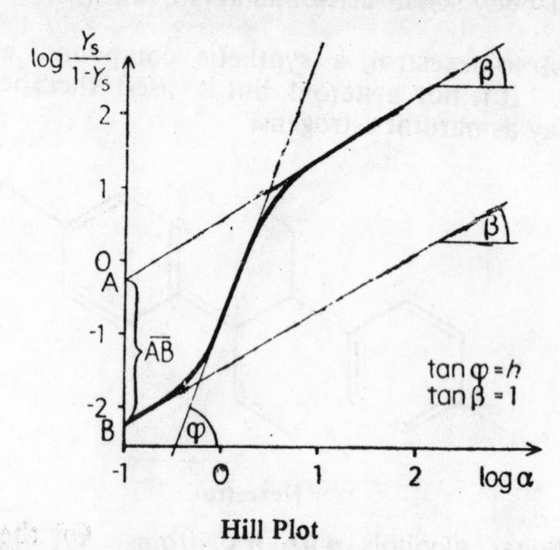

Hill Plot

Hill Reaction. The light-dependent production of oxygen by the photosynthetic system in the presence of an artificial oxidizing agent (electron acceptor). R. Hill first observed this reaction in illuminated isolated chloroplasts, in the absence of CO_2, and using iron(III) oxalate as oxidizing agent. Iron(III) oxalate (Fe^{3+} is reduced to Fe^{2+} in the reaction) can be replaced by potassium ferricyanide, quinone and other compounds (Hill reagents). Spinach chloroplasts catalyse the following H:

$$4K_3Fe(CN)_6 + 2H_2O + 4K^+ \rightarrow 4K_4Fe(CN)_6 + 4H^+ + O_2$$

The "natural" Hill reaction is the photolysis of water, and the "natural" Hill reagent is oxidized NADP.

Hippuric Acid: C_6H_5-CO-NH-CH_2-COOH. The N-benzoyl derivative of glycine. Mammalian herbivores detoxify benzoic acid by converting it to hippuric acid.

His. Abb. for L-Histidine.

Histamine: β-imidazol-4(5) ethylamine, a biogenic amine. M_r 114.14. It is formed by enzymatic decarboxylation of L-histidine. It stimulates the glands in the fundus of the stomach to secrete digestive juices, dilates the blood capillaries (important for increasing blood flow and decreasing blood pressure), increases the permeability (urtication and redening after local application of histamine), and causes contraction of the smooth muscles of the digestive tract, the uterus and the bronchia (in bronchial asthma). It is catabolized by diamine oxidases and aldehyde oxidases to imidazolylacetic acid.

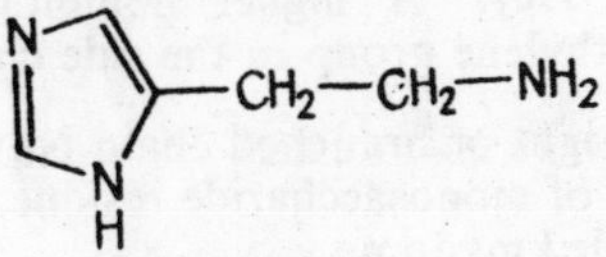

$CH_2-CH_2-NH_2$ (imidazole ring: N, N-H)

Histamine

L-Histidine Abb. His. Imidazolylalarine, a half-essential amino acid used in protein synthesis. The proportion of His in hemoglobin is especially high. It is also a component of carnosine and anserine. His is part of the active centers of many enzymes, and is an important buffer in the physiological pH range.

Histones. A group of simple, basic proteins of the cell nucleus. They are not tissue specific, and occur in all eukaryotic organisms. They have low molecular weights. H. form reversible complexes with DNA, called nucleohistones. They are the most important component of the chromatin-associated proteins and are thought to act less as specific regulators of gene expression than as nonspecific repressors of transcription, which, by changing the conformation of the chromosomes, limit the availability of the DNA.

hMG. Abb. for human menopausal gonadotropin.

Holarrhena Alkaloids, Kurchialkaloids. A group of steroid alkaloids which are the characteristic active substances in plants of the dogbane (Apocynaceae) genus Holarrhena. The representatives so far isolated (about 50) are formal derivatives of the hydrocarbon pregnane which is substituted in the 3 and 20 positions with amino or methylamino groups, *e.g.* holarrhimine and conessine.

These alkaloids are biosynthesized via cholesterol and pregnenolone.

Holosides. The compounds consisting of glycosidically linked sugar residues, *e.g.* oligo-and polysaccharides.

Holothurines. A group of highly toxic compounds from sea cucumbers (Holothurioidea). They are triterpene saponts which contain holothurinogenins as aglycons and D·glucose, D-xylose, 3(O)-methylglucose and D-quinose as sugars.

Homoarginine, Abb. Har. A higher homologue of arginine with an additional methylene group in the side chain.

Homocysteine, Abb. Hcy. A higher homologue of cysteine with an additional methylene group in the side chain.

Homoglycans. Straight or branched chain polysaccharides containing only one kind of monosaccharide residue. They are widespread in the vegetable kingdom.

Homologous Proteins. Refers to the proteins which have arisen through divergent evolution from a common ancestor. They usually have very similar primary and tertiary structures. Examples are the cytochromes, hemoglobin and myoglobin, the ferredoxins (non-heme iron proteins), fibrin peptides, immul noglobulins, peptide hormones (*e.g* insulin and hypophysca-hormones), snake venom toxins and enzymes like the serine proteases of the pancreas (trypsin, chymotrypsin, elastase) or the blood-clotting enzymes (*e.g.* plasmin) thrombin and lactate dehydro-genase.

Homopolymer. A polymer built up of identical monomeric units, for example, amylose and polyphenylalanine. In a narrower sense, Homopolymers are synthetic polynucleotides in which all the nucleotides contain the same base, for example polyadenylic acid. Polyuridylic acid, polydeoxyadaylic acid Homopolymers in the narrower sense are synthesized in vitro from nucleosides di-or triphosphates using the appropriate polymerases without a matrix.

Hordenine, Anhaline: N, N-dimethyltramine. One of the biogenic amines, widely distributed in nature. As a derivative of phenylethylamine of is one of the amines which increase blood pressure, but it has a low physiological activity.

Hormones. The organic compounds in the plant and animal kingdoms are synthesized in cells and glands specialized for this function. Very low concentrations of hormones usually produce large metabolic responses in another tissue of the same organism, or in another organism. As a rule, they are not species-specific. Together with the nervous system, they serve to transmit information between cells. In contrast to the nervous system, however, the endocrine system cannot store information.

HP. Abb. for Haptoglobin.

HSK Cycle. Abb. for Hatch-Slack-Kortschak cycle.

Human Menopausal Gonadotropin, HMG, Castration Gonadotropin. A glycoprotein of the anterior lobe of the pituitary. Carbohydrate content 30%. HMG has a similar action to follicle stimulating hormone. It primary structure is not yet known.

Humulene. Isomeric monocyclic sesquiterpene hydrocarbons found in various aromatic oils.

Hyaluronic Acid. An unbranched mucopolysaccharide. The basic subunit is a diaccharide, N-acetyl-D-glucosamine glycosidically linked β-1, 4 to D-glucuronic acid. The latter is β-1, 3 linked to the next disaccharide unit. This acid occurs in various animal tissues and joint fluids. It is synthesized from D-glucose in the fibroblasts.

Hyaluronic acid

Hybridization. Hybridization of nucleic acid molecules is a molecular biological technique for comparison of the nucleotide sequences and characterization of the information contents of differing nucleic acids. The filter, or gel technique is important.

Hydrocarbon Degradation, Microbial Hydrocarbon Degradation. Some microorganisms are able to degrable hydrocarbons and use them as their sole source of carbon and energy. The degradation of hydrocarbons depends highly on their structure. The most important pathway is the oxidation of one end to the corresponding fatty acid via the alcohol and aldehyde as intermediates. Further degradation of the fatty acids is achieved by β-oxidation. Ring cleavage is always preceded by the formation of a phenol by an oxygenase. Further degradation proceeds via pyrocatechol, cis, cis muconic acid and α-ketoadipic acid to the components of the tricarboxylic acid cycle, acetic acid and succinic acid.

Hydrocyanic Acid, Hydrogen Cyanide, HCN. A highly poisonous compound found widely in nature in the form of Cyanogenic glycosides. It is released from these by β-glucosidases (such as emulsin) and oxinitrilases. A number of plants, especially those which contain cyanogenic glycosides, can metabolize HCN, usually by binding it to serine or cysteine to form cyanoalanine. Addition of water converts the latter to asparagine.

Hydrogen Metabolism

1. Metabolic redox reactions, involving pyridine nucleotide and flavin coenzymes;
2. All metabolic reactions involving hydrogen, *i.e.* hydrogenation, dehydrogenation, transhydrogenation, activation and formation of molecular hydrogen.

Hydroxamic Acids. Derivatives of carbonic acid containing the tautomeric group

$$R-\underset{\underset{O}{\|}}{C}-NHOH \rightleftharpoons R-\underset{\underset{OH}{|}}{C}=N-OH$$

They are especially important in the iron metabolism of many organisms. Well-known examples of hydroxamic acids are aspergillic acid synthesized from the amino acids leucine and isoleucine by Aspergillus flavus, and the Siderochromes.

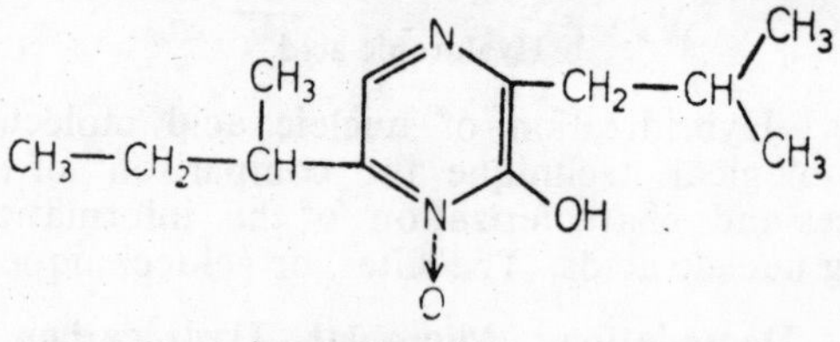

Aspergillic acid

Hydroxyacid. A carboxylic acid, in which one or more hydrogen atoms of the alkyl moiety is replaced by an hydroxyl group. The position of the OH-group in the alkyl chain is indicated by α, β, γ, δ, etc., or by 2, 3, 4, 5, where the C-atom of the COOH is No. 1; thus lactic acid is 2-hydroxypropionic, or α-hydroxypropionic acid. Important biological hydroxyacids are glyceric, malic, lactic and citric acids.

N-2-Hydroxyethylpiperazine-N-2-ethanesulfonicacid, Hepes. Used for the preparation of buffers in the pH range 6.8—8.2.

5-hydroxymethylcytosine. A pyrimidine compound, one of the rare nucleic acid bases. It is not synthesized as a modification of cytosine already incorporated in the nucleic acid but is formed de novo in the course of pyrimidine biosynthesis as 5-hydroxymethyldeoxcytidylic acid. It is found in the DNA of bacteriophages of the T2, T4, T6 series in place of cytosine.

Hydroxynervonic Acid: Δ^{15}-2-hydroxytetracosanoic acid, CH_3-$(CH_2)_7$-CH=CH-$(CH_2)_{12}$-CHOH-COOH. An hydroxylated, unsaturated fatty acid. An important component of cerebrosides.

3α-hydroxy-5α-pregnan-20-one. A catabolite of progesterone. Like its stereoisomers, 3β-hydroxy-5α-pregnan-20-one, m.p. 194°C, $[\alpha]_D+91°$ (alcohol), and 3α-hydroxy-5α-pregnan-20-one, m.p. 149°C, $[\alpha]_D+106°$ (alcohol), 3-αH. appears in the urine of pregnant women.

Hyocholic Acid: 3α, 6α, 7α-trihydroxy-5β-cholan-24-oic acid, one of the bile acids, a trihydroxylated steroid carboxylic acid. It is a component of pig and rat bile and, like the main component of pig bile, hyodeoxycholic acid (3α, 6α-dihydroxy-5β-cholan-24-oic acid). It is important as a starting material for the synthesis of steroid hormones.

Hyp. Abb. for hypoxanthine and hydroxyproline.

Hyperchromic Effect. Increase in the extinction of a solution at a particular wavelength due to structural changes in the solute molecules. It is important in the denaturation of DNA when the double helix held together by hydrogen bonds is transformed into a disordered random coil.

Hypochromic Effect. Optical phenomenon in molecules with several chromophores, in which the sum of the optical densities of the individual components is greater than the optical density of the whole molecule. This effect depends on the content of adenine and thymine and is therefore greater in DNA than in RNA. Double-stranded polynucleotides have a greater effect than single-stranded, because the effect is intensified by hydrogen bonds.

Hypophysis, Pituitary Gland. A vertebrate organ for hormone production. In humans, it weights 0.7 g and lies at the base of the brain. It is connected to the midbrain by the hypophyseal stalk. It is composed of two parts with different ontogenies.

Hypothalamus. The lowest part of the midbrain, which also includes the thalamus and epithalamus. The hypophyseal stalk with the neurohypophysis arises from the underside of the hypothalamus. As part of the limbic system, the hypothalamus is a "gateway to consciousness". Afferent (incoming) stimuli from the breast and abdominal areas and from the circulatory system are processed by the thalamus and passed on to the cerebrum. Conversely, efferent stimuli from the cerebrum flow to the thalamus and hypothalamus.

Hypoxanthine Abb. Hyp: 6-hydroxypurine. A purine derivative. Hyp occurs in the course of aerobic Purine catabolism through deamination of adenine compounds or through hydrolysis of inosine compounds. It is found as a rare base in certain transfer RNAs. It is widely distributed in the plant and animal kingdoms.

I

IAA. Indolo Acetic Acid.

Ibotenic Acid. α-Amino-3-hydroxy-5-isoxazoleacetic acid. It is a psychotropic, weekly insecticidal substance which, like its decarboxylation product, Muscimol belongs to the group of amanita poisons. It is pharmacologically very active, but less so than musimol in most tests. Another derivative of this acid, is the erythro-dihydroibotenic acid, Tricholomic acid.

IDP. Abb. for Insoine 5′-diphosphate.

I.E.P. Abb. for Isoelectric point.

Ig. Abb. for Immunoglobulins.

Ile. Abb. for L-Isoleucine.

Imidazole Alkaloids. Alkaloids of sporadic occurrence, possessing the imidazole ring system. The biosynthesis of these alkaloids is coupled to histidine metabolism.

Immobilized Enzymes. Soluble enzymes bound to an insoluble organic or inorganic (*e g.* porous glass) matrix, or encapsulated within a membrane in order to increase their stability and make possible their repeated or continual use.

Immune Response. A uppecific protective or defense reaction against foreign substances. These foreign substances, known as antigens, release from the surface of the lymphocytes the signal for cellular or humoral immune response.

Immune Respression. The specific suppression of the Immune response by corticosteroids, antimetabolites (purine analogs, folic acid antagonists), alkylating substances like cyclophosphamide ionization radiation or antilymphocyte serum. It is an imporitant form of therapy for autoimmune diseases and for preventing the rejection of transplanted organs.

Immunization. The artificial stimulation of antibody production for protection against pathogens and other antigens.

Immunofluorescence. A sensitive technique for detection of antigens or antibodies in which the specific antibody (direct I.) or the anti-antibody (indirect I.) is labelled with fluorochromes, for example fluorescein isothiocyanate. After the antigen-antibody reaction has occurred, the complex can be located by fluorescence microscopy.

Immonoglobulins, abb. **Ig., Antibodies.** Specific defense proteins found in blood plasma, lymph and many body secretions of all vertebrates.

Immunology. The science of the biological and chemical bases of immunity, or the defense mechanisms of the human and animal organism which are activated by the invasion of antigens.

Immunotolerance. The lack of an immune response to :

(a) the body's own substances or antigens with which the body has had contact since before or shortly after birth (natural I) and

(b) larger or smaller amounts of certain antigens (tolerogens), which the body does not recognize as foreign and therefore tolerates (acquired I). The acquired immunotolerance can only be maintained when the tolerogens are constantly present; otherwise the immune response to this antigen arises again. Immunosuppressive measures facilitate the induction of an I. in adults.

IMP. Abb. for Inosine 5′-monophosphate.

Indican. A glucoside composed of indoxyl and one molecule of glucose. It is found in Indigofera species, for example the indigo plant (Indigofera tinctoria), dyer's woad (Isatis tinctoria) and several other higher plants. It is the natural precursor for natural Indigo.

Indicaxanthin. A yellow pigment from the prickly pear (Opuntia ficusindica) belonging to the group of betaxanthines. It differs from betanin by the substitution of the cyclodopa residue by L-proline. It is the best known representative of the betaxanthines.

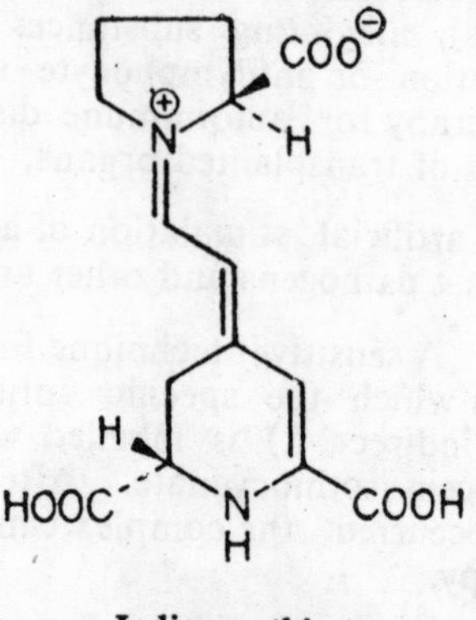

Indicaxanthin

Indigo. A dark blue vat dye, known in ancient times and particularly valued in the middle ages. It was formerly extracted from a few Indigofera species, like Indigofera tinctoria- and dyer's woad (Isatis tinctoria).

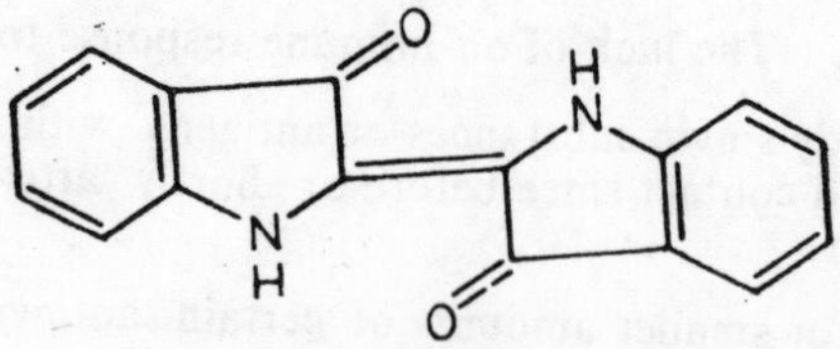

Indigo

Indole Alkaloids. One of the largest groups of alkaloids, with more than 600 representatives. Biogenetically, almost all indole alkaloids are derived from tryptophan; the majority also contains a monoterpene component which is responsible for the large variety of forms. It is advantageous to classify the indole alkaloids according to their sources (Table), because many plant genera are characterized by particular structural forms of the alkaloids.

TABLE

Classification of the Indole Alkaloids

Structural type	Typical representative
Carboline	Harman alkaloids
Pyrrolidinoindole	Physostigmine
Ergoline	Ergot alkaloids
Iridoid indole alkaloids	Rauwolfia, Curare (Calabash curare), Vinca Strychonos, Cinchona alkaloids

Inducer. A chemical compound which stimulates the synthesis of enzymes (Enzyme induction). Inducer may be the substrate of the enzyme or another, effector, such as a hormone.

Industrial Biochemistry, Biochemical Technology. A branch of biochemistry concerned with the application of biochemical processes and principles in industry. Industrial biochemistry is of particular importance in the chemical, pharmaceutical and food industries.

Industrial Fermentation. A process in which products are synthesized or improved using micro-organisms. Discontinuous fermentation processes consist of the following steps : production of the nutrient medium, sterilization, innoculation, monitoring of the fermentation, separation and work-up of the product.

Industrial Microbiology, Technical Microbiology. A main area of technical biochemistry. It is concerned with the application of microbial production and processes in industry. The microbial production of products in industry is called product synthesis. Industrial microbiology has an important role in the following branches of industry : fermentation industry, pharmaceutical industry (production of antibiotics, vitamins and enzymes), chemical industry (production of proteins, enzymes and biochemicals), and food industry (baker's yeast). Industrial microbiology is also very important for sewage treatment and composing of garbage.

Infectious Nucleic Acids. Viral nucleic acid which has entered the host cell to replicate and to have its genetic information translated by the host cell's synthetic apparatus into viral products.

Informofers. Protein particles which are generated when the RNA component of nucleoprotein particles (containing mRNA or its precursor) is cleaved off.

Informosomes. According to Spirin, nucleoprotein particles containing mRNA which can be isolated from the cytoplasm of eukaryotic cells. They can be separated from ribosomes and polyribosomes by density-gradient centrifugation. They are thought to be transport particles for mRNA. They pick up the mRNA formed in the cell nucleus, probably at the nuclear membrane, and make it available in the cytoplasm for the formation of polyribosomes.

Inhibin. The name given in 1932 to a hypothetical hormone from the male gonads. The existence of inhibin was proposed in order to account for certain aspects of the feedback regulation of gonadotropin secretion, particularly that of FSH. According to modern interpretation inhibin is a nonandrogenic molecule and a feedback inhibitor of FSH secretion, produced by the seminiferous tubules when they are stimulated by FSH. Existence of inhibin, is not universally accepted. The current search for a male contraceptive pill has resulted in renewed interest in the possible existence of inhibin.

Inhibitor Peptides Low molecular weight oligopeptide-fatty acid compounds of microbial origin which irreversibly inactive plant and animal proteases. The inhibition is stoichiometric.

Inhibitor Proteins For the most part, low molecular weight, resistant proteins with compact structures and a lack of species specificity. Inhibitor proteins inhibit reversibly or irreversibly either, at the molecular level, certain anabolic or catabolic metabolic processes, or, at the cellular level, growth and maturation processes of normal and malignant cells. The best known group or inhibitor proteins are the protease inhibitors, a group of protease-insensitive, disulfide-rich polypeptides widely distributed in the animal and plant kingdoms. They are especially frequent in the nutritional protein (egg white) of many eggs and in plant seeds. A large number of animal proteinase inhibitors are secretory proteins, like the trypsin inhibitor of the mammalian pancreas, blood and seminal plasma, milk, colostrum, salivary, and snail mucous.

Inhibitors, Antagonists. Substances which slow down or prevent chemical or biochemical reactions from occurring. A competitive inhibition is often involved, in which of the inhibitor competes with the Effector for the active site (receptor) according to the law of mass action. Many inhibitors are known which specifically inhibit particular enzyme reactions. The Antibiotics are a special class of inhibitors.

Initial Rate Technique. A graphic method to determine the initial reaction rate from the reaction kinetic curves. The slopes of the tangents passing through the origin t = O for the reaction curves at various substrate concentrations are the initial rates at those concentrations.

Initiation. The beginning of synthesis of biopolymers; Deoxyribonucleic acid; Ribonucleic acid; Protein biosynthesis.

Initiation. Codon, Start Codon. A sequence of three nucleotides in the mRNA which is recognized by the anticodon of formylmethionyl-tRNA and thus serves as the start signal for polypeptide synthesis. It has the sequence 5′-AUG and is apparently localized in a sterically favorable position of the mRNA.

Initiation Factors. Catalytic proteins required for the initiation of tRNA synthesis and of Protein synthesis. At least three structurally and functionally different initiation factors have been identified in bacterial protein synthesis.

Initiation tRNA, Starter tRNA, Formylmethionyl-tRNA. Abb. **F-met-tRNA$_F$**. A tRNA specific for methionine which differs in its primary and tertiary structure from the tRNA (Met-tRNA$_M$) which supplies the methionine incorporated into the middle of a polypeptide chain. The Met is formylated after esterification to tRNA$_F$ by a formyltransferase which transfers the formyl group from formyltetrahydrofolic acid. This enzyme is apparently absent from the cytoplasm of eukaryotes, so the Initiation tRNA for protein synthesis on 80S ribosomes is Met-tRNA$_F$. F-Met or Met is removed from most proteins during their synthesis (Post-translational modification of proteins).

Ino. Abb. for inosine.

Inosine Phosphates. Nucleotides, phosphoric acid esters of inosine :

1. Inosine 5′-monophosphate, abb. IMP : inosinic acid, hypoxanthine riboside 5′-phosphoric acid is made up of the purine base hypoxanthine, D-ribose and phosphoric acid. Inosine phosphate is the precursor of all the other purines. It is the first intact purine compound in the course of Purine biosynthesis and all other purine nucleotides arise from it. Together with guanylic acid, IMP serves as a flavoring principle and is isolated for this purpose either from meat extract or from hydrolysates of yeast nucleic

acids, or it is produced on a large scale by mutants of certain microorganisms, for example Corybacterium glutamicum.

2. Inosine 5′-triphosphate, abb. ITP : Can, as an energy-rich phosphate, replace ATP in certain metabolic reactions (carboxylations). It is formed by phosphorylation of IMP via inosine 5′-diphosphate, abb. IDP.

3. Cyclic inosine 3, 5′-monophosphate, abb. cyclo-IMP, cIMP A structural analog of cyclic adenosine 3′, 5′-monophosphate (Adenosine phosphates) which like the adenosine derivative, inhibits the growth of certain transplantable tumors.

Inosine, Abb. **Ino, Hypoxanthinosine.** Hypoxanthine riboside, 9β-D-ribofuranosylhypoxanthine, a β-glycosidic nucleoside of D-ribose and hypoxanthine.

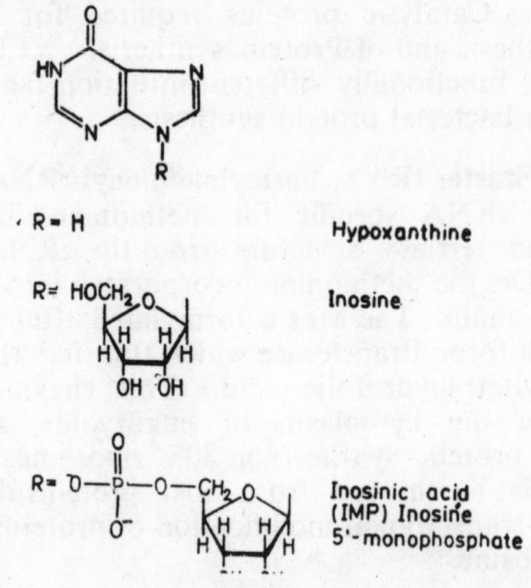

R = H — Hypoxanthine

R = $HOCH_2$ (ribose) — Inosine

R = $^-O-P(O)(O^-)-O-CH_2$ (ribose) — Inosinic acid (IMP) Inosine 5′-monophosphate

Structure of Hypoxathine, Inosine and Inosinic Acid

It occurs free, especially in meat and yeast, and is formed by dephosphorylation of inosine phosphates. It fulfills a specific function as a component of the anticodon of certain tRNA.

Insect Hormones. Substances responsible for directing the life cycle of insects, mostly of low molecular weight. Three hormones are involved in the direction of post-embryonic development of insects.

1. activation hormone (brain hormone, adenotropic factor),

2. molting hormone (Ecdysone), and
3. juvenile hormone.

Insulin. A polypeptide hormone, M_r 5780 (bovine). It is the only hormone which reduces the blood sugar concentration. Like thyroxin and somatotropin, it affects the entire intermediary metabolism, particularly of the liver, fat tissues and musculature. The sequence region A8 to 10 is species specific. Insulin tends to form aggregates.

The biosynthetic precusor of insulin is proinsulin, a single-chain polypeptide of known primary structure (81 to 86 amino acid residues, M_r 9000). It is formed in the B-cells of the islets of Langerhans. The C-peptide (connecting peptide) is enzymatically removed to form I. The physiological stimulus for synthesis and secretion is glucose in the form of hyperglycemia.

The total synthesis insulin was first achieved in 1963. Insulin increases the permeability of the cell to monosaccharides, amino acid and fatty acids, and accelerates glycolysis, the pentose phosphate cycle and, in the liver, glycogen synthesis. It promotes the synthesis of fatty acids and proteins. It is degraded

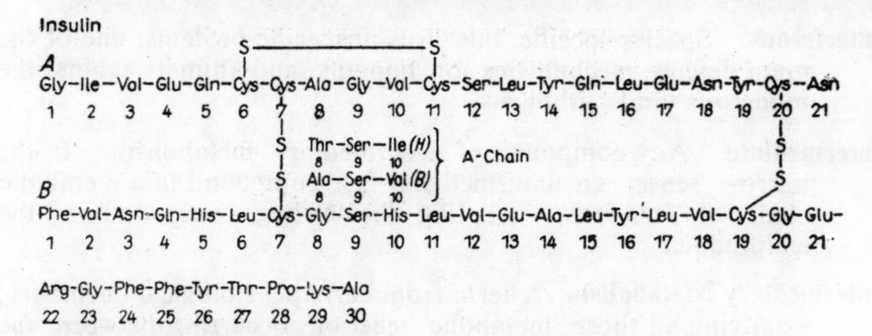

Primary Structure of Sheep Insulin. Human (H) and Bovine (B) Insulin Differ From Sheep Insulin in the Sequence Region A8 to 10; in Addition, in Human Insulin the C-terminal Alanine of the B-chain is Replaced by Threonine

in the liver by a thiol protein disulfide oxidoreductase and an insulin-specific protease. It is determined radioimmunologically. The normal level in human blood is 1 ng/ml. A disturbance in the synthesis, increased degradation or inactivation of I. leads to diabetes mellitus, the most common human endocrine disease, except for overweight.

Integrated Rate Equation. An equation which represents the concentration of the substrate or product as a function of time. The corresponding plots are called the progress curves, which can also be obtained by direct measurement (Enzyme kinetics). By integrating the Michaelis-Menten equation, one obtains for example, with $(S_o - S) = P$. the integrated velocity equation $P(t) = V_m t + K_m \ln S/S_o$, where P is the product concentration, S the substrate cancentration, so the substrate coneentration at t =o, and V_m and K_m are the maximal velocity and Michaelis constant, respectively.

Intercalation. A special interaction between dye moiecules (*e.g.* proflavin) and DNA in which the dye molecule inserts itself between two neighboring base pairs of the DNA double helix. This stretching of the DNA can give rise to errors in the transcription or translation of the DNA, causing either a mutation or a defective mRNA, and thus the wrong amino acid sequence of a protein.

Interconversion. In biochemistry in general, the changing of one intermediary metabolic product into another; in enzymology, in particular, transformation of the active form of an enzyme into the inactive form, and vice versa, as a possibility for physiological regulotion of metabolism.

Interferons. Species-specific, but virus-unspecific proteins, one of the main defense mechanisms of humans and animals against the numerous viral pathogens.

Intermediate. Any compound of Intermediary metabolism. In the narrow sense, an intermediate is a compound in a metabolic chain of reactions, excludifig the starting compound and the endproduct.

Intermediary Metabolism. A term from early physiological chemistry, signifying all those metabolic reactions occurring between the uptake of foodstuffs and the formation of excretory products. In modern usage, it is essentially identical with primary metabolism.

Interphase. The phase between two mitoses (G_1, S and G_2 phases) in the Cell cycle in which the chromosomes are completely unwound, as chromatin, in a membrane-bound nucleus.

Intracellular Digestion. The process of digestion within the cell, in which the lysosomal system plays an important role. It affects exogenous materials brought into the cell from outstde, and endogenous cell components.

Intron. An intervening sequence in eukaryotic DNA; the term is also applied to the intervening sequence in the RNA transcript

Intron, along with the introns coding sequences (exons) are transcribed, then removed to produce functional mRNA. Genes containing intron are known as split genes. The process of removal of an intron in an RNA transcript and the joining of the neighbouring exons is known as splicing. The base sequence of intron in mRNA begins with 5′GU and finishes with AG3′. These sequences act as recognition sites for splicing enzymes. The 5′GU AG3′ pattern is not found in tRNA precursor molecules, suggesting the existence of at least two splicing enzymes, one for mRNA, the other for tRNA.

Inulin. High molecular weight vegetable reserve carbohydrate. It is a fructan consisting of 20 to 30 fructofuranose units α-glycosidically linked. Probably the reducing end of the chain terminates with glucose It is found as a reserve substance in the tubers and roots of many members of the Compositae, like dahila and Jerusulem artichoke tubers. It is used in food for diabeties.

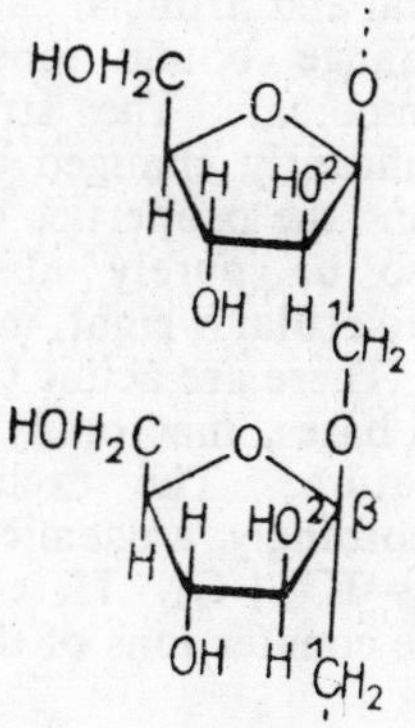

Inulin

Invertase, β-h-fructosidase A saccharide-cleaving hydrolase from yeast, fungi and higher plants. Yeast invertase is a dimeric glycoprotein, Neurospora invertase a tetrameric glycoprotein.

Invertebrate Hormones, Hormones of Invertebrate Animals. The best studied of these are the insect and crustacean hormones. Many of these have not been chemically characterized, but the structures and biochemical effecfs of two are known : Icdysone and Juvenile hormone.

Invert Sugar. A mixture of equal parts D-glucose and D-fructose which, in contrast to dextrorotatory sucrose, is levorotatory. Invert sugar is generated by acid or enzymatic hydrolysis of sucrose.

Ion-exchange Chormatography. A liquid-solid phase chromatographic method for analytic and preparative separation and purification of mixtures of substances. It is based on electrostatic binding between cations or anions of the substance to be studied or separated and the corresponding Ion exchanger. (The components of an electrolyte solution are separated by successive adsorption and desorption through ion exchange. It can be run as a front, replacement or elution process. Ion exclusion and ion retardation are special techniques of ion exchange chromatography. It can be used to separate ionized from nonionized substances or substances ionized to different extents, *e.g.* inorganic from organic acids. Ions appear in the eluate sooner than nonions, because the latter penetrate the interstial and the interior resin phases and thus pass through the column more slowly.

Ion Exchangers. All natural and artificial substances, mostly solids, which are able to exchange bound ions for ions from the surrounding liquid medium. The structure of the solid ion exchangers is not significantly changed in the process. The exchange depends on the properties of the ions involved; in addition, there can also be purely absorptive binding. Ion exchangers are high molecular weight, insoluble polyelectrolytes capable of swelling. There are acidic (solid acids, macropolyacids) and basic (solid bases, macropolybases) types of widely varying chemical nature. The exchange of ions occurs stoichiometrically; accordingly, a chemicale quilibrium can be established : $IG_1+G_2 \rightleftharpoons IG_2+G_1$. Here I is the ion exchanger, and G_1 and G_2 are the counter ions of the exchanger and the milieu.

Ionophore. A compound which increases the permeability of membranes to ions. Ionsphores act by delocalizing the charge of the ion and shielding it from the hydrophobic region of the lipid bilayer. They possess both hydrophobic (confers lipid solubility) and hydrophilic (binds the ion) regions.

Ion Pumps. Metabolic cycles within cell membranes which can transport ions against the prevailing concentration gradient. The bioelectric membrane potential of the nerves, for example, is based on different distributions of Na^+ and K^+ ions.

IP. Abb. for isoelectric point.

IPA. 6-Δ^2-isopentenylaminopurine; 6-(3-methyl-2-butenyl) aminopurine N^6-γ, γ-dimethylallyladenine; also known as bryokinin. IPA is a Cytokinin.

Ipecacuanha Alkaloids. Iridoid Isoquinoline alkaloids. The most important representative of the group is Emetine.

Ipomoeamarone. A Phytoalexin with a sesquiterpene structure. It is formed when sweet potatoes are infected by phytopathogenic microorganisms. Biologically, it is synthesized from mevalonic acid via farnesol.

Iridodial. A representative of the Iridoids. The dialdehyde form is in equilibrium with the half-acetal form (Fig.) I. was first identified in the defensive secretion of various ants.

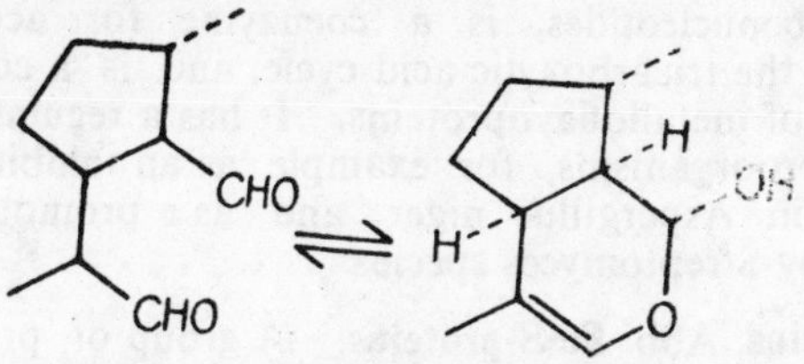

CHO CHO ⇌ H OH H O

Iridodial

Iridoids. A group of natural products characterized by a methylcyclopentanoic monoterpene skeleton. Iridoids used to be called, incorrectly, pseudoindicans, because some of them can be transformed into intensely blue compounds in the presence of air and acids. This process has notbe en elucidated, but it is not chemically related to the transformation of indican glycosides into indigo.

Iridomyrmecin. A monoterpene lactone, one of the iridoids. It was first isolated from a pheromone mixture from ants of the genus Iridomyrmex. Treatment with alkali produces the isomeric isoiridomyremecine, which is a pheromone for other ant species.

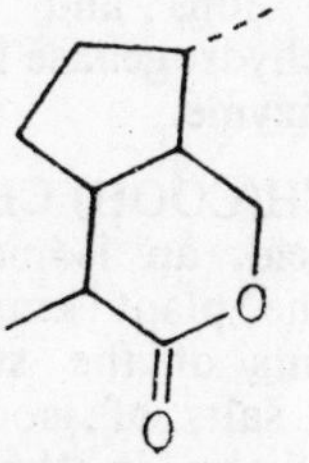

O O

Iridomyrmecin

Iron, Fe. A bioelement found in all living cells. The human body contains 4 to 5 g, of which 75% is in hemoglobin. The Fe in organisms occurs in the II and III oxidation states; in higher animals it is stored bound to protein. It is transported in the blood as a complex with transferrin from which it is transferred enzymatically to metal-free porphyrin molecules. Non-heme iron is also found in a number of compounds, for example, Iron-sulfur proteins. The Fe metabolism of microorganisms is mediated by a group of natural products called **Siderochromes.**

Fe catalyses most redox reactions in the cell (Cytochromes; Chlorophyll). It is involved in the reduction of ribonucleotides to deoxyribonucleotides, is a coenzyme for aconitase (EC 4.2.1.3) in the tricarboxylic acid cycle, and is a component of a number of metalloflavoproteins. It has a regulatory role in many microorganisms, for example as an inhibitor of citrate synthesis in Aspergillus niger, and as a promoter antibiotic synthesis by Streptomyces species.

Iron-sulfur Proteins, Abb. **Fe-S-proteins.** A group of proteins found in all organisms. They contain iron-sulfur centers (iron-sulfur clusters) and take part in electron transfer processes. They are involved in H_2 metabolism, nitrogen and carbon dioxide fixation, oxidative and photosynthetic phosphorylation, mitochondrial hydroxylation and nitrite and sulfite reduction.

Isocitrate Dehydrogenase (E.C. 1.1.91 or 1.1.1.92). An enzyme belonging to the class of dehydrogenases which reduces isocitrate at the secondary hydroxyl group during the operation of the Tricarboxylic acid cycle. It simultaneously catalyses the reversible decarboxylation of the resulting oxalosuccinate to 2-oxoglutarate. It transfers the hydrogen to NAD^+ or $NADP^+$. The NAD^+ -specific form is found only inside the mitochondria, is allosterically activated by ADP, inhibited by ATP, and catalyses the reaction only in the direction of 2-oxoglutarate. The $NADP^+$-dependent cytoplasmic enzyme requires manganese(II) ions and produces hydrogen for syntheses. Isocitrate dehydrogenase is allosteric and therefore serves as a regulatory enzyme.

Isocitric Acid: $HOOC\text{-}CH_2\text{-}CH(COOH)\ CHOH\text{-}COOH$. A monohydroxy tricarboxylic acid, an isomer of citric acid, which is widely distributed in the plant kingdom and occurs in free form especially in plants of the stone-crop family (Crassulaceae), and fruits. The salts of isocitric acid are important metabolically as intermediates in the Tricarboxylic acid cycle where they are formed from citrate by the enzyme aconitase, then oxidized to 2-oxoglutarate. In the Glyoxylate cycle isocitrate is cleaved to succinate and glyoxylate.

Isoelectric Point, Abb. I.P. or I.E.P. The pH value of a solution at which the net charge on the dissolved ampholyte is zero, *i e.* the sum of the cationic charges is equal to the sum of the anionic charges. The isoelectric point of electrolytes, *e.g.* amino acids, peptides or proteins, may lie in the range from pH 1 (pepsin) to pH 11.8 (protamine), and is characteristic for each ampholyte.

Isoenzymes, Isozymes. Multiple forms of an enzyme with the same substrate specificity, but genetic differences in their primary structures. If there are no differences in primary structure, one speaks of Pseudoisoenzymes.

Isoflavones. Phenolic natural products belonging to the group of flavonoids. They are derived from 3-phenylchromane and differ from the isomeric flavones and neoflavones (4-phenylchromanes) in the 3-position of their B rings.

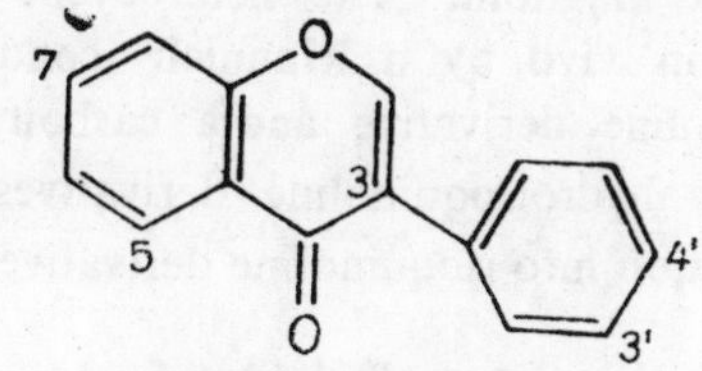

Isoflavone ring system

L-isoleucine, Abb. **Ile:** L-α-amino-β-methylvaleric acid, CH_3-CH_2-$CH(CH_3)$-$CH(NH_2)$ COOH. An aliphatic, neutral amino acid found in proteins. Ile is found in relatively large amounts in hemoglobin, edistin, casein and serum proteins, and in sugar beet molasses, from which it was first isolated in 1904 by F. Ehrlich. It is an essential amino acid, and is both glucoplastic (degradation via propionic acid) and ketoplastic (formation of acetate).

Isomaltose. A reducing disaccharide, composed of two molecules of D-glucopyranose linked 1, 6-glycosidically. It is formed by the enzymatic degradation of branched polysaccharides, *e.g.*, amylopectin.

Isopentenylpyrophosphate. An intermediate in the biosynthesis of the Terpenes.

Isopenthylaetate: $(CH_3)_2CH$-CH_2-CH_2-O-CO-CH_3. The most effective alarm pheromone (Pheromones) of the boney bee. It is synthesized in the glandular tissue of the sting. palps, and is released when the bee stings. Its odor attracts other bees.

Isopeptide Bond. A covalent cross-linking bond between the ϵamino group of lysine and the side-chain carboxyl group of glutamate or aspartate, formed by condensation :

$$H_2N\text{-}CH(COOH)\text{-}CH_2\text{-}CH_2\text{-}COOH + H_2N\text{-}CH_2\text{-}CH_2\text{-}CH\text{-}CH_2\text{-}CH(NH_2)\text{-}COOH \xrightarrow{H_2O} N\epsilon$$

γ-glutamyl)-lysine. It has been found in polymerized fibrin and native wool. It is not hydrolysed by the body's own digestive proteases, but only by the bacteria in the large intestine. Its presence in nutritional proteins therefore reduces their nutritional value.

Isoquinoline Alkaloids. A large group of alkaloids occurring widely in the plant kingdom. The heterocyclic skeleton is usually synthesized in vivo by a Mannich condensation between a phenylethylamine derivative and a carbonyl component. The resulting tetrahydroisoquinoline derivatives are converted by dehydrogenation into isoquinoline derivatives.

Isorubijervine. A Veratrum alkaloid of the Jervatrum type. M_r 413.65, m.p. 237°C, $[\alpha]_D + 6.5°$ (alcohol). It occurs in yeasts (Veratrum album, Veratrum eschscholtzii and Veratrum viride) and differs (α-solanine) in having an additional 18-hydroxyl group. In the glycoalkaloid isorubijervosine, from Veratrum eschscholtzii it is linked to D-glucose.

Isotope Technique, Tracer Technique. The use of radioactive and stable isotopes (more exactly, nuclides) in biological, chemical and physical research, and in technology. Since atoms, groups of atoms (functional groups) or molecules are labelled by addition (in the case of elements) or chemical incorporation of indicator atoms, this technique is regarded as one of the indicator methods.

TABLE

Selected Nuclides of the Bioelements

Nuclide	Symbol	Half-life	Type of radiation
Hydrogen			
(Deuterium)	^{2}H	Stable	—
(Tritium)	^{3}H	10.46 years	β, very soft
Carbon	^{13}C	Stable	
	^{14}C	5568 years	β, soft
Nitrogen	^{13}N	10.05 min	β
	^{15}N	Stable	
Phosphorus	^{32}P	14.3 days	β
Sulfur	^{35}S	87.1 days	β soft

Isovaleraldehyde: $(CH_3)_2CH\text{-}CH_2\text{-}CHO$. A naturally occurring aldehyde. p=0.7977, m.p. 92.5°C. It is a colorless, sharp-smelling liquid which is very reactive and polymerizes easily in the presence of acid. It occurs in many aromatic oils, especially in eucalyptus oils, and is obtained synthetically by oxidation of isoamyl alcohol.

ITP. Abb. for Inosine 5′-triphosphate.

IUB. Abb. for International Union of Biochemistry.

IUPAC. Abb. for International Union of Pure and Applied Chemistry.

J

Jervine. A Jeveratrum type of Veratrum alkaloid with a C-nor-D-homo structure. It is the main alkaloid of white and green hellebore (Veratrum album and Veratrum viride).

Juniperic Acid: 16-hydroxypalmitic acid, $HOCH_2\text{-}(CH_2)_{14}\text{-}COOH$. A fatty acid. It is a typical wax acid in the waxes of many gymnosperms, *e.g.* juniper (Juniperus communis).

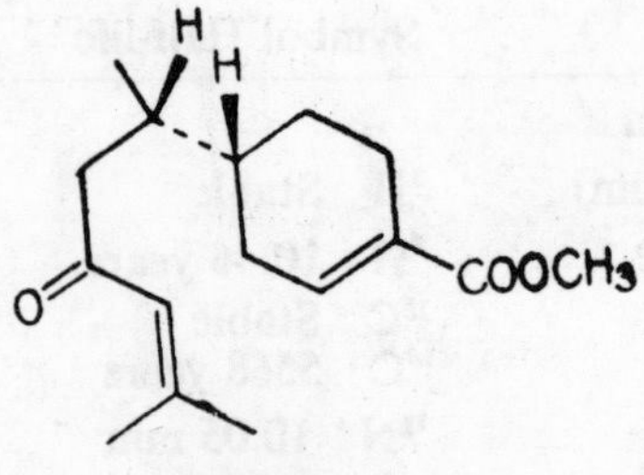

Juvabione

Juvabione, Paper Factor. A monocyclic sesquiterpene ester from the wood of the North American balsam fir (Abies balsamea). (+)-J., an oil, M_r 266, $[\alpha]_D^{20}+79.5°$ (c=3.5, chloroform). It is specific for Pyrrhocoris apterus and only the dextrorotary form is biologically active.

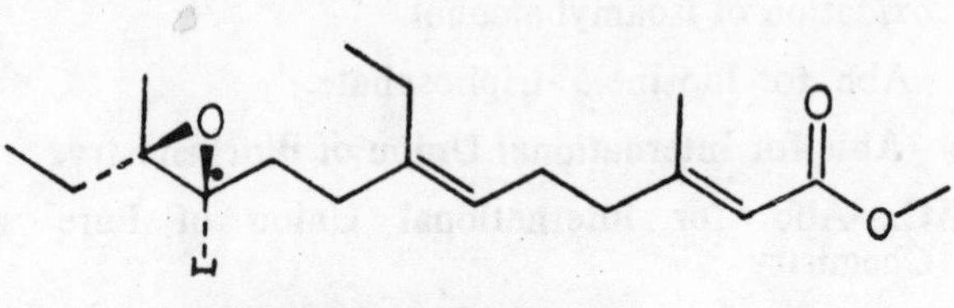

Cecropia juvenile hormone

Juvenile Hormone, Larval Hormone, Status-quo Hormone. Insect hormone responsible for the control of molting. The first J.h. to be isolated and to have its structure elucidated was obtained from the abdomens of the male silk worm moth (Hyalophora cecropia) in 200 μg quantities. Homologs of this compound were discovered later, and other J.h. were postulated. Juvabione and farnesol derivatives are among the natural compounds with J.h. activity. Some synthetic substances are more active than the J h., and they must be used in increased measure in the future as part of the integrated attack on insects.

K

Kallikrein, Kininogenin, Kininogenase. A proteolytic enzyme (EC 3.4 21.8) which preferentially cleaves Arg and Lys peptides. There are at least three types.

Kanamycin. An aminoglucoside antibiotic Streptomycin.

Kasugamycin. An aminoglucoside antibiotic streptomycin.

Kauren. It is ent-kaur-16-ene, a tetracyclic diterpene which is found in the plant kingdom in both the (+) and (—) forms. (—)-K. M, 272, m.p. 50°C, $[\alpha]_D^{20}$—75°.

Kava. A narcotic drink made from the roots of the kava plant (Piper methysticum L.) in the Pacific islands. It has a mild pain-killing and euphoric effect.

Keratins. Insoluble, cystine-rich intracellular structural proteins in the epithelial tissues of land vertebrates. They occur mostly in epidermis, fur hair, wool, claws, hoofs, horns, scales, beaks and feathers.

Keratan Sulfate. An acid mucopolysaccharide-composed of N-acetyl-D-glucosamine 6-sulfate and D-galactose, alternatingly β-1, 3 and β-1, 4 glycosidically linked. It is found in the cornea of the eye, in cartilage, in the aorta and in the intervertebral discs.

Kermesic Acid A bright red insect dye belonging to the group of anthraquinones, m p. 250°C (d). It is structurally closely related to Carminic acid; it possesses the identical structure of a tetrahydroxylated methylanthraquinone carboxylic acid, but the C-glycosidic glucose on C-2 is absent. It makes up 1 to 2% of kermes, the dried bodies of female scale insects Kermococcus ilicis.

Ketoacid. A carboxylic acid which contains the carbonyl group—C=O in addition to the carboxyl group COOH. Depending on the position of the carbonyl with respect to the carboxyl, the acid is referred to as an α, β, or γ-ketoacid.

Ketogenesis. The formation of ketone bodies. The primary product of ketogenesis is acetoacetate. It is synthesized in the liver

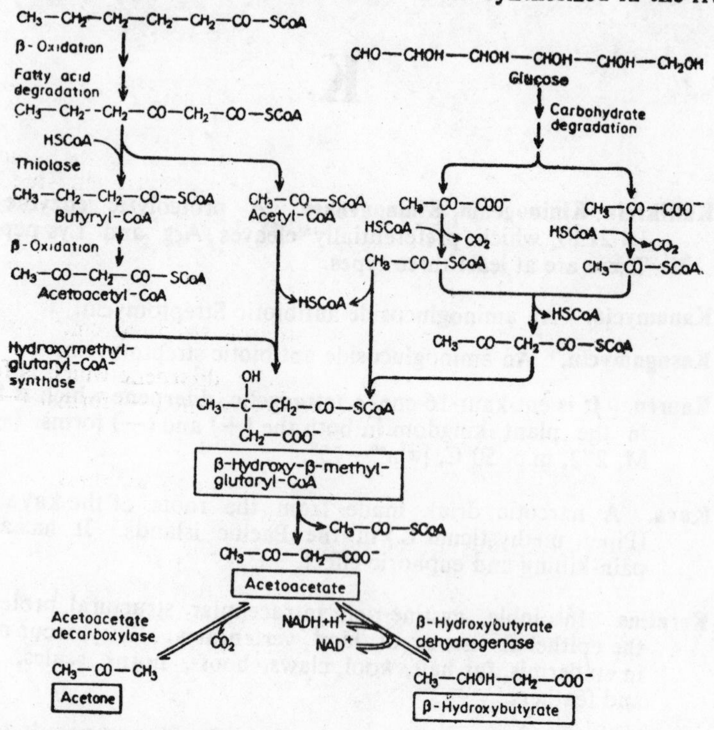

Biosynthesis of the ketone bodies acetoacetate, β-hydroxybutyrate and acetone. Alternate pathway for the formation of acetoacetate

from acetyl-coenzyme A via acetoacetyl-CoA and β-hydroxy-β-methylglutaryl-CoA. β-Hydroxybutyrate dehydrogenase catalyses the conversion of acetoacetate to β-hydroxybutyrate. The enzyme is located in the mitochondria. Acetone is formed by the spontaneous decarboxylation of acetoacetate. Acetoacetate is also produced by the degradation of the ketoplastic amino acids leucine, isoleucine, phenylalanine and tyrosine.

α-Ketoglutaric Acid, 2-oxoglutaric Acid: HOOC-CO-CH_2-CH_2-COOH. A keto-dicarboxylic acid which represents an important branching point in the tricarboxylic acid cycle. α-K.a. is formed as its anion (α-ketoglutrate) by oxidative decarboxylation of isocitrate, by transamination of glutamate in amino acid metabolism, and by degradation of lysine via glutarate and α-hydroxyglutarate. The oxidative decarboxylation of α-ketoglutarate yields succinyl-coenzyme A. Its reductive amination leads to glutamate.

Ketone Bodies Organic compounds produced by ketogenesis in the organism. Ketone bodies are acetoacetate and the compounds formed from it, β-hydroxybutyrate and acetone. The increased production of under certain pathological conditions, *e.g.* diabetes mellitus, leads to acidosis, because acetoacetate and β-hydroxybutyrate are present as anions, which reduce the concentration of HCO_3 in the blood. Other consequences are the excretion of the K b. by the kidneys, production of acid urine, and damage to the central nervous system.

Ketoses. The polyhydroxyketones, a subgroup of monosaccharides (the other is aldoses). The characteristic feature is their non-terminal—C=O group, which is given the lowest possible number in a systematic numbering.

Kidney. A paired mammalian organ with excretory and metabolic functions. Human kidneys are bean-shaped, weigh about 150g. each, and lie in the posterior part of the abdominal cavity, on either side of the vertebral column. The functional unit of the kidney is the nephron, consisting of a glomerular capsule and a long tubule. Each K. contains about a million nephrons. Kidneys are responsible for maintaining the constancy of the internal milieu of the body.

Kinases. Enzymes which catalyse the transfer of a phosphate residue from ATP to another substrate, especially to the alcoholic hydroxyl groups of monosaccharides. Some important kinases are hexokinase, which phosphorylates several hexoses at the C-6 position, glucokinase, which is responsible for the formation of glucose 6-phosphate formation in the liver, and phosphofructokinase.

Kinetic Data Evaluation. Data evaluation by computer is performed by a noninear regression analysis, using an objective procedure, in which the sums of the squares of the differences between calculated and experimental values are minimized (method of least squares).

Kinetic Equations. A system of differential equations which describes the changes with time of the concentrations of enzyme species as a function of the reaction rate constants, the concentrations of enzyme species and the concentration variables. The kinatic equation for the Michaelis-Menten scheme are :

$$dE/dt = -k_1 S.E + k_{-1} ES, \quad dES/dt = k_1 S.E - (k_1 + k_2) ES$$

Kinetin. 6-furfurylaminopurine, the model substance for cytokinins. Kinetin in conjunction with other factors, such as auxins, induces renewed cell division in resting plant tissue. It influences the nucleic acid and protein metabolism of the plant. In addition to many other physiological effects, it prevents yellowing of isolated leaves and creates an attraction centre for protein synthesis.

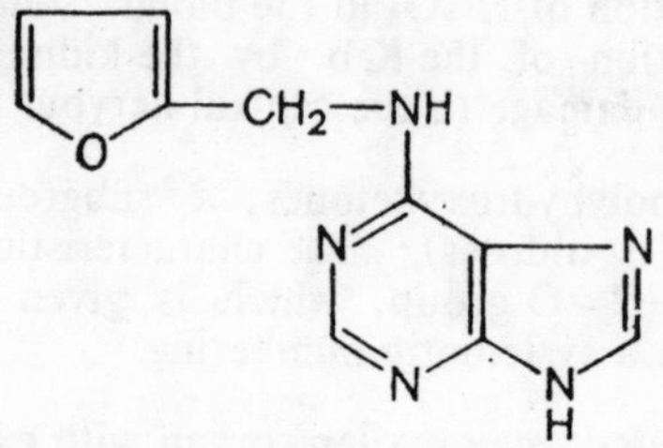

Kinetin

King-altman Method A method for derivation of rate equations according to simple rules. These rules come from the application of the determinant theory in the solution of inhomogeneous systems of linear equations. One first draws all possible geometric figures of the enzyme graphs which transform the various enzyme forms (enzyme species) into one another. The number of lines (edges) is 1 less than the number of enzyme forms, Circles and cycles are forbidden and are eliminated. The edges are assigned the appropriate reaction rate constants or the products of rate constants and concentration variables of the corresponding step of the reaction.

(*e.g.* $E \xrightarrow{k_{is}} ES$) (edge analysis). According to the King-Altman rules, the following distribution equation then holds: Enzyme form/E_t=sum of the products of the edge analysis of all pathways leading to this enzyme form, divided by Σ. Here E_t is the total enzyme concentration, Σ the sum of the numerators of all distribution equations of the enzyme graph. The rate equation is then obtained by multiplication of the product-producing enzyme forms, such as EP, with the associated catalytic constant: $v=k_{cat}$ EP. Here k_{cat} is the catalytic constant, EP the enzyme product complex. If there are several product-producing enzyme forms the partial rates are added.

Kynurenine: 3-Anthraniloylalanine. An intermediate in Tryptophan degradation.

L

Labdadienyl Pyrophosphate. Refers to an intermediate in the biosynthesis of the Diterpenes.

Lac Repressor Protein. The first repressor substance (product of a repressor gene) to be isolated and characterized. It is an acidic (I.P. 5.6) allosteric protein of M_r 152000. It consists of four identical subunits (M_r 38000,347 amino acids, sequence and spatial structure known). Its primary structure has recognizable similarities to histones or to β-galactosidase.

Lac System. The region of the genome of Escherichia coli and other enterobacteria which controls the ability to utilize lactose and other β-galactosides. It consists of the structural genes LacZ for β-galactosidase, LacY for galactoside permease and LacA for thiogalactoside transacetylase, an operator, a promotor and a regulator gene which is responsible for the synthesis of the Lac repressor.

Lactate Dehydrogenase, Abb. LDH, Lactic Acid Dehydrogenase. An oxidoreductase and a much studied isoenzyme. LDH catalyses an NAD-or NADH-dependent side reaction of glycolysis: lactic acid⇌pyruvic acid. LDH is absolutely specific for L(+)-lactate (salt of lactic acid), since D(−)-lactate is not dehydrogenated. The highest LDH activities are found in heart muscle and liver.

Lactic Acid: CH_3-CHOH-COOH. An aliphatic, optically active hydroxy acid, widely distributed in plants, especially seedlings.

Lactose, Milk Sugar. A reducing disaccharide. It crystallizes from water as the β-form above 93°C, and as α-lactose monohydrate below 93°C. It consists of galactose β-1,4 glycosidically linked to glucose, both monosacharide residues in the pyranose form.

It is the most important carbohydrate in the milk of all mammals. Human milk contains 6 to 8%, cow's milk 4 to 5%.

It is cleaved by β-galactosidase or acid hydrolysis. It is used as a starting material in pharmaceutical preparations, and as a carbon source in the culture of microorganisms, *e.g.*, in the production of penicillin.

Lamp-brush Chromosomes. Very large chromosomes, up to 1 mm long, which occur in the meiotic prophase in newts and a few other animals They form loops along the sides, which give them a brush-like appearance. The loops are not fixed structures. They represent unwound, individual chromomers, consisting of DNA. protein and RNA. They appear at the sites of active RNA synthesis, thus indicating increased physiological activity in the particular chromosome section.

Lanolin, Wool Fat, Wool Wax. The fatty or more correctly waxy substance secreted by the skin of the sheep, m.p. 36—42°C. It is a complicated mixture of fatty acids, alcohols, fats and waxy substances

Lanosterol, Kryptosterol. 5α-lanosta-8(9), 24-dien-3β-ol A tetracyclic triterpene alcohol. It is also one of the zoosterols It occurs in large amounts in the wool fat of seep. It is derived from the hydrocarbon 5α-lanostane. It is biosynthesized from squalene, via 2,3-epoxysqualene, and as the primary product of this reaction, if important in the synthesis of all further tetracyclic triterpenes of the lanostane type and of the steroids.

Lanosterol

Lathyrinogenic Amino Acids. The nonproteogenic amino acids occurring in the seeds of some species of vetech (Lathyrus). They include diaminobutyric acid —

$$H_2N-(CH_2)-CH(NH_2)-COOH$$

(neurolathrinogenic effect), β-aminopropionitrile, which occurs as the glutamyl peptide in the seeds of Lathyrus odoratus, and presumably the N-oxaloyl α, β-diaminopropionic acid—

$$HOOC-CH(NH_2)CH_2-NH-CO-COOH.$$

The disease caused in humans and animals by lathyrinogenic amino acids is called lathyrism, and takes various forms, neuro(nerve) and osteo (bone) lathyrism β(Nγ-glutamyl) amino propionitrile, for example, causes skeletal abnormalities in rats.

Lauric Acid. N-dodecananoic acid, $CH_3-(CH_2)_{10}-COOH$. One of the most widespread fatty acids, a typical wax fatty acid. It is present in the seed fats of the laurel family (Lauraceae), and makes up 52% of the fatty acids in palm seed oil, 48% in coconut fat, 4 to 8% in butter. It is an acid component of spermaceti.

LDH. Abbreviation for Lactate dehydrogenase.

Lead Pb. A highiy toxic cumulative element in man and animals. It affects adversely nearly all setps in heme synthesis; it inhibits the mitochondrial enzyme 5-aminolevulinic acid synthase, but inhibits even more strongly 5-aminolevulinic acid dehydrase. The result of this inhibition is an increase in the blood level of 5-aminolevulinic acid, which is also detectable in the urine. Other enzymes inhibited by absorbed Pb are cytochrome P_{450} (liver), adenyl cyclose (brain and pancreas), enzymes of collagen synthesis, some ATPases, and lipoamide dehydrogenase.

Lectins, Phytohemagglutinins. A group of antibody-like proteins found in plant seeds. It can bind specifically to the surface of various cell types like erythrocytes, leukemia cells, yeast cells and several species of bacteria, causing them to agglutinate. The reaction occurs between the sugar-binding site site of the lectins and the terminal, non-reducing mamose, glucose and fructose residues of saccharides in the corresponding surface receptors of the cell.

Leghemoglobin, Legoglobin. An autoxidizable hemoprotein, present in the root nodules of leguminous plants. It is structurally and functionally related to hemoglobin and myoglobin Amino acid sequence of L, shows homology with that of animal myoglobin X-ray crystallography shows similar topology and three-dimensional structure of it and animal myoglobins. It is essential for symblotic nitrogen fixation in the root nodules of leguminous plants, where it is responsible for the rapid flow of oxygen to the bacteroids (cf. role of myoglobin in transport of oxygen to respiratory enzymes of muscle).

L-Leucine, Abb. Leu. L-2-amino-4-methylvaleric acid,

$$(CH_3)_2CH-CH_2-CH(NH_2)-COOH.$$

An aliphatic, neutral amino acid found in proteins. Leu is both an essential dietary amino acid and ketogenic. It is particularly abundant in serum albumins and globulins. It is degraded to isovaleric acid by deamination and decarboxylation, then further to acetic acid via acetoacetic acid. The biosynthesis of leu follows the scheme for branched amino acids (L-isoleucine) and branches off at the level of 2-oxo-3 methylvaleric acid, which undergoes condensation with an acetyl group The subsequent reactions are analogous to those in the tricarboxylic acid cycle, and result in 2-oxo 4-methylvaleric acid, which is transaminated to Leu.

Leucoanthocyanidins. A group of flavonoid natural products derived from flavan-3, 4-diol, and differing in their degree of hydroxylation They are common in plants, especially in wood, bark and the rinds of fruits. The dimers, oligomers and polymers of with cathechols (flavan-3-ols) form polyphenols which have the properties of taning agents.

Leucoplasts. Colorless plastids in plant cells, generally those which are not exposed to light. These include the starch-storing amyloplasts, the protein-storing aleuroylasts, the fat-storing elaioplasts and the etioplasts, which are the colorless chloroplasts of sprouts and stolons (for example, potato shoots) which have been kept in the dark etiolated)

Leucopterin. 2-amino-5,8-dihydro-4, 6, 7 (1H)-pteridinetrine. A white pigment found in the wings of cabbage white butterfles

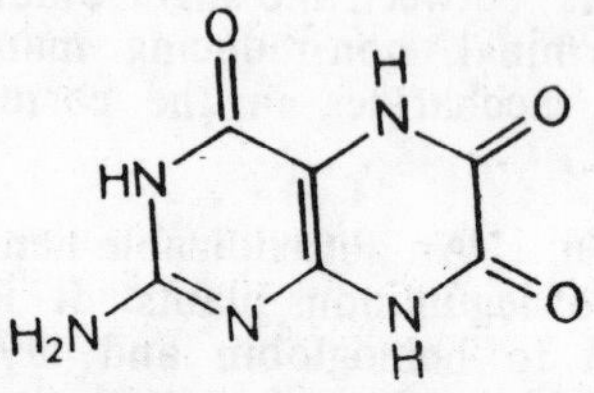

Leucopterin

and other butterfles. It is biosynthesized from guanine and two C-atoms of a pentose which serves as a precursor.

LH. Abbreviation for luteinizing hormone.

Liberin. A Releasing hormone; a suffix (liberin) used in the nomenclature of releasing hormones.

Licanic Acid Couepic acid, Ketoeleotearric Acid, Oxoeleotearic Acid. 4-oxo-9, 11, 13-octadecatrienoic acid,

$$CH_3(CH_2)_3CH{=}CH)_3\ (CH_2)_4\ CO\ CH_2)_2{-}COOH.$$

Isolated from the seed fat or oil of Licania rigida. It is the only unsaturated oxoacid that has been isolated from a natural fat, and it is also present in the seed fats of other Licania species. *e.g.*, L.arborea (Mexico). L. crassifolia (East Indies) and L. venosa (Guyana), and the seed fats of several species of Parinarium.

Leukotrienes. A class of peptolipids derived from arachidonic acid. These are slow reacting substances. Slow reacting substance. A (abbreviation SRS-A) is a L. involved in asthma and allergies.

Slow reacting substance A (a Leukotriene)

Lichenln, Moss Starch. A polysaccharide serving both a storage and as structural compound. It is composed of 150 to 200 D-glucose units linked β-1 4 with about 25% β-1,3 glucosidic linkages distributed at random in the molecule. It is found in many lichens and has antineoplastic properties.

Liebermann Burchard Reaction. A reaction used for colorimetric determination of sterols. The substance to be tested is dissolved in chloroform and treated with sulfuric acid and acetic anhydride. A color change from pink to blue to greedin dicates unsaturated sterols and is the basis for the quantitative determination of cholesterol in blood.

Life. The sum of all the chemical and physical processes occurring within an organism and enabling it to maintain its structural identity. Life includes the complex interactions and the integration in time and space of these processes.

Light Compensation Point. The light intensity at which the rate of photosynthesis (CO_2 incorporation) and the rate of respiration (CO_2 production) are balanced.

Lightening Hormone Pyr-Leu-Asn-Phe-Ser-Pro-Gly-Trp-NH_2. A neurohormone produced by the eye-stalk glands of crustaceans. The hormone is released from nerve endings in the gland in response to visual stimuli and controls the pigment granules in the hypodermal chromatophores. In this way the animal adjusts its color to match the surroundings.

Light-harvesting Protein. A strongly hydrophobic, integral membrane protein, isolated from the thylakoids of many angiosperms, gymnosperms and green algae. It is chiefly associated with photosystem II, but some activity with photosystem. I has also been demonstrated. Chlorophylls a and b are bound in equimolar amounts, together with lutein and β-carotene.

Lignin. A polymer responsible for the thickening and strengthening of plant cell walls. The properties associated with wood are due to the incrustation of plant cell walls with lignin Chemically lignin cannot be exactly defined. According to Freudenberg, lignin is a highly cross linked, macromolecular, branched polymer, formed irreversibly by dehydrogenation and condensation. According to Adler and Gierer, lignin is an essentially acid-resistant polymorphic, amorphous incrustation material found in wood, consisting of methoxylated phenyl-propane units linked by ether linkages and C—C bonds.

Lignoceric Acid n-tetracosanoic acid, $CH_3-(CH_2)_{22}-COOH$. A fatty acid. It occurs as a component of glycerides (usually less than 3%) in many seed oils, such as ground nut and rape seed oil. It is a component of certain cerebrosides (*e g*., kerasin), phosphatides and waxes.

Linalool, Coriandrol. A doubly unsaturated monoterpene alcohol found in essential oils. Both optical isomers occur naturally. The pure alcohol and its esters are used in perfumery.

Linoleic Acid. $\Delta^{9,12}$—octadecadienoic acid,

$$CH_3-(CH_2)_4-CH=CH-CH_2-CH=CH-(CH_2)_7-COCH.$$

An essential fatty acid. This acid is widely distributed in plant and animals. It occurs as a glyceride component in many fats and oils, and it is found in phosphatides. It is an essential dietary constituent formammals.

Linolenic Acid. $\Delta^{9,12,15}$—octadecatrienoic acid,

$$CH_3-CH_2-CH=CH-CH_2-CH=CH-CH_2-CH$$
$$=CH-CH_2-(CH_2)_6-COOH.$$

An essential fatty acid. This acid occurs in animals and plants, and it is especially common in plant fats and glycerophosphatides. This acid cannot be synthesized by mammals.

Lipases. A group of carboxylesterases, which preferentially hydrolyse emulsified neutral fats to fatty acids and glycerol or monoglycerides. Calcium ions are required for activity. Pancreatic lipase also requires taurocholate.

Lipids. A heterogeneous group of biological compounds, which are sparingly soluble in water, but very soluble in lipophilic solvents. They can be extracted from animal and plant tissues with a variety of organic solvents, *e.g.*, benzene, chloroform, trichloroethene. As a class, lipids are defined by their solubility; they include many chemically unrelated compounds. Thus the lipids include neutral fats (triacylglycerides), waxes, terpenes (monoterpenes, diterpenes, carotenoids, steroids, etc.) The more complex L, such as Glycolipids and Phosphotipids are also called lipoids, Glycerides and Waxes are known as saponifiable. lipids, whereas the Terpenes are calles nonsaponifiable L.

Lipoic Acid, 6-thioctic Acid, (+)-5[3-(1,2-dithi-oanyl)] Pentanoic Acid. A coenzyme of hydrogen transfer and acyl group transfer reactions. It is a component of the pyruvate dehydrogenase and the 2-oxoglutarate dehydrogenase complexes. Multienzyme complexes), which catalyse the oxidative decarboxylation of the corresponding 2-oxoacids.

S—S COOH

(3R)-1, 1-dithiolane-3-pentanoic acid
(the oxidized form of lipoic acid)

Lipoproteins. Conjugated proteins in which the prosthetic groups are lipid. They occur in blood plasma and cell cytoplasm, in the membranes of the cell and cell organelles, and in egg yolk Blood plasma lipo-proteins are responsible for the transport and distribution of lipids (lipid hormones; lipids absorbed by the intestine, such as neutral fats, phospholipids, free and esterified cholesterol, free fatty acids, and fat-soluble vitamins) via the lymph and blood system to the liver and other organs.

Liposome, Phospholipid Bilayer Vesicle. An aqueous compartment enclosed by a completely sealed lipid bilayer. Liposomes are formed by ultrasonic irradiation (sonication) of phosphoglycerides (or other suitable lipids) and water.

Lipotropic Substances. Compounds directly or indirectly involved in fat metabolism, which can prevent or correct fatty degeneration of the liver. They serve as substrates of phosphatide biosynthesis of these substrates. Thus, choline and any substance capable of contributing methyl groups for choline synthesis (*e g.* methionine lipotropic substances.

Lipotropin, Lipotropic Hormone, Adipokinetic Hormone, abb. **LPH.** A polypeptide hormone from the hypophysis. LPH promotes lipolysis, and it acts via the adenyl cyclase system. β-LPH contains 91 amino acid residues, Mr 9894 (porcine). γ-LPH contains 58 amino acid residues, identical in sequence to residues 1-58 of β LPH. Corticotropin, melanotropin, ACTH and LPH all contain an identical heptapeptide in their structures. They belong to the so-called ACTH family of peptide hormones, and they are derived from a common precursor (Peptides).

Lithocholic Acid. 3α-hydroxy-5β-cholanoic acid, or 3α-hydroxy-5β-cholan-24-oic acid. It is a monohydroxylated steroid carboxylic acid, and one of the bile acids. It has been isolated from the bile of man, cow, rabbit, sleep and goat. It is normally prepared from bovine bile. It is formed from chenodeoxycholate by intestinal bacteria. It is absorbed and returned to the liver for secretion. It is not readily conjugated, and it is relatively toxic to the liver. It may be important in the pathogenesis of liver damage following billary stains.

Liver. The largest metabolic gland of vertebrates. Liver and pancreas arise from the midgut in the course of embryonic development. The human L. weighs 1.5 kg and lies in the abdominal cavity in the right sulphrenic space. 1 to 2 blood per minute is supplied by the portal vein and the liver arteries, and leaves via the liver veins. The liver secretes bile, which flows out into the duodenum through a system of vessels, of which the gall bladder is a side arm. The functional unit of the liver is the hepatocyte, which is surrounded by blood vessels and bile capillaries. Metabolically, the liver is the most versatile organ in the body. Nearly all the products of digestion which are absorbed by the intestine are carried to the liver by the portal vein, there to be transformed or degraded.

Living Matter. In biochemistry, the body substance of living organisms and cells. It has been determined by elemental analysis that about 40 chemical elements occur in living matter, but of these, about 10 account for more than 99% of the body substance. The dominant inorganic compound is water, the milieu in which life processes take place. Of the organic molecules of living matter (biomolecules), the quantitatively most important are the carbohydrates, proteins and lipids, which are therefore the main classes of nutrients for humans animals.

TABLE 1

Water; Mineral and Organic Contents (in Percent of Fresh Weight) of Animal and Plant Body Substance

The values are rough approximations.

Class of substance	Animal (%)	Plant (%)
Water	60	75
Mineral	4.3	2.5
Carbohydrates	6.2	18
Lipids	11.7	0.5
Proteins	17.8	4.2

TABLE 2

Chemical Composition of the Human Body (Adapted from Rapoport)

The values have been rounded off.

Class of Substance	%	kg. Fresh weight)
Water	60	42.0
Minerals	4	2.8
Carbohydrates	1	0.7
Lipids	15	10.5
Proteins	19	13 3
Nucleic acids	1	0.7

Lobella Alkaloids An extensive group of 2, 6-disubstituted piperidine alkaloids of the genus Lobelia, especially from the medicinal plant, Lobelia inflata, cultivated in some European countries and in the USA. There are three structural types, depending on the functional groups of the substituents: lobelidiols, lobelionois and lobelidiones (Fig.). R_1 and R_2 may be $-CH_3$, $-C_2H_5$ $-C_6H_5$, and the configuration of the side chains may be cis or trans.

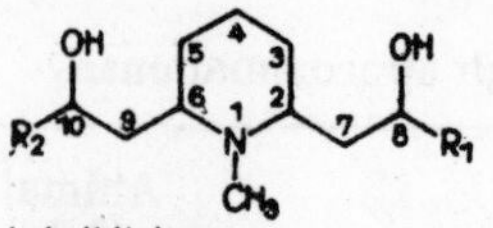

Lobelidioles

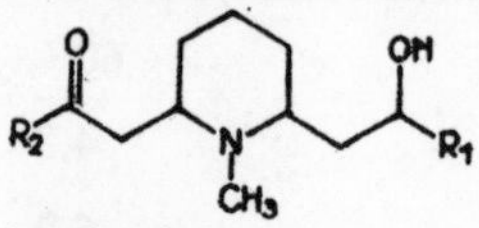

Lobelionoles

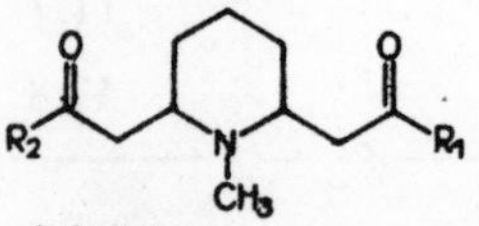

Lobelidiones

Lobelin. (—)-lobelin, cis-8, 10 diphenyl-lobelionol. The main Lobelia alkaloids. Structurally, it is a lobelionol, in which both R_1 and R_2 are phenyl ($-C_6H_5$) groups. L., crystallizes as colorless needles, m.p. 130—131°C, $[\alpha]_D^{15}$ $-43°$ (c = 1, ethonol). It is used medicinally as a respiratory analeptic. On account of its nicotine-like properties, it is also used in the treatment of smoking addiction.

Loganin. An ester glucoside belonging to the iridoid group. Cleavage with emulsin produces the aglycon, loganetin. L. and the free acid, loganic acid, are found in Strychnos and Menyanthus spp. It is a key compound in the biosynthesis of irids and many alkaloids.

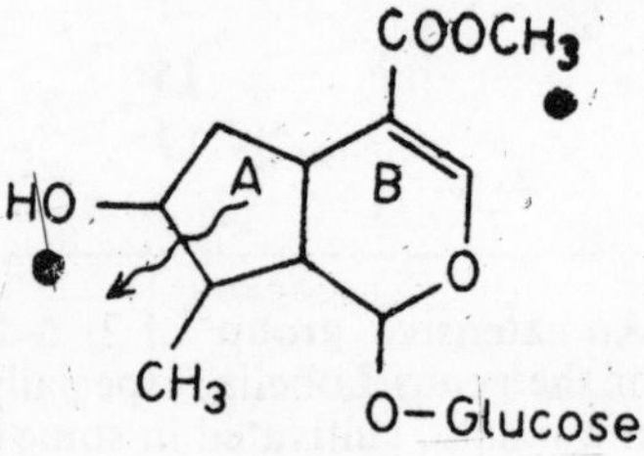

Loganin

Lophenol. 4α-methyl-5α-cholest.7-en-3β-ol. A zoo- and/or phytosterol, M_r 400.69, m.p. 151°C, $[\alpha]_D$ +° (chloroform). It has the structure of a tetracyclic triterpene with a 31, 32-bisdemethylianostane skeleton. Its structure is therefore intermediate between lanosterol and the sterols derived from lanosterol.

Luciferase. A low molecular weight oxidoreductase, which catalyses the dehydrogenation of iuciferin in the presence of oxygen, ATP and magnesium ions. During this process, 96% of the energy released appears as visible (mostly blue) light. This is the basis of Bioluminescence.

Luciferin. A collective name for the substrates of luciferases. By the action of the enzyme and in the presence of oxygen. It gives rise to bioluminescence. Electron-excited states of the oxidation product of L. (thought to be peroxides) are responsible for light emission. The structures of life L. are known.

Lupeol. A pentacyclic, triterpene alcohol. It has a 5α-Lupandering system. It occurs free, esterified and as the aglycon of triterpene. Saponins in many plants. It has been found, *e.g.* in the latex if Ficus spp., in the seed coats of the yellow lupin (Lipinus luteus) and in the leaves of misletoe. It has been detected in the cocoons of the silkworm, Bombyx mori.

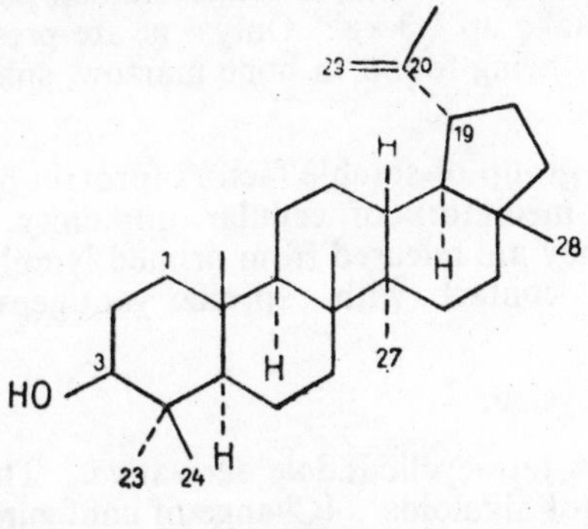

Lupeol

Lupin Alkaloids. A group of quinolizidine alkaloids containing a ring system variously known as quinolizidine, octahydrophyridocoline, norlupinane, or 1-azabicyclo [0,4,4] decane. This ring system may be further condensed with other N-containing ring systems, so that tri- and tetracyclic, as well as bicyclic lupins are known. Chief representatives are lupine, lupanine; sparteine and cytisine.

Lutein, Xanthophyll. (3 R, 3′R, 6′R)-β, ε-carotene-3, 3′-diol, or 3,3′-dihydroxy-α-carotene. A carotenoid of the xanthophyll group.

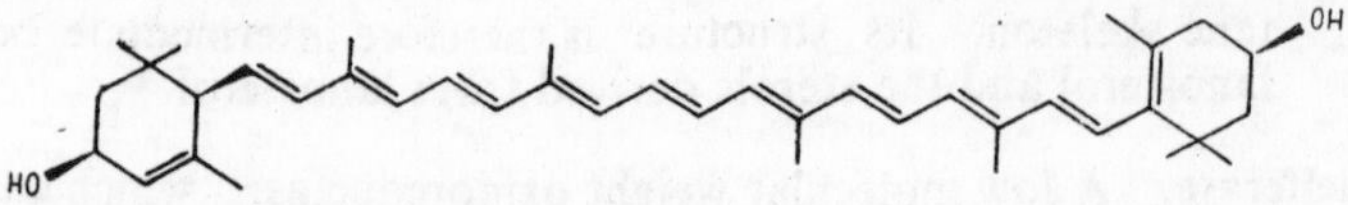

Lutein

It contains the same chromophore as α-carotene, and it is isomeric with zeaxanthin. It is a yellow pigment present, together with carotene and chlorophyll, in all green parts of plants. It is also present in many yellow and red flowers and fruits. It is also found in animals, *e.g.* in bird feathers, egg yolk and the corpus luteum. It may be present in free form, or as an ester; it has no vitamin A activity.

Lycopene. One of the carotenoids. It is an unsaturated, aliphatic hydrocarbon of isoprenoid origin. It is a red plant pigment, widely distributed and especially plentiful in fruits and berries.

Lymphocytes. Colorless, nucleated blood cells, approximately the same size as erythrocytes. They are immunologically competent, *i.e.* they recognize antigens, and they play a central role in the Immune response (An average human possesses about 10^{12} L., which make up 1.3 kg. Only 3 g are present in the blood, the majority being found in bone marrow, spleen, lymph glands and lymph.

Lymphokines. A group of soluble factors (protein or polynucleotide); which are mediators of cellular immunity, but are not antibodies. They are released from primed lymphocytes when they come into contact with specific antigens. M_r 5000 to 80000.

Lys. Abb. for Lysine.

Lysergic Acid. A tetracyclic indole derivative. The D-form, is one of the Ergot alkaloids. (Change of configuration at C-8 produces biologically inactive isolysergic acid. It is a close structural relative of LSD.

Lysergic Acid Diethylamide, LSD. A synthetic derivative of D-lysergic acid. It is the most potent psychotomimetic substance known. Hallucinogenic derivatives of lysergic acid are found in the Mexican ritual drug, ololiuhqui, these are lysergic acid amide (ergine) and lysergic acid hydroxyethylamide. LSD can be prepared semisynthetically from ergot alkaloids, which also occur in Rivea seeds (*e.g.* ergometrine). The hallucinogenic action of LSD, which is characterized by a state resembling schizophenia, was discovered accidentally by A. Hofmann during the recrystallization of a sample of LSD.

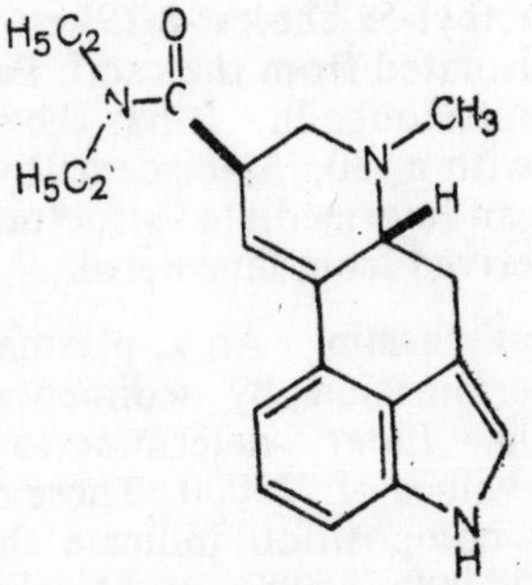

D-Lysergic acid diethylamide (LSD-25)

L-Lysine, Lys. 2, 6-diaminocaproic acid, H_2N-$(CH_2)_4$-$CH(CH_2)$-COOH. It is a basic proteogenic, essential amino acid. The proteins of cereals (wheat, barley, rice) and other vegetable foodstuffs are rather poor in Lys. Children and young growing animals have a particularly high requirement for Lys, since it is needed for bone formation. Like threonine, Lys does not take partinreversible transmination.

Lysosomes. Organelles, 0.2—2 nm diameter, found in the cytoplasm of eukaryotic cells. They are bounded by a single lipoprotein membrane, but otherwise show no fine structure. Under the light or electron microscope. They are markedly polymerphic. They can be characterized biochemically or histochemically, but not morphologically. They are sites of Intracellular digestion particularly of biological macromolecules, such as proteins, polynucleotides, polysaccharides, lipids, glycoproteins, glycolipids, etc. Approximately 40 different lysosomal hydrolases are responsible for this degradative activity; they all show optimal activity at acidic pH values.

Lysozyme, Endolysin, Muramidase, N-acetylmuramide Glycanohydrolase (EC 3.2.1.17). A widely occurring hydrolase, found in phases, bacteria, plants, invertebrates and vertebrates. In the latter, it is found particularly in egg white, saliva, tears and mucosas.

M

Macdougallin. 14α-methyl-5α-cholest-8 (9)-en-3β, 6α-diol, a phytosterol. It was isolated from the cacti, Peniocerius fosteriunus and Peniocerius macdougalli. It has the structure of a tetracyclic triterpene with a 30, 31-bisdemethyllanostane skeleton, and it represents an intermediate structure between lanosterol and the sterols derived from lanosterol.

α_2-Macroglobulin, α_2-antiplasmin. An α_2-plasma protein. The first reported Mr determination by sedimentation diffusion gave a value of 820,000. Later determination by sedimentation equilibrium gave values of 725000. These results, together with studies of subunit composition, indicate that the true Mr lies in the range 650,000-725000. α_2-M is a glycoprotein containing 8.2% carbohydrate. The carbohydrate moiety contains mannose, fucose, N-acetylglucosamine and sialic acid. Electron microscope studies of the protein reveal a structure resembling two beans facing each other; these two identical subunits are bound noncovalently. Each subunit consists of two peptide chains linked covalently by a disulfide bridge. α_2-M. binds tightly and inhibits a number of proteases of varying specificity and origin, *e.g.* trypsin, plasmin, thrombin, kallikrein and chymotrypsin. It is therefore a natural inhibitor of plasmin.

Macrolides, Macrolide Antibiotics. A group of antibiotics from various strains of Streptomyces, all having the same complex macrocyclic structure. They inhibit protein synthesis by blocking transpeptidation and the translocation on the 50S ribosomal subunit (similar to Chloramphenicol,). Examples of these are erythromycin (Fig.), spiramycin, oleandomycin, carbomycin, angolamycin, leucomycin, picromycin. Almost all M, are used therapeutically as broad spectrum antibiotics.

Madder Dyes. Refer to the plant dyes from madder (Rubia tinctorum) and other members of the madder family (Rubiaceae). Important representatives are the glycosidically bound components alizarin and purpurin.

Magnesium, Mg. A cation widely distributed in biological systems, with many different biological functions. Mg is the fourth most abundant cation in the vertebrate, and it has great biological significance as a constituent of the porphyrin system of chlorophyll. Mg ist an essential nutrient, and it plays an important part in metabolism by acting as a cofactor of many different enzyme system.

Maleic Hydrazide, MH. 1, 2-dihydro-3, 6-pyridazinedione. A synthetic plant growth retardant, which is used as a herbicide against grasses. It causes a unique depression of growth, inhibition being marked but temporary. It inhibits germination of seeds, suppresses growth of roots and terminal shoots, and retard flower and bud development. It prevents the formation of suckers in tobacco and tomatoes, and prevents the sprouting of onions and potatoes.

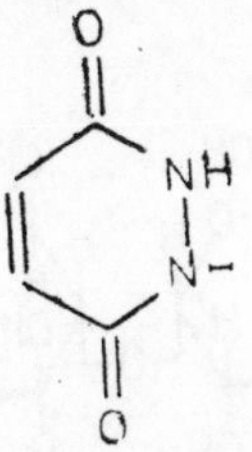

O, NH, N

Maleic Hydrazide

Malformin A. A heterodetic cyclic pentapeptide with antibiotic activity, from Aspergillus niger. It cause malformation of the roots of cereals.

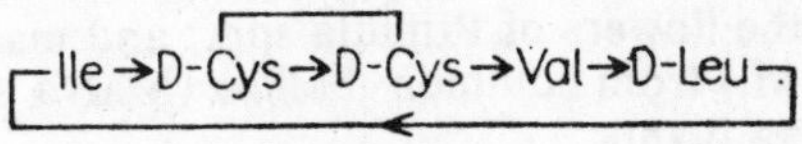

Ile→D-Cys→D-Cys→Val→D-Leu

Malformin A

Malic Acid. Monohydroxysuccinic acid, HOOC-COOH-CH_2-COOH. A dicarboxylic acid found in many plant juices. The malate ion is formed in the Tricarboxylic acid cycle and the Gloxylate cycle. Malate plays an important role in the Diurnal acid rhythm of the Crassulaceae (stone-crop family).

Malic Enzyme(s), L-malate-NADP Oxidoreductase, Decarboxylating. (EC 1.1.1.40). An important enzyme found in most organisms, which catalyses the decarboxylation of L-malate to pyruvate and CO_2, with concomitant reduction of $NADP^+$ to NADPH (or the synthesis of malate by the reverse reaction :

$$HOOC\text{-}CH_2\text{-}CHOH\text{-}COOH\text{-}+NADP^+ \rightleftharpoons CH_3\text{-}CO\text{-}COOH+CO_2+NADPH+H^+.$$

Malonic Acid. HOOC-CH_2·COOH. A dicarboxylic acid, which has been found in the free form in plants, but is of only sporadic occurrence. m.p. 135.6°C. A metabolically important derivative of malonic acid is malonyl-CoA, an intermediate of Fatty acid biosynthesis.

Maltose, Maltobiose, Malt Sugar. 4-O-α-D-glucopyranosyl-D-glucose. A reducing disaccharide. It consists of two molecules of D-glucose linked by an α-1, 4-glycosidic bond; both glucose residues are present in the pyranose form. It is a stereoisomer of cellobiose. It serves as a fermentable substrate in brewing, as a component of prepared bee food, as a substrate in microbiological growth media, and generally in food and pharmaceuticals as a nutrient and sweetner (one third as sweet as glucose).

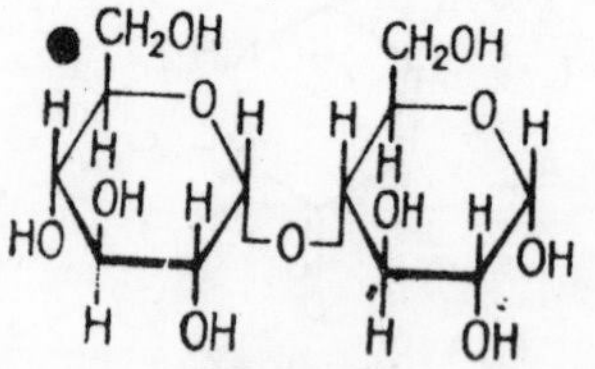

α-Maltose

Malvidin. 3, 5, 7, 4′-tetrahydroxy-3′, 5′-dimethoxyflavylium cation. The aglycon of various anthocyanins. 10 natural glycosides of malvidin are known, *e.g.* oenin (syn. primulin, ligulin), the 3 β-glucoside of malvidin, is the pigment in the skins of black grapes and the flowers of Primula spp.; and malvin (3, 5-di-β-glucoside of M.) from common mallow (Malva sylvestris) and other flowering plants.

Manganese, Mn. A bioelement present in all living cells, which usually conain less than 1 ppm. on a dry weight basis, or less than 0.01 mM in fresh tissue. Bone contains 3.5 ppm. Mn (II) is necessary for sporulation in Bacillus subtilis, and these bacteria can maintain an intracellular Mn concentration of 0.2 mM against an external concentration of 1 μM. Mn is an essential nutrient for animals and plants. Mn deficiency in animals leads to degeneration of the gonads and to skeletal abnormalities; the characteristic skeletal abnormality in chickens is called slipped tendon disease, or perosis. In plants, Mn deficiency results in chlorosis and mottling.

Mn may also be important in the regulation of enzyme activity, *e.g.* nonadenylylated glutamine synthetase requires Mg (II), but the adenylylated form binds Mn (II). Also the specificity of nucleases and DNA polymerases is changed when

Mg (II) is replaced by Mn (II). The physiological significance of these differences is not clear.

Mannans. Polysaccharides widely distributed in plants as reserve material, and in association with cellulose as hemicellulose. Mannans of plants consist of D-mannose predominantly in α-1, 4-glycosidic linkage. Yeast cell walls contain mannans, consisting of a backbone of α-1, 6-linked mannose with short (1-3 mannose units) branches attached by α-1, 2 and α-1, 3-linkages.

D-Mannitol, Mannite, Manna Sugar. A hexitol related to D-mannose. Mr 182.17, HOH_2C-CHOH-CHOH-CHOH-CHOH-CH_2OH. It is found widely in plants and plant exudates, and in fungi and seaweeds. Mannitol is used as a sugar substitude for diabetics.

D-Mannosamine. 2-amino-2-deoxy-D-manose, an amino sugar related to mannose, in which the hydroxyl group on C-2 of mannose is replaced by an amino group. It is a constituent of neuraminic acids, animal mucolipids and animal muco-proteins.

D-Mannose, Seminose, Carubinose. A monosaccharide hexose. It is a C-2 epimer of D-glucose. In the metabolism of D-mannose the activated form of the sugar is the GDP derivative (not the UDP derivative which occurs for many other sugars). It is degraded via its 6-phospho-derivative, which is converted to glucose 6-phosphate.

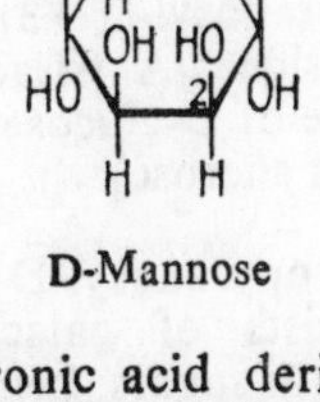

D-Mannose

D-Mannuronic Acid. A uronic acid derived from D-mannose. It is a constituent of the polyuronide, alginic acid.

Marker Synthesis. A laboratory synthesis developed by R.E. Marker for the conversion of diosgenin into progesterone.

Mavacurin. One of the Curare alkaloids.

Melanins. High molecular weight, amorphous polymers of indole quinone, empirical formula $(C_8H_3NO_2)_x$, containing 6-9%

nitrogen. These are natural pigments occurring predominantly in the animal kingdom in vertebrates and insects, and occasionally in microorganisms, fungi and higher plants. Mammalian colors are determined chiefly by two types of melanins the black or brown insoluble, nitrogenous Eumelanins and the lighter colored sulfur-containing, alkali-soluble Phaeomelanins.

Melanotropin, Melanocyte Stimulating Hormone, MSH. A polypeptide hormone produced by the polygonal cells of the pars intermedia of the hypophysis, except in species that lack this structure, *e.g.* chicken, porpoise, whale, where MSH is produced by the neurohypophysis. The production of MSH is under the control of MRH and MIH of the hypothatamus. MSH causes the melanophores to expand, thus producing a generalized darkening of the skin. The function of MSH in humans is not clear.

Melatonin. N-acetyl-5-methoxytryptamine. A biogenic amine. It is a hormone produced by the pineal gland. It inhibits the development of gonadal function in young animals, and the action of gonadotropins in mature animals. Its synthesis is suppressed by light, acting via the eyes and the nervous system, and it is increased in the dark.

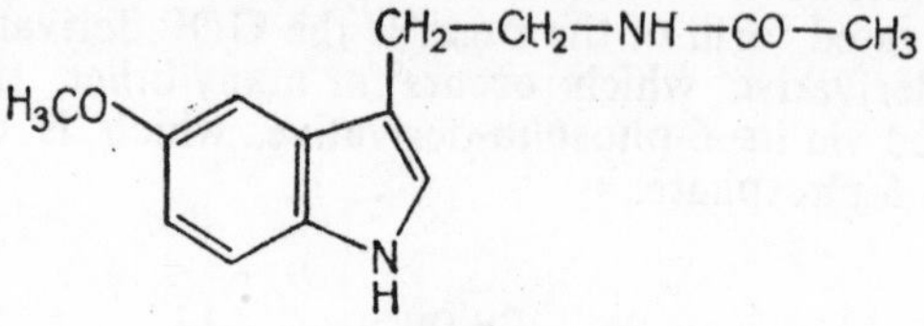

$CH_2—CH_2—NH—CO—CH_3$

H_3CO

N
H

Melatonin

Melezitose. 0-α-D-glucopyranosyl-(1→3)-0-β-D-fructofuranosyl (2→1)-α-D-glucopyranoside. A trisaccharide found in plants. It consist of a molecule of D-glucose linked to the disaccharide turanose (an isomer of sucrose).

Melibiose. 6-0-α-D- galactopyranosyl-D-glucose. A reducing disaccharide which consists of galactose and glucose linked α-1,4-glycosidically, both sugars being in the pyranoid form. It represents a disaccharide grouping in the trisaccharide raffinose.

Melissic Acid. n-triacontanoic acid, CH_3-$(CH_2)_{28}$-COOH. A higher monocarboxylic fatty acid. It is present as an ester in beeswax and montan wax.

Melissyl Alcohol, Myricyl Alcohol. 1-triacontanol, or 1-hydroxytriacontane, CH_3-$(HC_2)_{28}$-CH_2OH. A high molecular weight alcohol, which is present in plant cuticle waxes. Beeswax consists principally of the palmitate ester of melissyl alcohol (myricin).

Melittin. A linear, toxic (hemolytic) polypeptid-amide, and the chief component of bee venom (about 50% of dried venom, and at least in 50-fold molar excess over other venom constituents). The primary product of melittin biosynthesis is prepromelittin, consisting of 70 amino acid residues. Removal of the N-terminal signal sequence of 21 residues leaves promelittin. The protease required for removal of the signal sequence is present in many animal cells and is not species-specific, *e.g.* promelittin and not prepromelittin is the first detectable product when melittin mRNA is injected into oocytes of Xenopus laevis.

Membrane Enzymes. The enzymes present on the surfaces of the many different types of biological membranes. There are, *e.g.* glucose 6-phosphatase of the endoplasmic reticulum, galactosyl transferase of the Golgi apparatus, oligomycin-sensitive ATPase of inner mitochondrial membrane, monoamine oxidase and rotenone-insensitive NADA-cytochrome c-reductase of the outer mitochondrial membrane, and the sodium-independent ATPase and the 5-uncleotidase of the cell membrane. A very large number of reactions occur on membranes, and the membrane enzymes play a central part in cell metabolism.

Membranochromic Pigments. The plant pigments which impregnate the cell wall, *e.g.* the phenols and quinones that give color to heart wood.

Memory. The ability of the central nervous system to store information. Information enters the brain in the form of nerve impulses, and it is thought that temporary or short-term memory is also a matter of reverberations in nervous circuits. It can be completely erased by a brain concussion (retrograde amnesia). Long-term memory, in contrast, is not affected by concussion. Its physiological basis is thought to be changes in the synapses. Information can be transferred from short to long-term memory.

Menadione. Vitamin K_3 and provitamin of the vitamin K group.

Menaquinone-6. Vitamin K_2.

p-Menthadienes. Doubly unsaturated, monocyclic monoterpene hydrocarbons.

Menthol. It is p-menthane-3-ol, commercially the most important of the monocyclic, monoterpene alcohols. It contains 3 asymmetric carbon atoms. The naturally occurring form is (−) menthol, which is the chief constituent of peppermint oil.

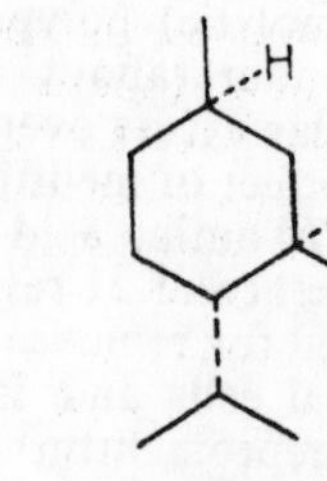

(—)-Menthol

Mescaline. It is 3, 4, 5-trimethoxy-phenylethylamine, the principal hallucinogenic compontent of the drug Peyotl or Peyote, used as a ceremonial intoxicant by Mexican Indians living near the Mexico-USA border.

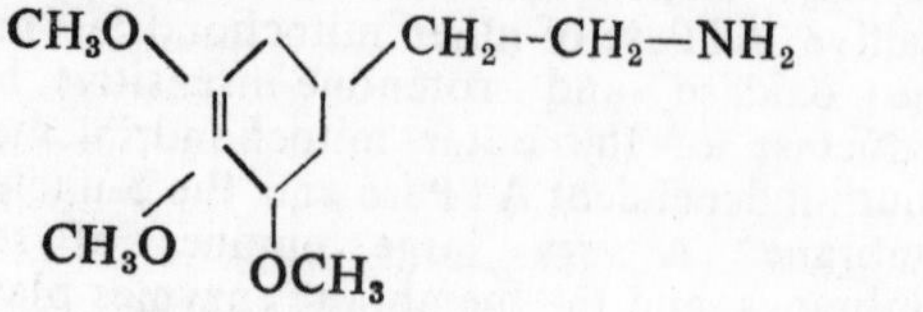

CH_3O CH_2--CH_2—NH_2

CH_3O OCH_3

Mescaline

Messenger RNA, mRNA, Template RNA. RNA that is transcribed from the codogenic strand of DNA; during this pocess, information encoded in the linear sequence of nucleotides in the codogenic DNA srtand acts as a template for the formation of a complementary code in the linear sequence of nucleotides in the mRNA. The functional unit of protein biosynthesis is the polyribosome (polysome), consisting of several ribosomes attached at regular intervals along the length of the mRNA. One molecule of mRNA may carry information for the synthesis of more than one protein, having been transcribed without interruption from several neighbouring cistrons of the DNA. This polycistronic mRNA has so far been found only in prokaryotic organisms. The polypeptides translated from polycistronic mRNA usually have related functions, *e.g.* 10 enzymes in the histidine biosynthesis pathway are encoded in and translated from a polycistronic mRNA containing about 12000 nucleotides, Mr 4×10^6. Viral RNA is functionally very similar to mRNA. The entire length of viral RNA or mRNA is not translated into protein. The start condon for protein synthesis (always AUG) is located some distance in from the 5′-end of the mRNA molecule. Thus, the 5′-end commences with a sequence of nucleotides, which is not translated, and appears to act as a site for the recognition by and of specific initiation factors. For example, in the Lac-mRNA of Escherichia coli, the start condon for β-galactosidase occurs at position 39.

Functional mRNA is single stranded.

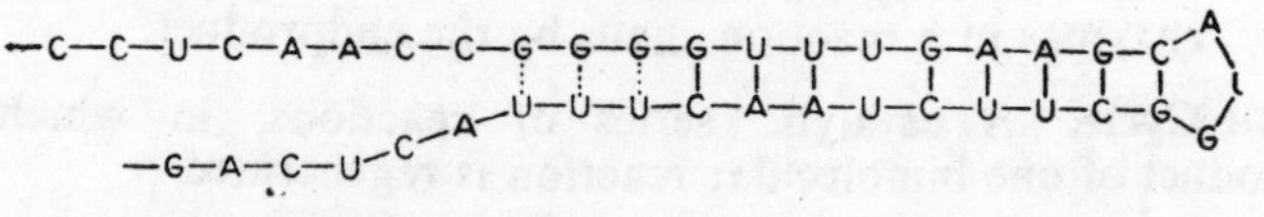

```
                                                              A
—C—C—U—C—A—A—C—C—G—G—G—G—U—U—U—G—A—A—G—C     \
                 :  :  :  |  |  |  |  |  |  |  |      U
                 U—U—U—C—A—A—U—C—U—U—C—G    /
—G—A—C—U—C—A                                    G
```

Secondary Structure in Region of the Coat Protein Cistron of R 17-phage RNA. A adenine, C cytosine, G guanine, U uracil

Mesterolone. 1 α-methyl-3β-hydroxy-5α-androstan-3-one, a synthetic androgen. It shows high activity when administered orally. Like Methyltestosterone. It is used for the therapy of male gonadal insufficiency and endocrine disorders.

Mestranol. 17 α-ethinyl-3-methoxyestra-1, 3, 5 (10)-trien-17 β-ol, a synthetic estrogen. It has high biological activity; when administered orally, and it is used as a component of Ovulation inhibitors.

Met. Abb. for L-methionine.

Metabolic Control, Metabolic Regulation. Metabolism is subject to control, and an analogy may be drawn with electronic and mechanical regulation processes used in technology. In a purely formal way, living systems can be regarded as cybernetic machines. Control is a fundamental principle in the organization of living organisms, and depending on the nature of the signal or method of information transfer, there are four broad types :

1. Neural (nervous) control. The nerve impulse is an electrical signal, and the regulatory response may also be electrical (*e.g.* further nerve impulse to a muscle) or chemical (*e.g.* production of a hormone). The nervous system may be considered as a physiological broadcasting system.

2. Hormonal (humoral) control. Hormones act as chemical signals in a regulation system that is superimposed on the more basic levels of M.c.

3. Differential gene expression. The signals (or triggers) of differential gene expression may be chemical (hormones)

or environmental (*e.g.* light). Differential gene expression is responsible for the regulation, at a molecular level, of differentiation and development.

4. Feedback and feedforward mechanisms, in which metabolites themselves act directly as signals in the control of their own breakdown or synthesis. Feedback is negative or positive. Negative feedback results in inhibition of the activity or synthesis of an enzyme or several enzymes in a reaction chain by the endproduct.

Mctabolic Cycle. A catalytic series of reactions, in which the product of one bimolecular reaction is regenerated :

$$A+B \longrightarrow C+A$$

Thus A acts catalytically and is required only in small amounts, and A can be considered as a carrier of B. The catalytic function of A and other members of the metabolic cycle ensure economic conversion of B into C; B is the substrate of the metabolic cycle and C is the product. If intermediates are withdrawn from the metabolic cycle, *e.g.* for biosynthesis, the stationery concentrations of the metallic cycle intermediates must be maintained by synthesis Replenishment of depleted metabolic cycle intermediates is called anaplerosis.

Metabolic cycles are anabolic, catabolic or amphibolic. The Calvin cycle is an anabolic (synthetic) cycle.

Metabolic Shunt, Metabolic Bypass. A metabolic pathway which bypasses some reactions and exploits others of a primary metabolic pathway.

Metabolism. The sum total of chemical (and physical) changes that occur in living organisms, and which are fundamental to life. Nutrients from the environment are used in two ways by living organisms; they may serve as constituents in the synthesis of components of the organism (assimilation), or they may be oxidatively degrated for purposes of energy production (dissimilation).

Metabolite. A substance produced or consumed by metabolism. Biopolymers are not included in this definition. The precursors and degradation products of biopolymers are, however, true metabolities. All small molecules produced or converted by enzymes during metabolism are metabolities.

Metalloproteins. The proteins containing complexed metals. In the metalloenzymes, the metals are functional components. In the metal-transporting M. (*e.g.* the blood proteins, transferrin and coeruloplasmin) and the metal-storage depot proteins (*e.g.* ferritin), the metal binding is reversible and the metal is a temporary component.

Metallothioneins. Cytoplasmic metalloproteins, inducible in mammalian liver and kidney in response to Zn, Cu, Hg and especially Cd. From any single source (*e.g.* horse kidney), there appears to be only one apoprotein (*i.e.* thionein), and the metallothioneins from that source differ only with respect to the bound metal.

Methemoglobin. A hemoglobin in which the iron is trivalent (Fe III). M are unable to transport oxygen.

L-Methionine, Met. α-amino-γ-methylmercaptoutyric acid, a sulfur-containing essential proteogenic amino acid.

$$\underset{\text{Homoserine}}{CH_2OH-CH_2-HCNH_2-COOH} + \underset{\text{Cysteine}}{SH-CH_2-HCNH_2-COOH} \longrightarrow \underset{\text{Cystathionine}}{CH_2-S-CH_2-HCNH_2-COOH} \longrightarrow$$

$$\longrightarrow \underset{\text{Homocysteine}}{CH_2SH-CH_2-HCNH_2-COOH} + CH_3COCOOH + NH_3$$

(+ Vitamin B_{12}) ← N^5-Methyltetrahydrofolic acid

$$\underset{\text{Methionine}}{CH_2-S-CH_3-CH_2-HCNH_2-COOH} + \text{Tetrahydrofolic acid}$$

Methionine Biosynthesis in Escherichia Coli

Methods of Biochemistry. Usually methods for the study of metablic processes, but in the widest sense including isolation, identification and characterization of natural substances. Methods for the study of metabolism are classified into 3 types :

1. In vivo methods employ whole organisms, their organs or cells, or populations of cells of micro-organisms. In balance studies, substances are administered to the organism and the time course of their conversion to various products is determined by analysis of body

materials or excretory products (feces, urine, expired gases).

2. In vitro methods are performed outside the whole organism. They are essentially "test tube" methods, employing crude cell homogenates, sub-cellular fractions thereof, or purified enzymes. For methods of cell disruption, subcellular fractionation and enzyme purification Proteins and Density gradient centrifugation.

The techniques of histochemistry and cytochemistry are also in vitro methods: the sites of metabolites, enzymes or metabolic reactions are identified in organ or tissue slices and in cells by characteristic chemical reactions, *e.g.* specific color reactions.

Methyl-accepting Chemotaxis Proteins, MCP. The cell membrane proteins in Escherichia coli involved in the iniation and control of chemotactic behavior. There are at least 3 different MCP in E. coli, each responsive to stimuli from different types of chemoreceptor, including both attractants and repellants.

Methylated Xanthines. N-Methyl derivatives of xanthine, biosynthesized by the enzymatic methylation of free xanthine (N-1, 3 and 7) with S-adenosyl-L-methionine. Caffeine, theobromine and theophylline occur in certain plants and are known as purine alkaloids.

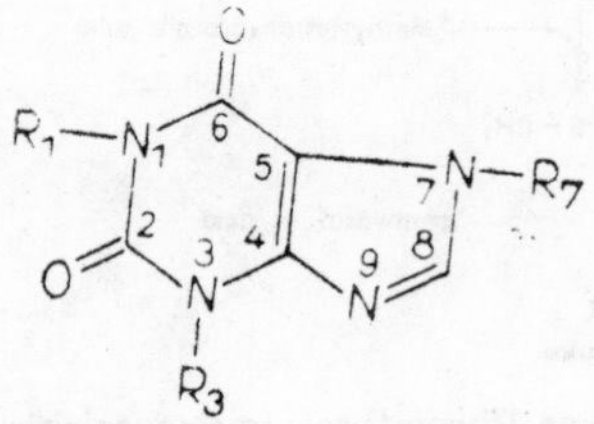

	R_1	R_3	R_7
Xanthine	H	H	H
Theophylline	CH_3	CH_3	H
Theobromine	H	CH_3	CH_3
Caffeine	CH_3	CH_3	CH_3

Structures of methylated xanthines

O-Methylbufotenin. A toxin from the toad, Bufo alvarius which has also been found in plants. In addition to its general properties as a toad poison, it also has a psychotropic effect.

Methylglyoxal. An intermediate of carbohydrate degradation in certain orgainsms. In some bacteria (Pseudomnas spp.) Methylglyoxal which is converted to lactate (precursor of pyruvate) by a catalytic cycle involving gluthione.

N^6 cis-γ Methyl-γ-hydroxymethylallyl adenosine. 6-(4-hydroxy-3-methyl-but-cis-2-enyl)-aminopurine. It is isomer of Zeatin and the free compound, zeatin, shows cytokinin activity.

Methyltestosterone. 17 α-methyltestosterone, 17α-methyl-17β-hydroxyandrost-4-ene-3-one. A synthetic androgen. It is shows high biological activity when administered orally, and it is used especially for the therapy of hypogenitalism, hormonal impotence, and peripheral circulatory disturbances. It is the 17α-derivative of Testosterone.

Mevaldic Acid. An intermediate in Terpene biosynthesis.

Mevalonic Acid. An intermediate in Terpene (see) biosynthesis.

MF. Abb. of maize factor (Zeatin).

MH. Abb. of Maleic hydrazide.

Michaelis-menten Kinetics

1. The Michaelis-Menten stoichiometric model shows the relationship between free enzyme (E), substrate (S), enzyme-substrate complex (Michaelis complex, ES) and product (P) :

$$E + S \underset{k_{-1}}{\overset{k_1}{\rightleftharpoons}} ES \xrightarrow{k_2} E + P$$

   where k_1 and k_2 are rate constants (k_2 is also known as the catalytic constant, or k_{cat}) $(k_{-1}+k_2)/k_1$ is a kinetic constant, known as the Michaelis constant and represented by k_m. If $k_2 \ll k_{-1}$, then $K_m = k_s = k_{-1}/k_1$ (Michaelis condition), and the equilibrium constant k_s is known as the substrate constant. If $k_{-1} \ll k_2$, then $K_m = k_2/k_1$ (Briggs-Haldane condition).

2. The Michaelis-Menten rate equation shows the relationship between v (rate of reaction), V_m (maximal rate when enzyme saturated with substrate), and S (substrate con-

centration : $v = V_m S/(K_m + S)$. When $S \ll K_m$, the reaction is first order, and $v = (V_m/K_m)S$. When $K_m \leqslant S$, the reaction is zero order, and $v = V_m = \text{constant}$. V_m and K_m are enzyme kinetic parameters. The Michaelis-Menten equation is often valid for other cases, where the derivation of the kinetic parameters from the rate constants is more complicated.

Microbiological Conversions, Microbiological Transformations. Conversions of materials occurring is one or more stages, and catalysed by micro-organisms. These conversions are the result of microbiological enzyeme action, and often have no importance for the microbial cell.

Microbiological Industry, Fermentation Industry. A branch of industry in which materials are produced or converted by the action of micro-organisms. It includes fermentation industry (production of alcoholic beverages and organic acids), production of antibiotics, production of enzymes and biochemicals, production of banking yeasts and yeasts for foodstuff manufacture.

Microbiological Preservation. The preservation of plant material by lactic acid produced by lactic acid fermentation. The lactic acid fermentation is used in the preparation of silage and sauerkraut.

Microbiological Processing of Ores, Microbial Mining. The use of acidophilic thiobacilli to leach metals from ores, where their concentration is too low for economic extraction by smelting. The most important organism for this purpose is Thiobacillus ferrooxidans, whose natural growth environment is a solution of metals, iron in particular, in 0.05 M sulfuric acid. It obtains energy by oxidizing sulfur, sulfides and ferrous iron.

β_2-Microglobulin. The smaliest known plasma protein; 100 amino acid residues of known primary structure. Increased quantities of β_2-M. are found in the urine in Wilson's disease and in cadmium poisoning.

Micronutrients. Trace nutrients.

Microsomes. A heterogeneous fraction of sub-microscopic vesicles, 20-200 nm diameter, formed during disruption of the cell by the resealing of fragments of the endoplasmic reticulum (and to some extent of the plasma membrane). Under the electron microscope, Ribosomes can be seen attached to the outside of the microsomes. Preparation of microsomes is by differential centrifugation of disrupted, homogenized cells; following the sedimentation of larger fragments, microsomes are sedimented at 100 000 g.

MIH. Abb. for melanotropin release inhibiting hormone.

Milk Proteins. Soluble proteins present in milk, consisting of caseins and whey proteins. The chief caseins are αS-, β- and Kappa casein. The most important whey proteins are β-lactoglobulin, α-lactalbumin and lactoferrin. In addition, milk contains several enzyme proteins, *e.g.* lactoperoxidase, xanthine oxidase, and immunoglobulins IgG, IgA and IgM. etc.

Mineral Nutrients, Major Inorganic Elements, Inorganic Bulk Elements, Macroelements. Inorganic nutrients required by living organisms in greater quantity than the trace elements. They are absorbed as cations (Na^+, K^+, Ca^{2+} etc.) and anions (Cl^-; I^-, NO^-_3, $SO^2_3{}^-$ etc.).

Minimum Protein (or Nitrogen) Requirement. The amount of complete protein required daily to compensate the nitrogen lost by excretion. Adults require 25 to 35 g complete (containing the optimal amounts of essential amino acids) protein per day. The absolute minimum protein requirement is the amount of nitrogen excreted on a protein-free but calorically adequate diet, about 2.4 g N=15 g protein per day for adults.

Miraculin. A taste-modifying glycoprotein from the miraculous berry (Synsepalum dulcificum, family Sapotaceae) native to West Africa.

Miraxanthins. Yellow Betaxanthins found in Mirabilis jalapa.

M-I : betalamic acid conjugated with methionine sulfoxide.

M-II : betalamic acid conjugated with aspartic acid.

M-IV : betalamic acid conjugated with tyramine.

M-VI : betalamic acid conjugated with dopamine.

Mirestrol, Miroestrol. A potent plant estrogen from the tubers of Pueraria mirifica found in Thailand. The tubers of the plant are known in folk medicine for their rejuvenating and oral contraceptive properties.

Mirestrol

Mitochondria, Chondriosomes (obsolete). Organelles present in all eukaryotic cells. These are 0.3-0.5 μm long, and they show a wide range of shapes and sizes. The average mitochondria is rather elongated and about the same size as a cell of Escherichia coli.

Electron microscopy of mitrochondria shows a characteristic internal structure. There are 2 concentric membranes, each 5-7 nm thick. Between the outer and the inner membranes lies the intermembrane space (also called external matrix or outer mitochondrial space) (Figs. 1 and 2). These 2 membranes have different submicroscopic structures, their biogenesis is different, and they are functionally distinet. The outer membrane can be removed by osmotic rupture.

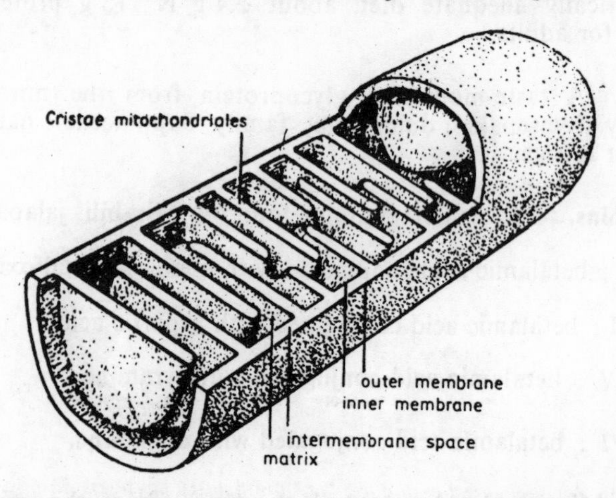

Diagram of a Cristae-type Mitochondrion. Each membrane consists of a bilayer.

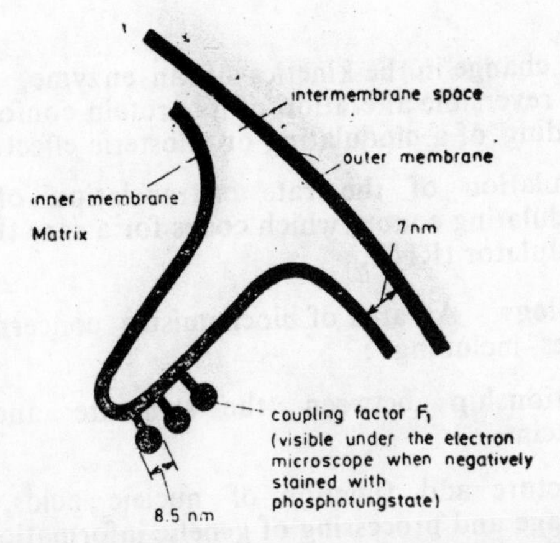

Diagrammatic Representation of part of a Mitchondrion, showing an infolding of the Inner membrane to form a Crista, and the arrangement of knob like coupling factors (F_1) Facing into the Matrix.

Mitomycin C. An aziridine antibiotic. It was first isolated in 1956 from Streptomyces caespitosus. It leads to an inhibition of DNA synthests. In the natural, oxidized state, M. is inactive. In the cell it undergoes intramolecular rearrangement and reduction to form an unstable bifunctional alkylating agent. The accompanying formation of carbonium ions at positions 1 and 10 enables the compound to bind covalently to DNA and irreversibly cross link the two strands of the double helix, thus preventing the action of DNA polymerase. It is bacteriocide and a cytotoxin. The aziridine antibiotics include other mitomycins and porfiromycins.

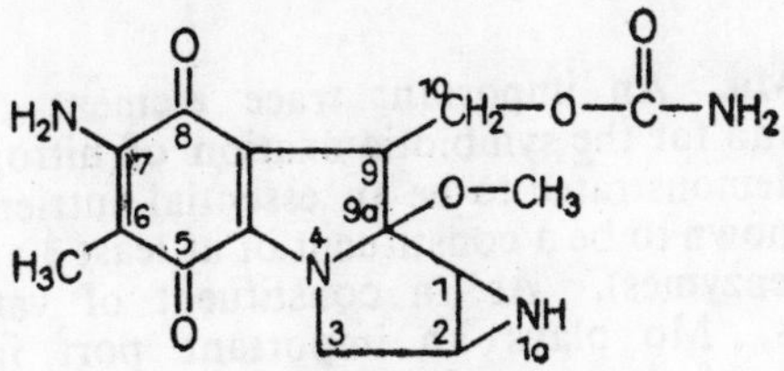

Mitomycin C

Mitosis. Nuclear division in the somatic cells of eukaryotic organisms, producing two identical daughter cells with the same genetic constitution. M. occurs in 4 stages, called prophase, metaphase, anaphase and telophase. In mitosis the daughter cells each contain the same number of chromosomes as the parent cell.

mlRNA. Abb. of giant messenger-like RNA.

Modulation

1. The change in the kinetics of an enzyme, resulting from the reversible alteration of its protein conformation by the binding of a modulating or allosteric effector.

2. Regulation of the rate of translation of mRNA by a modulating codon, which codes for a rare tRNA known as modulator tRNA.

Molecular Biology. An area of biochemistry concerned with large molecules, including :

1. relationship between the structure and function of proteins,

2. structure and function of nucleic acids, including the storage and processing of genetic information,

3. molecular basis and mechanism of biological regulation,

4. "molecular evolution".

Molecular biology attempts to elucidate the relationship between the structure and function of macromolecules. In a very narrow sense, molecular biology is the biochemistry of nucleic acids.

Molecular Genetics. A component of genetics and of Molecular biology. Classical genetics defined the gene, on the basis of its phenotypic expression and its mathematical distribution, as a hypothetical unit of inheritance. M.g. recognizes genes as defined segments of DNA, and shows the molecular basis for the causal relationship between the gene and its phenotypic expression.

Molybdenum, Mo. An important trace element for plants and bacteria, and for the symbiotic fixation of nitrogen. Mo has not been demonstrated to be an essential nutrient in animals, but it is known to be a constituent of at least 3 animal enzymes (Molybdoenzymes). As a constituent of various molybdoenzymes, Mo plays an important port in the nitrogen metabolism of plants and bacteria. In biological systems there is a partial antagonism beteen Mo and Cu.

Monellin. An intensely sweet tasting dimeric protein; the A-chain contains 44, the B-chain 50 amino acid residues. It tastes 3000 times (weight basis), or 90000 times (molar basis) sweeter than sucrose. It was discovered in the fresh fruits of an African berry, Dioscoreophyllum cumminsii ("wild red berry") of the Menispermaceaea. Another sweet protein is **Thaumatin.**

Monoterpenes. Aliphatic mono-, di-. or tricyclic terpenes, formed from two isoprene units ($C_{10}H_{16}$). Their occurrence is limited chiefly to plants. Most mono-terpenes are volatile liquids, and they are present in volatile oils and balsams. M. are biosynthesized from geranyl pyrophosphate via its isomer neryl pyrophosphate.

Montanic Acid. N-octacosanoic acid, CH_3-$(CH_2)_{26}$-COOH, Mr 424.73, m.p. 91°C. Esterified montanic acid is present in various waxes, *e.g.* montan wax, beeswax, Chinese wax.

Morindone. 1,2,5-trihydroxy-6-methylanthra-quinone. An orange-red anthraquinone. It occurs in the roots, bark and wood of Morinda species. In same Asian countries, particularly India, it was earlier used extensively as a natural dye.

Morphactins. A group of highly potent synthetic plant growth regulatos. These are derivatives of flurenol (9'-hydroxy-flurene-[9]-carboxylic acid, R=H), chiefly by esterification of the carboxyl group and/or substitution in the the ring system, *e.g.* chloroflurenol (R = Cl).

R
HO COOH

Flurenol (R=H)
Chloroflurenol (R=Cl)

Morphine. Medically the most important, and quantitatively the chief member of the opium alkaloids. It is isolated in large quantities from opium. It acts as an anesthetic wihout decreasing consciousness, and it is the most powerful analgesic known. High doses cause death by respiratory failue.

Morphogenesis. The genesis of form and shape; the ability of a molecule, group of molecules, an organ or organism to assume a certain shape.

MRH. Abb. for melanotropin releasing hormone.

Mucopolyosaccharides. Heteroglycans of animal connective tissue. Each of these acidic polysaccharides consists of an acetylated hexosamine (N-acetylglucosamine, or N-aceylgalactosamine) and a uronic acid (usually glucuronic acid, sometimes iduronic acid as in dermatan sulfate), which form a characteristtc repeating disaccharide unit.

Multienzyme Complex. A heteropolymeric protein, consisting of an ordered association of functionally and structurally different enzymes, which catalyse successive steps in a chain of metabolic reactions.

Multisubstrate Enzymes. Enzymes that require two or more substrates in order to catalyse a particular reaction. Accordingly, the enzyme forms a ternary (two substrates), quaternary (three substrates), etc. complex.

Muramic Acid. N-acetyl-3-O-carboxyethyl-D-glucosamine, or O-lactyl-N-acetylglucosamine. An amino sugar derived from glucose. The N-acetylated form of muramic acid is present in the mucopolysaccharide-peptide complxes of bacterial cell walls.

Murein, Peptidoglycan. A cross-linked polysaccharide-peptide complex of indefinite size, of the inner cell wall layer of all bacteria. It constitutes 50% of the cell wall in Gram negative, and 10% in Gram positive bacteria. It is tough and fibrous, and it enables the cell to withstand high internal osmotic pressure. It forms an immense bag-shaped molecule, or "sacculus", which gives shape and rigidity to the bacterial cell. In Gram positive bacteria the murein layer may be up to 10nm thick, consisting of about 20 layers of cross-linked peptidoglycan. In Gram negative bacteria, murein is probably rather more flexible and less tough than in Gram positive bacteria. Murein consists of linear parallel chains of up to 20 alternating residues of β-1,4-linked residues of N-acetyl-glucosamine and N-acetylmuramic acid, extensively cross-linked by peptides.

Muscaaurins. Orange Betalains present in the cap of the fly agaric. All muscaaurins are derivatives of Betalamic acid.

Muscarin. A biogenic amine and one of the amanita toxins. It is a quaternary ammonium base. (+)-muscarin from the fly agaric (Amanita muscaria) is a salt of 2S,4R,5S-(4-hydroxy-5-methyl-tetrahydrofurfuryl)-trimethyl-ammonium. It is concentrated in the skin of the cap of the fruiting body. (+)-muscarin is presnt in high concentrations in species of the fungal genera Inoybe and Clitocybe.

Muscimol. The enol betaine of 5-aminomethyl-3-hydroxyisoxazole. It is strongly polar, and it is probably not a native constituent of the agaric (Amanita muscaria). It is pharmacologically very potent, and in most tests it is more active than ibotenic acid. The primary pharmacological effect is an inhibition of motor function.

Muscle Proteins. The proteins present in muscle and constituting 20% of muscle tissue. The insoluble proteins (also known as contractile or myofibrillar proteins) are all intimately involved in the process of contraction; they are located in the myofibrils, which consist of organized bundles of thick myofilaments and thin myofilaments. Thick filaments (about 1.5 μm long; 12-16 nm diam.) contain 200–400 molecules of the protein, myosin Thin filaments (8 nm diam.) consist of a double helical arrangement of the protein, actin (300-400 actin molecules 1.0 μm length), each chain being accompanied by thread-like tropomyosin molecules and globular molecules of troponin. In intact muscle, thick and thin filaments lie parallet to one another and in intimate contact, forming the long, thin acto-myosin complex.

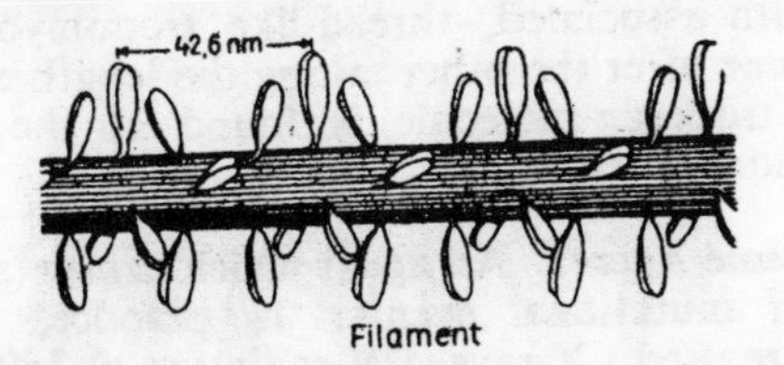

Diagram of a thick filament, consisting of numerous parallel myosin molecules.

During contraction, the glopular heads of the myosin molecules, protruding from the side of the filament, make contact with actin molecules of the neighboring thin filaments.

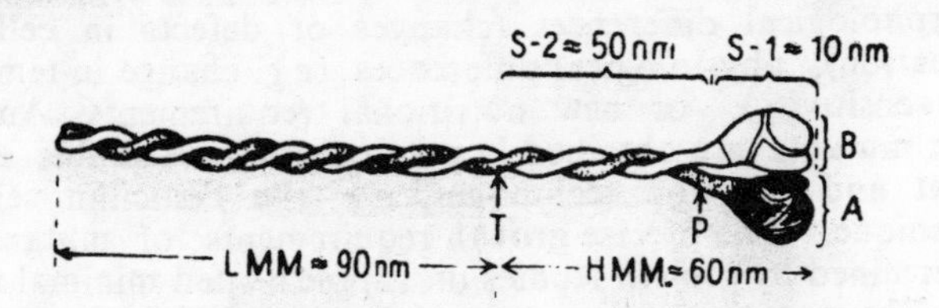

Diagram of a myosin molecule,

T site of attack by trypsin and high concentrations of papain. P site of attack by low concentrations of papain. LMM S-2 tail, S-1 two heads, consisting of 4 (A) and 3 (B) components.

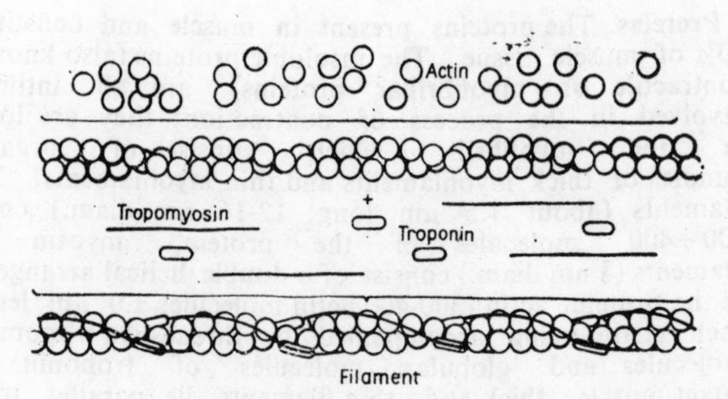

Diagram of a thin Filament.

The actin monomers form a double heclix (fibrillar or F-actin), with associated, thread-like tropomyosin molecules arranged one after the other along the length of the filament. A globular troponin molecule is found at the end of each tropomyosion molecule.

Mutagen, Mutagenic Agent. An agent which causes an increase in the rate of mutational events. In practice, the following mutagens are used : X-rays, UV-radiation at 260 nm (absorption maximum of DNA), nitrosomethyl-guanidine, methyl- and ethylmethane sulfonic acid, nitrite, hydroxylamine, base analogues, acridine dyes like proflavin, etc. The various physical and chemical mutagens have different mechanisms of action, some of which are imperfectly understood.

Mutants. Organisms, which as a result of Mutation show characteristic differences form the parent or wild type organism, *e.g.* morphological differences (changes or defects in cell wall fomation), physiological differences (*e g.* change in temperature sensitivity). or new nutritional requirements. Auxotrophic mutants are obtained in pure culture by suitable enrichment and selection techniques, *e g.* the Penicillin selection technique. The precise growth requirements of mutants are determined by growth studies on supplemented minimal media.

The mutation rate represents the probability of a mutation occurring in the generation of one cell; its value lies between 10^{-6} and 10^{-10}.

Mutant Technique. An important area of biochemical methodology, employing Mutants in particular Auxotrophic mutants of microorganisms. An auxotrophic mutant has a metabolic block at a certain point in the biosynthetic pathway of an essential product (*e.g.* an amino acid or coenzyme). Such a block is caused by an enzyme.

Mμ. Abb. for mlllimicron.

Mutarotation. A change in the optical rotation of an optical isomer, usually a carbohydrate, in aqueous solution.

Mutation. Chemical or physical changes in the genetic material of a cell or organism, which lead to a change in the genetic information. Mutation results in Mutants which, as a result of mutation, differ from the parent organism. The genetic information is stored in the DNA, which is the site of mutation. A single mutation repersents a change in a gene, which is a defined segment of DNA.

Mycobactin. A siderochrome synthesized by Mycobacterium spp.

Mycorrhiza. Symbiotic association between the roots of higher plant (forest trees, orchids, etc.) and a fungus. The fungal mycelium covers the root tips, and the root hairs become reduced or disappear. The fungal hyphae take up nutrients from the soil. The hyphae may penetrate the intercellular spaces of the root or actually penetrate the plant cells.

Mycotoxins. The metabolic products of certain fungi and other microorganisms, which are harmful to other organisms, especially vertebrates, including man.

Myoglobin. A single chain heme protein of skeletal muscle; 153 amino acid residues. Function of myoglobin is oxygen storage and transfer (*i.e.* from hemoglobin to respiratory enzymes). The affinity of myoglobin for oxygen is higher than that of hemoglobin. Muscle has a high content of myoglobin especially the cardiac muscle of diving mammals.

Myo-inositol. The earlier equivocal name, mesoinositol, should be avoided. It is a cyclitol found free and combined in animals and plants. It is biosynthesized from D-glucose, and the configuration cf C-atoms 1-5 of glucose is preserved during the conversion. It is an essential yeast growth factor and a constituent of the yeast growth factor mixture, "bios". It is present in brain cells, lens, thyroid gland, muscle, lung and liver. The total quantity of myo-inositol in the human body is about 40 g.

Myrcene. A triply unsaturated acyclic monoterpene hydrocarbon. It is a component of many essential oils, and it is prepared for the perfumery industry by pyrolysis of β-pinene (from oil of turpentine).

Myristic Acid. n-tetradecanoic acid, CH_3-$(CH_2)_{12}$-COOH. A fatty acid. C. Glyceral esters of myristic acid occur naturally in nearly all plant and animal fats, *e.g.* coconut oil, groundnut oil, linseed oil, rapeseed oil, milk fat and fish oil.

N

NAD. Abbreviation of Nicotinamide adenine dinucleotide. NAD^+ represents the oxidized form, NADH the reduced form.

NADP. Abbreviation of Nicotinamide adenine dinucleotide phosphase $NADP^+$ represents the oxidized form, NADPH the reduced form.

Nalidixic Acid. An antibiotic used therapeutically against Gram negative bacteria. It inhibits DNA replication in growing bacteria. It is a naphthyridine derivative.

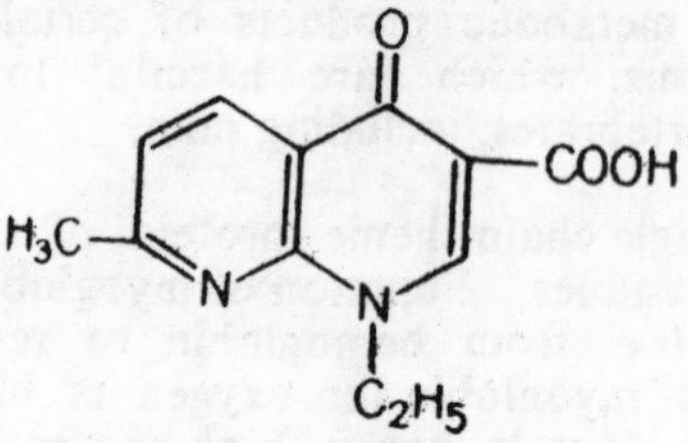

Nalidixio acid

Naphthoquinones. Naturally occurring derivatives of 1,4-naphthoquinone, *e.g.*. alkannin, eleutherin, juglone, lapachol, lawsone, lomatiol, plumbagin and shikonin.

Narcotios. Narcotic Drugs. Substances which act predominantly on the central nervous system. Depending on the dose, these show different phases of activity; small dose have a sedative action, somewhat higher doses are stimulatory, while even higher doses cause loss of consciousness (narcosis).

Natural Product Chemistry. A branch of chemistry concerned with the identification, isolation and structural elucidation of natural compounds.

Nearest neighbor Frequency The frequency at which a given pair of bases are immediately adjacent in the sequence of a polynucleotide chain. Since there are 4 bases. there are 16 possible pairs of nearest neighbors. Although it is now possible to determine the sequence of a short stretch of DNA. the parameter still gives useful information about the structure of the molecule.

Nebularine. 9-β-D-ribofuranosylpurine. A purine antibiotic synthesized by the mushroom Agaricus (Clitocybe) nebularis, and by Streptomyces spp. It is selectively active against mycobacteria. It has marked cytostatic properties, and in animals it is among the most poisonous of the purine derivatives.

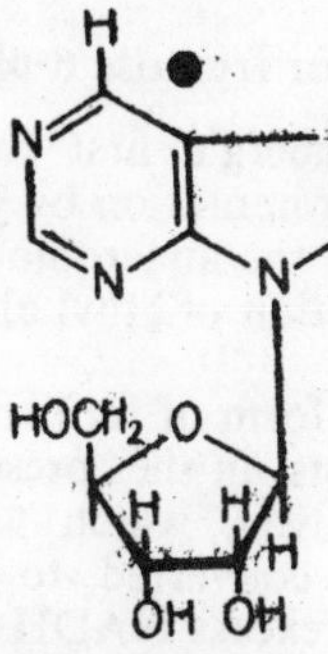

Nebularine

Negative Control. Inhibition of transcription by repression or inhibition of enzyme activity by metabolites.

Neomycin. An aminoglucoside antibiotic.

Neoxanthin. A xanthophyll, containing an allene group, two secondary and one tertiary hydroxy group. The chiral centers

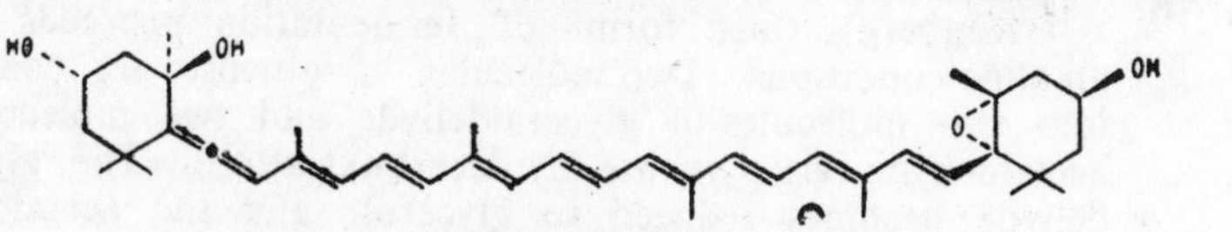

Neoxanthin

of N, are 3S, 5R, 6R, 3′S, 5′S, 6′S. It is present in the green parts of all higher plants

Neral. The cis-isomer of Citral.

Nerol. A doubly unsaturated acyclic monoterpene alcohol It is an oil with an odor of roses. It is a structural isomer of linalool and a cis-trans-isomer of Geraniol. It is the most valuable acyclic monoterpene alcohol used in perfumery. The pyrophosphate ester of N, neryl pyrophosphate is an intermediate in the biosynthesis of the Monoterpenes.

Nerve Growth Factor, NGF. A protein containing 118 amino acid residues of known primary sequence, Mr 13259 (mouse), NGF tends to form dimers.

Nervonic Acid. Δ^{15} tetracosenoic acid,

$$CH_3-(CH_2)_7-CH=CH-(CH_2)_{13}-COOH$$

A fatty acid, Mr, 336.6 m.p. 42°C. It is an acidic component of cerebrosides.

Neuberg Ester. Old name for fructose 6-phosphate.

Neuberg Fermentation. Neuberg's first form of fermentation is the normal "unsteered" fermentation by yeasts, *i.e.* all the reactions and side reactions of the anaerobic degradation of carbohydrate, with the production of ethyl alcohol.

Neuberg's second form of fermentation (or sulfite fermentation) occurs in yeasts in the presence of sodium hydrogen sulfite. The acetaldehyde, which would normally act as a hydrogen acceptor, is converted to its bisulfite addition compound. The resulting excess NADH reduces dihydroxyacetone phosphate to glycerol 1-phosphate (or 3-phosphate), a reaction catalysed by glycerol 1-phosphate (or 3-phosphate) dehydrogenase (Baranowski enzyme) EC 1.1.1.8). The oxido-reduction cycle (*i.e.*, reduction of NAD and its regeneration from NADH) is thus maintained. The glycerol 1-phosphate is dephosphorylated by a phosphatase (EC 3.1.3.21), and fee glycerol is produced. The process is therefore also known as a glycerol fermentation.

Neuberg's third form of fermentation proceeds under alkaline conditions. Two molecules of glucose are converted into two molecules of glyceraldehyde and two molecules of acetaldehyde (via pyruvate by decarboxylation). The glyceraldehyde becomes reduced to glycerol, and the acetaldehyde undergoes a dismutation to produce equivalent amounts of ethanol and acetate

Neuraminic Acid 5-amino-3,5-dideoxy-D-glycero-D-galactononulosonic acid. A C_9-compound which is formed from mannosamine and pyruvate, widely distributed in the animal kingdom in mucolipids, mucopolysaccharides, glycoproteins, and the oligosaccharides of milk. The N-and O-acetyl derivatives are called sialic acids.

Neuraminidase (EC 3.2.1.8): A hydrolase which cleaves N-acetylneuraminic acid from the non-reducing end of the heterosaccharide chain, producing glycoproteins and ganglioside. It occcurs in mixoviruses, various bacteria, blood plasma, and the lysosomes of many animal tissues. Its richest source ls the culture filtrate of the cholera organism, Vibrio cholerae.

Neuroendocrinology. The study of the interaction of nervous and hormonal systems in the regulation of metabolism, and in the adjustment of the individual organism to its environment.

Neurohormones. A group of hormones which, in conjunction with the nervous system and endocrine systems, play an important part in the regulation of somatic function, and in the adjustment of the individual organism to its environment. Examples

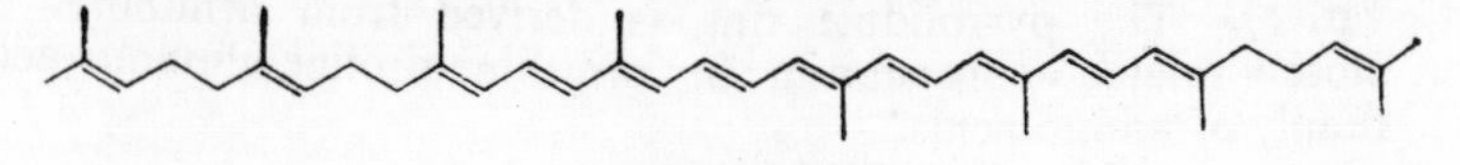

Neurosporene

are Releasing hormones Neurophypophyseal hormones and Neurotransmitters.

Neurohypophsial Hormones. A group of hormones produced in the hypothalamus (not, as the name suggests in the neurohypophysis.

Neurophysin. A carrier protein, which transports the neurohypophysial hormones within the nerve axon. N. I, II and III are cysteine-rich, linear polypeptides with a tendency to aggregate.

Neurosporene. A hydrocarbon carotenoid, containing 12 double bonds, nine of them conjugated. It is a direct biosynthetic precursor of lycopene, and it is found in Neurospora crassa.

Neurotoxins. Toxins that act specifically on nervous tissue. They may act as antagonists of neurotransmiters, *e.g.*, snake venom neurotoxine block acetyl choline receptors.

Neurotransmitters. Small, diffusible molecules, such as acetyl cholin, noradrenalin, amino acids, amino acid derivatives and peptides (*e.g.*, substance P) enkephalin, etc. which are released from synaptic vesicles (storage vesicles) at nerve endings synapses and motor end plates in response to electrical stimuli. They act as chemical messengers, which transmit a signal from a nerve cell to its target organ. Adrenergic N, *e.g.*, Noradrenalin and Adrenalin are formed in sympathetic postganglionic syapses.

NGF. Abbreviation of Nerve growth factor.

Niacin. A member of the vitamin B_2 complex.

Niacinamide. A member of the vitamin B_2 complx.

Nickel, Ni. Ni occurs only in traces in living system. In particular, it seems to be associated with RNA. A nickel metalloprotein named "nickeloplasmin" has been isolated from human and rabbit serum, but its function is not known.

Nicotiana Alkaloids, Tobacco Alkaloids. A group of pyridine alkaloids occuring mainly in the tobacco plant (Nicotiana). These contain a pyridine ring sysrem substituted at position 3 by pyrrolidyl or piperidyl residues.

The pyridine ring system of these alkaloids is derived biosynthetically from nicotinic acid (or a closely related derivative). The pyrrolidine ring is derived from ornithine or closely related compound in the ornithine-proline-glutamic acid family of amino acids.

Nicotinamide. A member of the vitamin B_2 complex.

Nicotinamide-adenine-dinucleotide, Abb. **NAD, Diphosphopyridine Nucleotide, Abb. DPN, Codehydrogenase I, Coenzyme I, Cozymase.** A pyridine nucleotide coenzyme involved in many biochemical redox processes. It is the coenzyme of a large number of oxidoreductases, which are classified as pyridine nucleotide-dependent dehydrogenases. Mechanistically, it serves as the electron acceptor in the enzymatic removal of hydrogen atoms from specific substrates.

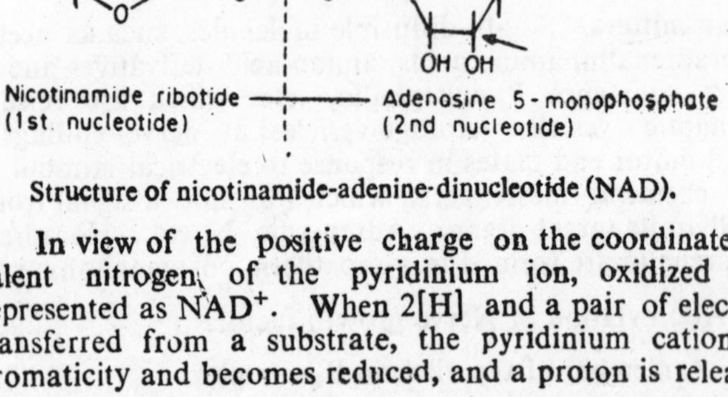

Structure of nicotinamide-adenine-dinucleotide (NAD).

In view of the positive charge on the coordinated pentavalent nitrogen of the pyridinium ion, oxidized NAD is represented as NAD^+. When 2[H] and a pair of electrons are transferred from a substrate, the pyridinium cation loses its aromaticity and becomes reduced, and a proton is released.

Substrate—H_2+NAD^+ $\rightleftharpoons$ Substrate+NADH+H^+. Thus of the hydrogens transferred, one becomes covalently bound to the NAD, while the other becomes a proton in equilibrium with the protons of the aqueous medium.

Nicotinamide-adenine-dinucleotide Phosphate, Abb. NADP, Triphosphopyridine Nucleotide, Abb. TPN, Coenzyme II. A pyridine nucleotide coenzyme, which differs from NAD by the presence of an extra phosphate residue in the 2′-position of the adenosine moiety.

NADP is the coenzyme of dehydrogenases and hydrogenases. It is important as the agent of hydrogen transfer in reductive syntheses, *e.g.*, in Photosynthesis and in Ammonia assimilation.

Nicotine. Chief representative of the Nicotiana-alkaloids Oxidative degradation of nicotine produces nicotinic acid. It is present in all parts of the tobacco plant (Nicotiana).

It has many different physiological effects Owing to its toxicity, it is not used therapeutically.

Nicotinic Acid. Pyridine 3 carboxylic acid. It is biologically important as a member of the vitamin B_2 complex. Isonicotinic hydrozide is antituberculosis agent, which shows antivitamin properties against both nicotinic acid and vitamin B_6.

NIH-shift. An anionotropy, whereby the cation intermediate produced in the hydroxylation of an unsaturated or aromatic substance by an oxygenase is stabilized. A group (usually a hydride ion) in the position of the incoming hydroxyl group is shifted to the neighboring position.

Nitrate Dissimilation, Dissimilatory Nitrate Reduction. Microbial reduction of nitrate to various products, which are not assimilated.

Nitrate Reductase. Enzymes which catalyse the reduction of nitrate to nitrate. All nitrate reductase studied so far contain iron and molybdenum. In the sequence of electron transfer, molybdenum appears to be the ultimate acceptor, which then transfers electrons to nitrate; during this process the molybdenum alternates between Mo(V) and Mo (VI).

Nitrate Reduction. In the narrow sense, nitrate reduction is the reduction of the nitrate ion (NO_3^-) to nitrite (NO_2^-), catalysed by Nitrate reductase. In the broad sense, it is the reduction of nitrate to gaseous products (nitrous oxide or nitrogen), or to ammonia.

Nitrate reduction may be assimilatory or dissimilatory. Assimilatory nitrate reduction, is found in green plants, green algae and fungi, where it involves the reduction of nitrate to ammonia, and operates in conjunction with the assimilation of the ammonia. It occurs in two stages : reduction of nitrate to nitrite by Nitrate reductase and the reduction of nitrite to ammonia by Nitrite reductase.

Nitrite Reductase. Dissimilatory nitrite reductase found in various bacteria, catalyses the reduction nitrite, probaby to nitric oxide; some authors claim that the reduction product is nitrous oxide. Nitrite reductase from Pseudomonas aeruginosa, Paracoccus denitrificans and Alcaligines faecalis contains both a c- and a d-type heme. The Pseudomonas nitrite reductase consists of two similar polypeptides, each of M_r 63000, and each containing two heme groups. Electrons are donoated by either cytochrome c_{551} or the blue copper protein, azurin. N.r. from Achromobacter cycloclastes and Pseudomonas denitrificans are both copper proteins.

Nitrogenase (EC 1.18.2.1). The enzyme system responsible for biological nitrogen fixation N consists of two proteins, both of which are required for activity. One of these proteins contains iron, molybdenum and acid-labile sulfur (Mo-Fe protein, component I; molybdoferredoxin, abb. azofermo) and the other contains iron and labile sulfur (Fe protein; component II; azoferredoxin abb. azofer.) The two components are present in the ratio I Mo-Fe protein: 2 Fe-proteins.

Nitrogenase catalyses the ATP-dependent reduction of several substrates that are of similar molecular size to, and are isoelectronic with molecular nitrogen, *e.g.* azide, nitrous oxide, cyanide, acetylene.

The activity of cell-free preparations of nitroaenase is usually measured by colcrimetric assay of ammonia, following incubation with molecular nitrogen.

Nitrogen Balance. Difference between the total nitrogen intake of an organism and its total nitrogen loss. Young growing animals are in positive nitrogen balance *i.e.* they retain more nitrogen (as protein added during growth) than they excrete. Mature, healthy adults show a zero nitrogen balance *i.e.* nitrogen intake is exactly balanced by nitrogen excretion. A negative nitrogen balance results from a defficiency of an essential amino acid; a decrease in the concentration of any proteogenic amino acid in the body's amino acid pool impairs total protein synthesis, so that the concentration of all other free amino acids ln the pool is increased; this leads to an exaggeration of degradative pathways and an increase in urea formation.

Nitrogen Catabolite Repression. Repression by ammonia of the synthesis of various enzymes of nitrogen metabolism.

Nitrogen Cycle. See Nitrogen fixation.

Nitrogen Fixation. A process in which atmospheric nitrogen is converted into ammonia. Activation of molecular nitrogen and its reduction to ammonia depend upon the catalytic activity of the enzyme Nitrogenase. The ammonia is then incorporated into the various nitrogenous compounds of the cell by the processes of Ammonia assimilation. Certain free living soil microorganisms, especially of the genera Clostridium and Azotobacter, are capable of nitrogen fixation. Other microorganisms fix nitrogen in symbiosis with higher plants, notably the Leguminosae.

Nonheme Iron Proteins, NHI-proteins. Proteins containing iron that is not bound in a heme system. In these proteins, the iron is bound by the sulfur of cyteins residues, and it is often also associated with inorganic sulfur. They are also called iron-sulfur proteins.

Nonhistone Chromatin Proteins. The "acidic chromain proteins", a highly heterogeneous group of tissue-specific proteins, which are bound to certain DNA sequences. Their M_r in detergents is 30000—70000, *i.e.* markedly higher than the Mr of Histones. As gene derepressors, they play a part in the regulation of gene expression in mammalian cells, especially in cell proliferation.

Nonsense Codon. An Amber codon, or Ochre codon.

Noradrenalin, Norepinephrine, Arterenol. Dihydroxyphenylethanolamine, a hormone with activity on the nervous system and the vascular system. It is a catecholamine and a biogene amine. Together with Adrenalin it is synthesized from L-tyrosine via Dopa (Dopamine) in the adrenal medulla and in the nervous system. In the sympathetic nervous system, acts as an adrenergic Neurotransmitter. It causes contraction of blood vessels, with the exception of the coronary vessels; it therefore causes an increase in the peripheral vascular resistance, with a consequent rise in blood pressure. The effect of noradrenalin on carbohydrate metabolism and blood sugar is similar to, but weaker than that of adrenalin.

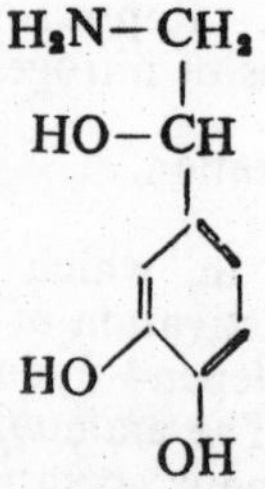

H_2N-CH_2

HO—CH

HO

OH

Noradrenalin

Norgestrel. A synthetic gestagen. It is used inoral contraceptives, and it is the most potent of the orally active gestagens.

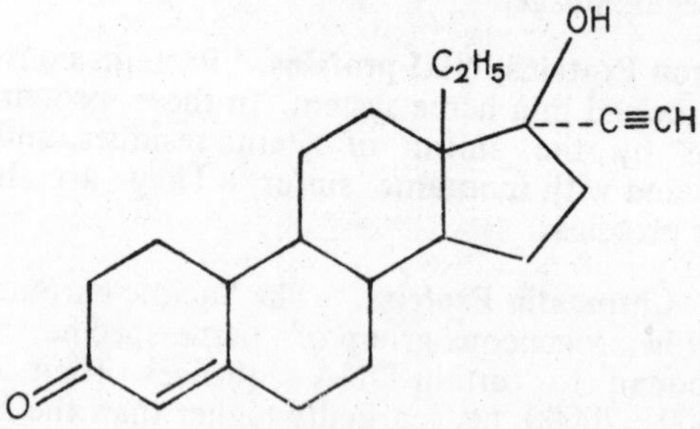

OH

C_2H_5

---C≡CH

O

Norgestrel

Nu Bodies, Nucleosomes. Subunits of Chromatin containing 180 to 200 base pairs of DNA and approximately an equal weight of histone. They get released by partial digestion of chromatin by staphylococcal nuclease.

Nucleases. A group of hydrolytic enzymes, which cleave nucleic acids. Exonucleases attack the nucleic acid molecule at its terminus, whereas endonucleases are able to catalyse a hydrolytic cleavage within the polynucleotide chain. Deoxyribonucleases (DNAases) are specific for DNA, and ribonucleases (RNAases) for RNA. All N. are Phosphodiesterases they catalyse the hydrolysis of either the 3′ or 5′ bond of the 3′, 5′-phosphodiester linkage.

Nucleic Acid Bases. Constituent bases of nucleic acids. These are fundamental to the function of nucleic acids in the storage and transfer of genetic information. They are Adenine Guanine, Cytosine, Thymidine, Uracil, and others that occur less frequently (see Rare nucleic acid components).

Nucleic Acid. Polymerized nucleotides; among the most essential components of all living cells, viruses and bacteriophages.

Nucleic acids are Polynucleotides with M_r between 20000 and several million. They contain three characteristic structural components :

1. the purine and pyrimidine bases, adenine, guanine, cytosine and uracil (in RNA), thymine (in DNA); in addition, there are over 30 other Rare nucleic acid components, which occur in various nucleic acids. These are components are formed by modification of existing structures or within the N.a., *e.g*. by methylation, hydrogenation rearrangement of normal bases;
2. the pentose monosaccharide, D-ribose (in RNA), or 2-deoxy-D-ribose (in DNA);
3. phosphoric acid.

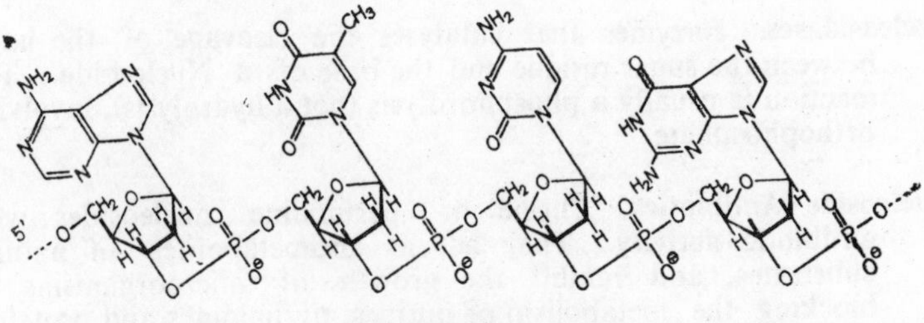

Polynucleotide structure

Nucleocidin. A purine antibiotic synthesized by Streptomyces calvus. It is active against bacteria and fungi, and it is used therapeutically against trypanosomes. It inhibits protein synthesis.

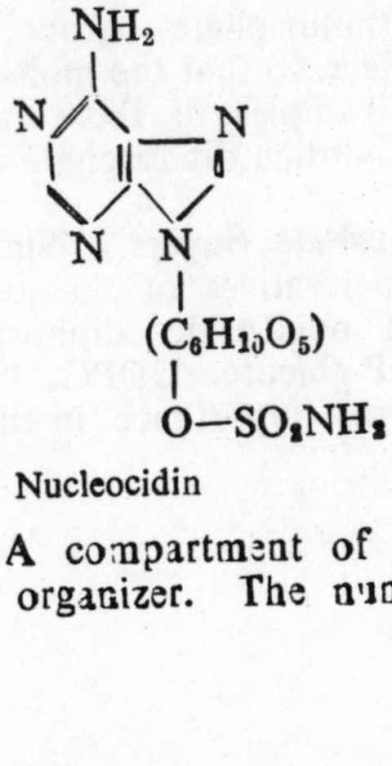

NH_2

N N

N N

$(C_6H_{10}O_5)$

$O{-}SO_2NH_2$

Nucleocidin

Nucleolus (plur. Nucleoli). A compartment of the nucleus. It contains the nucleolus organizer. The number of nucleolus per nucleus varies widely. The main components of the nucleolus are proteins (over 80% of the dry weight), RNA (over 50%) and DNA.

The nucleolus is the site of biosynthesis of ribosomal RNA. During nuclear division, the nucleolus temporarily disappears as a visible structure.

Nucleolus Organizer. A specific region on one or more eukaryotic chromosomes, where the nucleolus is formed. The DNA in this region contains genetic information for the synthesis of ribosomal RNA.

Nucleoprotein. Heteropolar complexes of nucleic acids (in particular, nuclear DNA) with basic, acid-soluble proteins (histones or protamines), and with acidic, base-or detergent-soluble non-histone-chromatin proteins. They occur mainly in the chromatin of the cell nucleus in its quiescent state, and in the chromosomes when the nucleus is active, *i.e.* dividing.

Nucleosidases. Enzymes that catalyses the cleavage of the bond between the sugar residue and the base of a Nucleoside. The reaction is usually a phosphorolysis (not a hydrolysis), involving orthophosphate.

Nucleoside Antibiotics. Purine or pyrimidine nucleosides with antibiotic activity. They act as antimetabolites of natural substrates, and inhibit the growth of microorganisms by blocking the metabolism of purines, pyrimidines and proteins. Some nucleoside antibiotics (*e.g.*) showdomycin) contain an analog base, others (*e.g.* gougerotin) contain an analog sugar, or both moieties may be modified (*e.g.* puromycin).

Nucleoside Diphosphate Compounds. Compounds containing a nucleoside diphosphate grouping. This grouping has an activating effect, so that the molecule has a high group transfer potential. Examples of these are Nucleoside diphosphate sugars and Cytidine diphosphate choline.

Nucleoside Diphosphate Sugars, Nucleotide Sugars. Energy-rich nucleotide derivatives of monosaccharides. The activating group is a nucleoside diphosphate. Uridine diphosphate glucose UDP-glucose, UDPG, "active glucose") is of wide-spread general importance in carbohydrate metabolism. It is synthesized from glucose 1-phosphate by reaction with uridine triphosphate.

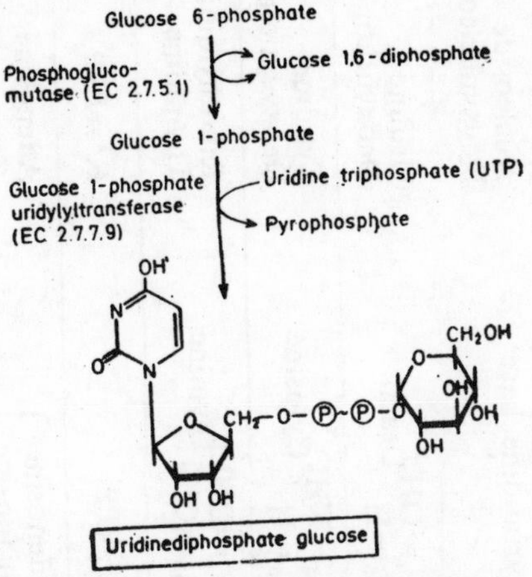

Synthesis of uridinediphosphate glucose

Nucleosides. The N-glycosides of heterocyclic nitrogenous bases. The N-glycosides of purines and pyrimidines with pentoses are of particular biological importance. The sugar component is either D-ribose or D-2-deoxyribose, both in the furanose form. C-1 of the pentose residue is linked to N-9 of the purine or N-1 of the pyrimidine by an N-glycosidic linkage (C-N bond).

Nucleosides and deoxynucleosides can be synthesized via a Salvage pathway. They are also produced by the hydrolysis of nucleic acids and nucleotides.

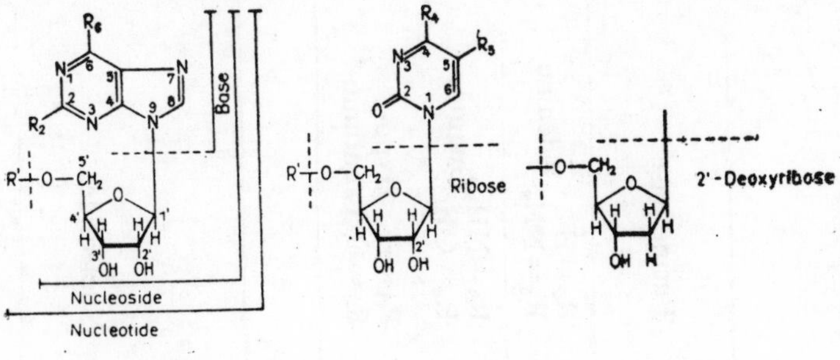

Purine base	$R^1=H$ Nucleoside or Deoxynucleoside	Pyrimidine base	$R^1=H$ Nucleoside or Deoxynucleoside
$R_2=H$, $R_6=NH_2$ Adenine	Adenosine Deoxyadenosine	$R_4=OH$, $R_5=H$ Uracil	Uridine Deoxyuridine
$R_2=NH_2$, $R_6=OH$ Guanine	Guanosine Deoxyguanosine	$R_4=NH_2$, $R_5=H$ Cytosine	Cytidine Deoxycytidine
$R_2=H$, $R_6=OH$ Hypoxanthine	Inosine (Deoxyinosine)	$R_4=OH$, $R_5=CH_3$ Thymine	Ribothymidine Thymidine
	$R^1=P$	$R^1=PP$	$R^1=PPP$
Nucleotides:	Adenosine, Guanosine, Inosine, Uridine, Cytidine, Thymidine } monophosphate	Adenosine, Guanosine, Inosine, Uridine, Cytidine, Thymidine } diphosphate	Adenosine, Guanosine, Inosine, Uridine, Cytidine, Thymidine } triphosphate

The sugar component of thymidine is 2′-deoxyribose

Nucleotide Coenzyme. A coenzyme containing a nucleotide structure. Nucleotide coenzymes are Pyridine nucleotide coenzymes the nucleoside diphosphate moieties of Nucleoside diphosphate sugars and Coenzyme A.

Nucleotides, Nucleoside Phosphates The phosphoric acid esters of Nucleosides. o-Phosphoric acid is esterified with a free OH-group of the sugar. If the sugar is D-ribose the nucleotide is called a ribonucleotide or ribotide. If the sugar is D-2-deoxyribose, the N. is a deoxyribonucleotide, deoxynucleotide or deoxyribotide. The phosphate may be present on position 2′, 3′ or 5′ (in deoxy-ribonucleotides, only the 3′-and 5′-phosphate are possible). The 5′-nucleoside phosphate are metabolically very important; they may be mono-, di-or triphosphorylated *e.g.* guanosine 5′-monophosphate, cytidine 5′-diphosphate and adenosine 5′-triphosphate.

Nucleus. A large structure (−5 μm diam.) in eukaryotic cells, containing the bulk of the cellular DNA, and representing the chief site for the storage, replication and expression of genetic information. In the period between cell divisions, *i.e.* at interphase, the nucleus is densely and uniformly packed with DNA and shows few distinct structures, even under the electron microscope. Distinguishable features are Chromatin nucleoplasm, nuclear membrane and the Nucleoli. At nuclear division, the highly structured Chromosomes are formed from the chromatin.

Nuphara Alkaloids. A group of alkaloids possessing a piperidine or quinolizidine ring system, which are found in various species of water lily (Nuphar app.). All nuphara alkaloids contain a sesquiter pene skeleton, which is cyclized by the inclusion of other atoms (nitrogen, oxygen or sulfur). The chief nuphara alkaloids are nupharidine nupharamine and thiobinupharidine.

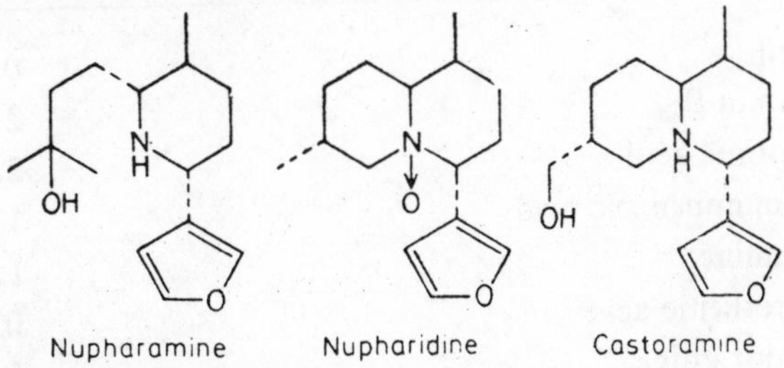

Nupharamine Nupharidine Castoramine

Nutrient Medium, Growth Medium. A medium, liquid or solid, for the cultivation of microorganisms, cells, tissues or organs.

Solid media are prepared from liquid media by the addition of a gelling agent, *e.g.* gelatine (nowdays used only in special cases), silicic acid (when exclusion of organic compounds is necessary), or Agar-agar (used widely in bacteriology, usually at a concentration of 1.5—2%).

TABLE

A Simple Synthetic Culture Medium for Microorganisms (after Schlegel)

Glucose	10.0 g
NH_4Cl	1.0 g
K_2HPO_4	0.5 g
$MgSO_4.7H_2O$	0.2 g
$FeSO_4.7H_2O$	0.01 g
$CaCl_2$	0.01 g
Trace element solution	1 ml
Water	1000 ml

TABLE 2

Vitamin Solution used for the preparation of culture media for soil and water bacteria (after Schlegel)

2—3 ml of this solution are added to 1000 ml of culture medium.

Biotin	0.2 mg
Vitamin B_{12}	2.0 mg
Nicotinic acid	2.0 mg
p-Aminobenzoic acid	1.0 mg
Thiamine	1.0 mg
Pantothenic acid	0.5 mg
Pyridoxamine	5.0 mg
Distilled water	1000 ml

TABLE 3

The A—Z solution, or trace element solution of Hoagland

$Al_2(SO_4)_3$	0.055 g
KI	0.028 g
KBr	0.028 g
TiO_2	0.055 g
$SnCl_2.2H_2O$	0.028 g
LiCl	0.028 g
$MnCl_2$ $4H_2O$	0.389 g
$B(OH)_3$	0.614 g
$ZnSO_4$	0.055 g
$CuSO_4.5H_2O$	0.055 g
$NiSO_4.(7H_2O)$	0.059 g
$Co(NO_3).6H_2O$	0.055 g
Distilled water	1000 ml

Nutritional Physiology of Microorganisms. The terms autotrophy and heterotrophy are too broad to distinguish between all the different forms of microbial nutrition. Different types of nutritional physiology are classified according to :

1. the nature of the carbon source;
2. the source and mechanism of formation of ATP;
3. the source of reducing power for the synthesis of cell constituents.

O

Oat Coleoptile Test, Avena Test. A biotest for the quantitative determination of auxins, which is carried out as follows :

The tip of the coleoptile is cut off, and the cotyledon within it is removed by pulling on its base. An agar block with the test substance is set on one side of the coleoptile stump.

The auxin diffuses into the side of the coleoptile covered by the agar and, due to the one-sided stimulation of growth, causes it to bend. The angle of the bend is a function of the auxin concentration.

Ochratoxins. Mycotoxins produced by Aspergillus ochraceus during food spoilage. In rats and mice these cause pronounced liver damage. The three known O, are designated A, B and C.

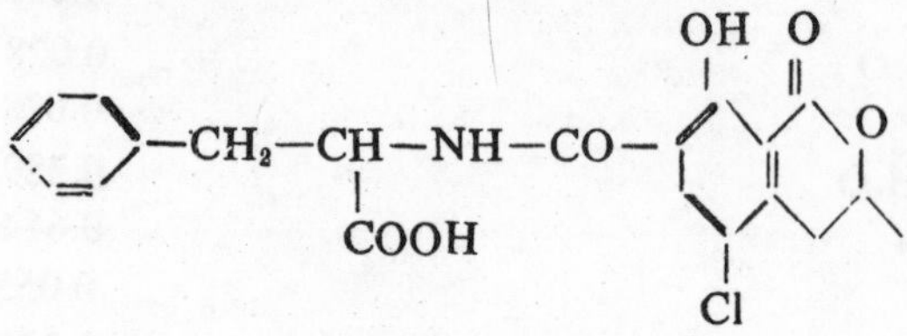

$$C_6H_5-CH_2-CH(COOH)-NH-CO-$$ (isocoumarin ring bearing OH, C=O, O, CH_3 and Cl)

Ochratoxin

Ochre Codon. The UAA sequence in mRNA. Like the amber codon, it signals the end of protein biosynthesis. The synthesized polypeptide chain is released after the incorporation of the amino acid encoded immediately before the ochre codon and the one most widely employed by all living systems.

Ochre Mutants. Bacterial mutants, which, as a result of a point mutation, possess a UAA codon in their mRNA. There are specific Suppressors of ochre mutations.

Ocimene. A triply unsaturated monoterpene hydrocarbon. It is a double bond positional isomer of Myrcene and a component of many essential oils.

D-Octopine. N-α(1-carboxyethyl)-arginine, N^2-(D-1-carboxyethyl)-L-arginine. It is found in the muscles of certain invertebrates, *e.g.* Octopus. Pecten maximus, Sipunculus nudus, where it serves as a functional analog of lactic acid *i.e.* the NADH produced by glycolysis is oxidized to NAD during the synthesis of D O., and NAD can be reduced to NADH by the reversal of the same process. It is biosynthesized from pyruvate and arginine by a reductive condensation, catalysed by an unspecific NADH-dependent dehydrogenase.

Oils. Water-insoluble, liquid organic compounds. They are combustible, lighter than water, and soluble in ether, benzene and other organic solvents. Naturally occurring oils may be glycerides, *e.g.* the oils stored in certain seeds, or fish liver oils; or they may be nonsaponifiable lipids, *e.g.* Essential oils.

Oiticica Oil. The seed fat of Licania rigida, a tree native to the semiarid areas of northeast Brazil. Licanic acid represents 50–80% of the total esterified fatty acids of Oiticica oil.

Oleandomycin. A Macrolide (see) antibiotic.

Oleanolic Acid. A monosaturated, pentacyclic Triterpene, with a carboxylic acid group. It differs structurally from β-amyrin by the presence of a carboxyl group in place of the 28-methyl group.

Oleic Acid. Δ^9-octadecenoic acid, $CH_3-(CH_2)_7-CH=CH-(CH)_7-COOH$. The most widely distributed naturally occurring unsaturated fatty acid. The double bond has the cis conformation. The trans isomer is called elaidic acid. O.a. is present in practically all the glycerides of depot and milk fats, and it is a component of phospholipids.

Oligonucleotides. Linear sequences of up to 20 nucleotides, joined by phosphodiester bonds. Position 3′ of each nucleotide unit is linked via a phosphate group to position 5′ of the next unit. In the terminal units, the respective 3′ and 5′ positions may be free (*i.e.*-OH groups) of phosphorylated.

Ommochromes. A class of natural pigments, which contain the phenoxazone ring system. Their colors range from yellow through red to violet. They are especially common in, but not limited to the Arthropoda, and were named from their occurrence in the ommatidia of the insect eye.

OMP. Abb. of Orotidine 5′-monophosphate.

Open System :

1. A system in dynamic equilibrium with its surroundings, *i.e.* there is a continual exchange of material, energy and information with the environment.
2. In biology, plants are considered to be O.s., whereas animals are closed systems. The plant is theoretically capable of unlimited growth; certain cells remain embryonal and able to divide and differentiate, so that growth occurs from vegetative sites, such as the meristematic regions of the shoot and root tips, intercalary meristems, etc.

Operon A group of neighboring genes, which represent a functional unit. A operon contains :

1. structural genes (S_1 to S_4) which code for thep rimary structures of enzyme proteins catalysing successive steps

in a metabolic pathway, *e.g.* the enzymes in the biosynthesis of an amino acid. The primary transcription product of this group of genes is a polycistronic mRNA; therefore, during the control of transcription, all the structural genes are affected equally.

2. The promotor (P in the Fig.) is the starting point of transcription. This section of the DNA is "recognized" by RNA polymerase with the aid of the sigma factor. The affinity of the promotor for RNA polymerase (apparently determined by the structure of the promotor) is one of the factors that regulate the transcriptional frequency of the operon. When the RNA polymerase has bound to the promotor, it must then pass through the operator region in order to reach the structural genes.

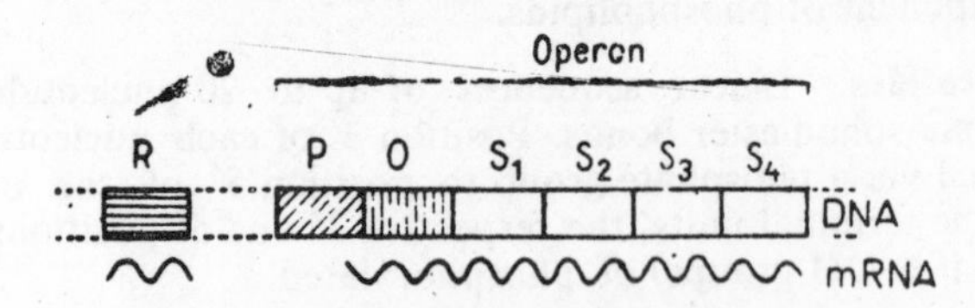

Schematic Representation of an operon

3. The operator (O in the Fig.) is the control gene for the function (*i.e.* the transcription) of the structural genes. It is able to bind a repressor protein, which is the product of the regulator gene (R in the Fig.). R. is not a part of the operon, and it is located in a different region of the chromosome. If the specific repressor protein binds to the operator, transcription of the structural genes is blocked, *i.e.* the RNA polymerase cannot pass to the structural genes. If the operator is unoccupied, transcription of the structural genes can proceed.

Opium. The congealed or dried latex of Papaver somniferum. The chief active constituents of opium are the Opium alkaloids. The action of opium is essentially similar to, but weaker than that of its constituent Morphine.

Opium Alkaloids. The Papaveraceae alkaloids that occur in Opium There are about 40 different O.a.; the chief representatives are Morphine (opium alkaloids) Codeine and Papaverine.

Opsonin. The name given by Wright and Douglas in 1903 to the thermostable material present in serum, which stimulates phagocytosis of bacteria. It is probably identical to C3b of th e complement system, although other complement components may also be active to a lesser extent. In addition to its role in the opsonization of immune complexes, C3b can bind to various structures, such as foreign erythrocytes and bacterial cells, and render them more readily phagocytosed by immune adherence to phagocytes.

Opsonization. The ability of serum to render an immune complex more readily phagocytosed. It is a property of the Complement system. Although immune complexes are subject to phagocytosis, interaction with complement greatly increases the rate. Component C3b (activated C3 of the complement system) has a labile binding site(s), which permits it to bind tightly to antigen-antibody complexes (opsonic adherence).

Optical Test. A method introduced in 1936 by Otto Warburg for the determination of the enzymatic activity of NAD and NADP-dependent dehydrogenases. Absorbance at 340 nm (or another suitable wavelength in that region) is measured as an index of the degree of reduction of NAD or NADP. The principle of this method is now widely used for measuring enzyme activities and the concentration of metabolites. Determinations of dehydrogenase activities by this test are now conveniently performed in an automatic recording spectrophotometer, so that the change in absorbance at 340 nm is monitored continuously with time, *e.g.* the activity of malate dehydrogenase is measured from the rate of increase of absorbance at 340 nm due to the production of NADH in the reaction : $NAD^{+} + Malate \rightarrow Oxaloacetate + NADH + H^{+}$.

Orchinol. 9, 10-dihydro-2, 4-dimethoxy-7-phenanthrol, A phytoalexin, m.p. 168-170°C. It is formed by the orchid, Orchis militaris, as a defense against infection by the fungus, Rhizoctania repens. The structurally related hircinol (9, 10-dihydro-4-methoxyphenanthrene-2, 5-diol, m.p. 1625°C) is formed by the orchid,

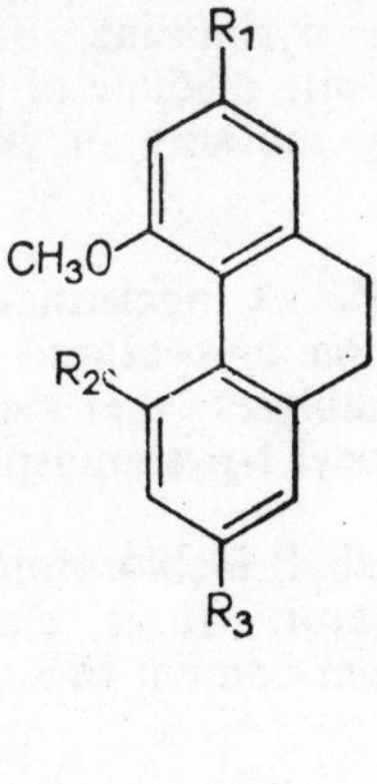

	R_1	R_2	R_3
I	OCH_3	H	OCH_3
II	OH	OH	H

I = Orchinol

II = Hircinol

Himantoglossum (Loroglossum) hircinum, against infection by Rhizoctonia.

Ord. Abb. of Orotidine.

ORD. Abb. of Optical Rotatory Dispersion.

Orn. Abb. of L-Ornithine.

L-Ornithine Abb. **Orn**. α, δ-diaminovaleric acid, 2, 5-diaminovaleric acid, $H_2N\text{-}CH_2\text{-}CH_2\text{-}CH_2\text{-}CH(NH_2)\text{-}COOH$. A nonproteogenic amino acid. In mammals, Orn. is an intermediate in the Urea cycle, and it is an intermediate in the biosynthesis of Arginine in all arginine-synthesizing organisms.

Oro. Abb. of Orotic acid.

Orosomucoid, **α_1-seromucoid**. **α_1-acid Glycoprotein**, abb. α_1**AGp**. A plasma protein of mammals and birds. It contains about 38% carbohydrate, and is the most carbohydrate rich and water soluble of all the plasma proteins. The seromucoid fraction of blood consists of orosomucoid together with two other carbohydrate rich proteins (C1-inactivator and hemopexin). Increased blood levels of orosomucoid and of the total seromucoid fraction are associated with inflammation, pregnancy and various disease states, such as cancer, pneumonia, rheumatoid arthritis.

Orotic Acid, Abb. **Oro**. Uracil 4-carboxylic acid. It is an intermediate in Pyrimidine biosynthesis, and it accumulates in large quantities during growth of mutants of Neurospora crassa, which require uridine, cytidine or uracil.

Orotidine, abb. **Ord**. Orotic acid-3-β-D-ribofuranoside, a β-glycosidic Nucleoside of D-ribose and the pyrimidine, orotic acid. It does not appear to be a normal intermediate of pyrimidine metabolism, but it is produced by mutants of Neurospora crassa.

Orotidine 5′-Monophosphate, Abb. **OMP**. A nucleotide of orotic acid. It is an intermediate in Pyrimidine biosynthesis. Orotidine 5′-phosphate pyrophosphorylase catalyses the synthesis of OMP from orotic acid 5-phosphoribosyl 1-pyrophosphate.

Orthonil, **PRB 8**. 2-(β-chloro-β-cyanoethyl)-6-chlorotoluene. A synthetic, stimulatory growth regulator. It is claimed that orthonil causes an increase in the sugar content of sugar beet.

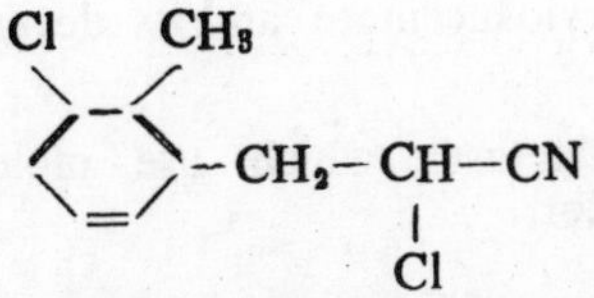

Cl CH$_3$

(benzene ring)—CH_2—CH—CN

|

Cl

Osmosis. A phenomenon associated with semi-permeable membranes, especially Biomembranes. When two solutions are separated by a membrane which is permeable only to selected components of the solution (*e.g.* water), the component which can cross the membrane will flow from the side on which its partial pressure is higher to the side on which its partial pressure is lower.

Ovosiston. An oral contraceptive consisting of a mixture of the progestin, Chlormadinone acetate, and the estrogen, Mestranol.

Ovulation Inhibitors. A group of steroids, which inhibit ovulation by feedback inhibition of the production of Luteinizing hormone and/or Follicular stimulating hormone. They are used extensively as oral contraceptives (otherwise known as "the pill").

Oxalic Acid: HOOC-COOH. It occurs widely in plants as its calcium, magnesium and potassium salts. Owing to its ability to bind calcium, large amounts of oxalic acid are poisonous. By forming insoluble calcium oxalate in the intestine, it hinders the absorption of calcium. Animals cannot metabolize oxalic acid. Higher levels may lead to kidney damage (formation of oxalate kidney stones).

Oxaloacetic Acid: HOOC-CO-CH_2-COOH. An oxodicarboxylic acid present in fairly high concentrations in plants, *e.g.* red clover, peas. The anion (oxaloacetate) is an intermediate in the Tricarboxylic acid cycle, and it is the oxoacid derived from Aspartic acid by Transamination. The enol form of O.a. may be cis or trans: hydroxymaletic acid (cis, m.p. 152°C), and hydroxyfumaric acid (trans, m.p. 184°B[d]).

Oxalosuccinic Acid. A β-ketotricarboxylic acid (2-oxotricarboxylic acid):

COOH

|

HOOC—CH_2—CH—CO—COOH

The anion (oxalosuccinate) is an intermediate in the isocitrate dehydrogenase reaction of the Tricarboxylic acid cycle. Isocitrate dehydrogenase catalyses both the oxidation of isocitrate to oxylosuccinate, and its decarboxylation to 2-oxo-glutrate.

Oxidases. Oxidoreductases which use molecular oxygen as an electron acceptor.

Oxidation. The loss of electrons. Classically, O. was defined as combination with oxygen or removal of hydrogen. The electrons are transferred to the oxidizing agent, which becomes reduced. Therefore oxidation is always coupled to Reduction, so that any oxidation or reduction is part of an oxidoreduction process. In metabolism there are different mechanisms of enzyme catalysed O., *i.e.* dehydrogenation, electron transfer, introduction of oxygen, and hydroxylation (Table).

Type of reaction	Enzymes	General reaction
Dehydrogenation	Oxidoreductases	$SH_2 \rightleftharpoons S$; $D \rightarrow DH_2$
Electron transfer	4 and 2 electron transferring oxidases	$SH_2 \rightarrow S$; $\frac{1}{2}O_2 \rightarrow H_2O$ or $O_2 \rightarrow H_2O_2$
Oxygen transfer	Dioxygenases	$S + O_2 \rightarrow SO_2$
Hydroxylation	Hydroxylases	$S + DH_2 + O_2 \rightarrow SOH + D + H_2O$

S = substrate. D = hydrogen carrying cofactor

Oxidation Reactions in Metabolism

Oxidative Phosphorylation, Respiratory Chain Phosphorylation. Formation of ATP coupled with the operation of the Respiratory chain. Energy available from the flow of electrons from substrate to oxygen via the respiratory chain drives the synthesis of ATP from ADP and inorganic phosphate. Oxidation of one molecule of reduced nicotinamide adenine dinucleotide ($NADH+H^+$) generates three molecules of ATP, while reduction of one molecule of reduced flavin adenine dinucleotide ($FADH_2$) yields two ATP. Complete oxidation of one molecule glucose yields 38 ATP, 2 from glycolysis and 36 from O.p.

The mechanism of oxidative phosphorylation *i.e.* the nature of the energy transducing mechanism that converts the energy of electron flow into the chemical energy of ATP, has always been a controversial area ot biochemistry.

Oxygenases. Enzymes that catalyse the incorporation of the oxygen of molecular oxygen into their organic substrates, *i.e.* the oxygen atom(s) appearing in the product is (are) derived from atmospheric O_2, and not from water. Dioxygenases (oxygen transferases), catalyse the introduction of both atoms of molecular oxygen. Monooxygenases or hydroxylases catalyse the introduction of one atom from molecular oxygen ; the other atom becomes reduced to water. Monooxygenases therefore require a second substrate, which serves as an electron donor; for this reason they are also called mixed function oxygenases. They catalyse the following type of reaction : $AH+O_2+DH_2 \rightarrow AOH+H_2O+D$, where AH is the substrate, AOH the hydroxylated substrate, DH_2 an electron donor, and D the oxidized electron donor.

Oxygen Metabolism. Metabolic reactions involving oxygen, including :

1. General oxidation of metabolites by Dehydrogenation, removal of electrons (oxidases, Flavin enzymes), and addition oxygen to the substrate.
2. Oxidation of reduced coenzymes via the Respiratory chain.

Oxytocin. A neurohypophysial, peptide hormone. It causes contraction of the smooth muscle of the uterus and of the mammary gland (milk ejection). It is structurally related to the other neurohypophysial hormone, Vesopressin; these two hormones have a common ancestry and there is an overlap in their physiological activities. In lower vertebrates, four other neurohypophysial hormones have been identified, which are variants of oxytocin and vasopressin, with different amino acid residues in positions 4 or 8, or both.

Cys—Tyn—Ile—Gln—Asn—Cys—Pro—Leu—Gly—NH_2
(disulfide bridge between the two Cys residues)

Oxytocin

It is synthesized in the hypothalamus in Nucleus paraventricularis, and transported to the posterior lobe of the pituitary (hypophysis) via the Tractus paraventriculo-hypotheseus, in combination with the transport protein, Neurophysin I. It is released into the blood stream in response to psychological and tactile stimulation of genitalia, or suckling stimulation of the mammary gland. It acts on its target organ via the adenylate cyclase system. Methods for the assay of O, are chiefly biological, based on the milk ejecting activity in lactating animals, or the behavior of perfused sections or strips of mammary tissue. Radioimmunoassay is also possible.

P

P_i. Inorganic phosphate, the anion of ortho-phosphoric acid, PO_4^{3-}.

Paeonidin: 3, 5, 7, 4′-tetrahydroxy-3-methoxyflavylium cation. The 3, 5-β-diglucoside (paeonin) is the chief pigment of peony.

Pahutoxin. The main toxin of the tropical box fish Ostraction lentiginosus.

$$CH_3-(CH_2)_{12}-\underset{\substack{|\\ O-\overset{\overset{\displaystyle O}{\|}}{C}-CH_3}}{CH}-CH_2-COO-CH_2-CH_2-\overset{\oplus}{N}(CH_3)_3$$

Pahutoxin

It is a choline derivative, highly poisonous for fish, but not for warm blooded animals.

Paleoproteins. A group of proteins from fossils, in particular the exoskeleton of molluscs (snail and mussel shells, and cuttle bone).

Palindrome: A nucleic acid sequence that is identical to its complementary strand (when both are read in the same 5′—3′ direction). In the region of a palindrome there is therefore a two fold rotational symmetry.

Palmitic Acid: n-hexadecanoic acid, $CH_3-(CH_2)_{14}-COOH$. A fatty acid. Together with stearic acid, it is one of the most widely distributed natural fatty acids, and is present in practically all natural fats, *e.g.* 36% in palm oil, 29% in bovine carcass fat, 16% in olive oil; it is also found in phosphatides and waxes.

Palmitoleic Acid: Δ^9-hexadecenoic acid, $CH_3(CH_2)_5-CH=CH-(CH_2)_7-COOH$. An unsaturated fatty acid. It is present in the glycerides of many plant and animal fats, and in phosphates.

Pancreas. A vertebrate organ producing a digestive secretion which enteis the adjacent duodenum in response to the hormones Secretin and Cholecystokinin. It also contains about 1 million islets of Langerhans, which have a diameter of 150 μm, a rich supply of blood and are innervated with unmyelinated nerve fibres. The islets contain various types of hormone-producing cells, the A cells which produce Glucagon the B cells which make Insulin and the D cells which manufacture Gastrin.

Pancreatic Enzymes. A group of at least 12 digestive enzymes, including some of the most intensively investigated of all enzymes. Autolysis of the pancreas does not occur, because the proteolytic enzymes, trypsin, chymotrypsin A and B, elastase and carboxypeptidase A and B, and phosyholipase A_2 are synthesized and stored in the pancreas as inactive zymogens.

Pancreatin. Defatted, powdered preparations of pancreas. It contains all the pancreatic enzymes in active form. It serves as a starting material for the laboratory purification of pancreatic enzyme. It is also used in the pharmaceutical industry for the preparation of enzyme tablets, which are used in cases of secretory malfunction of the pancreas.

Pantetheine 4′-phosphate, Phosphopantetheine. The phosphate ester of N-(pantothenyl)-β-aminoethanthiol. Its residue is present in the molecule of Coenzyme A. It is the prosthetic group of acyl carrier proteins in certain multienzyme complexes, *e.g.* fatty acid synthetase and gramicidin S synthetase, where it serves as a "swinging arm" for the attachment of activated fatty and amino acid groups.

Pantothenic Acid. A vitamin of the B_2 group.

PAP. Abb. for 3′-phosphoadenosine 5′-phosphate.

Papain (EC 3.4.22.2). A thiol enzyme from the latex and unripe fruit of Carica papaya (tropical melon or papaw). It is unusually stable to high temperatures and to high concentration of denaturing agents, *e.g.* 8M urea or organic solvents such as 70% ethanol or 15% dimethylsulfoxide. It is a carbohydrate-free, basic, single chain protein 212 amino acid residues; methionine absent; with 4 dissulfide bridges, and catalytically important cysteine (position 25) and histidine (position 158) residues. The molecule is a rotational elipsoid, divided by a cleft, and containing a predominance of antiparallel β-structures.

It is used medically for the treatment of necrotic tissue and eczema, and in protein chemistry for cleaving proteins into large peptides.

Papaveraceae Alkaloids, Papaver Alkaloids, Poppy Alkaloids. A group of Benzylisoquinoline alkaloids occurring especially in species of poppy (Papaver). They include the important Opium alkaloids.

Papaverine. An opium alkaloid occurring with morphine in various species of poppy. Physiologically it acts peripherally, and causes relaxation of smooth muscle; it is therefore used for the treatment of spasms in the gastrointestinal tract.

Paper Chromatography. A chromatographic separaction method, which employs a high quality filter paper (chromatography paper) as the carrier. It is almost pure partition chromatography on cellulose. The stationary phase is a film of water or adsorbed hydration layer on the cellulose fibres of the paper. The mobile phase, an organic solvent or mixture of solvents, migrates over the stationary phase.

In a defined solvent system, each substance has a rate of migration, which is expressed as the R_f value (ratio to the solvent front):

$$R_f = \frac{\text{migration distance of substance}}{\text{migration distance of the solvent}}$$

The migration distance of the solvent is the distance from the point of application of the substance to the front of the advancing solvent. Under standardized conditions (composition of solvent system; temperature; degree of vapor saturation in the chromatography tank), the R_f value is a constant.

In practice, paper chromatography consists of the physical separation, followed by special detection methods for the location of the individual substances. It may be performed on an analytical or preparative scale. With respect to the movement of the solvent system and position of the paper, paper chromatography can be ascending, descending or horizontal. Single of two dimensional procedures, and radial P c. (circular filter paper technique) are commonly used. Additional useful techniques are multiple development (the first solvent is removed by drying and a different solvent is run in the same dimension) and flow through or run off (the solvent is allowed to run for a prolonged period by dripping from the edge of the paper).

P APS. Abb. for Phosphoadenosinephosphosulfate.

Paraproteins. Abnormal immunolobulins, or normal proteins which are found in increased quantities in blood plasma in various hematological disturbances, known as paraproteinemias.

Parapyruvate. A dimer which accumulates during the storage of pyruvate.

$$^{-}OCC-\underset{CH_3}{\overset{OH}{\overset{|}{\underset{|}{C}}}}-CH_2-CO-COO^{-}$$

Parapyruvate

Parasitism. A close coexistence of two species, beneficial to one partner (the parasite) and at the expense of the other (the host). A facultative parasite (*e.g.* Claviceps, can live and grow without a host, and it can be easily cultivated in vitro. An obligate parasite (*e.g.* rickettsias, viruses, rust and smut fungi) cannot survive separately from its host, to which has become adapted by evolution.

Parathormone, Parathyrin. A hormone produced by the parathyroid gland, which influences the metabolism of calcium and phosphate. It is a single chain proteohormone with 84 amino acid residues of known primary structure.

Parathyroids, Epithelial Bodies. Two to six endocrine glands, about the size of peppercorns, on the back of the thyroid gland, or in some animals, inside the thyroid gland. They produce Parathormone.

Paromomycin. An aminoglucoside antibiotic.

Pasteur Effect. An inverse relationship between the rate of glucose utilization and the availability of oxygen.

This effect represents the regulation of glucose consumption to match the energy needs of the cell. Naturally the effect is only observed in facultative cells, *i.e.* cells which can adjust their metabolism to either acerobic or anaerobic conditions, *e.g.* yeast cells, muscle cells.

P 700 Chlorophyll A-protein. A strongly hydrophobic, integral membrane protein isolated from the thylakoids of many angiosperms, gymnosperms and green algae. It is a component of photosystem I.

Pectins. High M_r polyuronides, consisting of α-1, 4 glycosidically linked D-galacturonic acid residues. Some of the carboxyl groups are present as their methyl esters. The free acids are called pectinic acids.

They are especially plentiful in fieshy fruits, in roots, leaves and green stems. They are prepared from sliced sugar beet and from apple and citrus residues (following juice extraction) by gentle extraction with hydrochloric, lactic or citric acids.

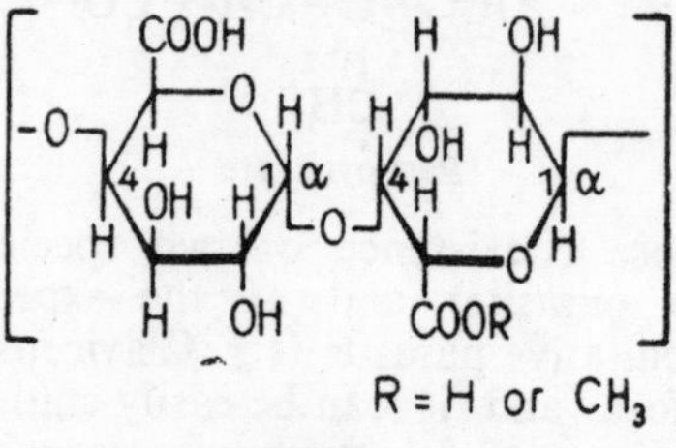

R = H or CH_3

Pectin

Pelargonidin; 3, 5, 7, 4'-tetrahydroxyflavylium cation. An aglycon of many Anthocyanins. It is widely distributed in higher plants as its various glycosides, which are responsible for the rose, orange-red and scarlet colors of petals and fruits.

Penicillinases (EC 3.5.2.6). Enzymes which catalyse the hydrolysis of the β-lactam ring of penicilin; the resulting penicilloic acids have no antibiotic activity.

Penicillins. The sulfur-containing antibiotics produced by fungi of the genera Penicillium, Aspergillus, Trichophyton and Epidermophyton. All P. contain a condensed β-lactam-thiazolidine ring system, whereas the acyl group (R) is variable.

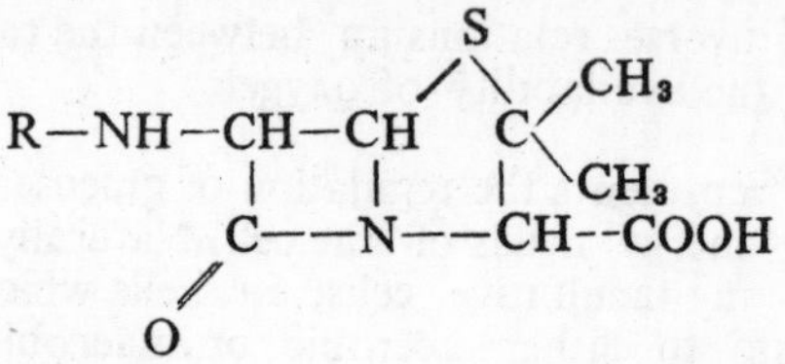

Name	R
Benzylpenicillin (penicillin G)	C_6H_5-CH_2-CO-
Pentenylpenicillin (penicillin F)	CH_3CH_2-CH=CH-CH_2-CO-

n-Heptylpenicillin (penicillin K)	$CH_3(CH_2)_6$-CO-
Penicillin N	HOOC-CH(NH_2)$CH_2)_3$-CO-
6-Aminopenicillanic acid ("6 APS")	H-
Penicillin X	HO-C_6H_4-CH_2-CO-
Penicillin V	C_6H_5O-CH_2-CO
Ampicillin	C_6H_5-CH(NH_2)-CO-

They are biosynthesized from the amino acids α-amino-adipic acid, cysteine and valine, which become linked by peptide bonds. The residue of α-aminoadipic acid is usually subsequently lost and replaced by a different acyl residue, but it is retained in the molecule of penicillin N.

They act by inhibiting cross linkage of the muropeptide in the Murien layer of the cell wall. The cell wall is thus weakened and cannot withstand the high internal pressure of the bacterial cell (about 30 atmospheres). They are the most widely and intensively used antibiotics; they are well tolerated by the animal organism and have a relatively broad spectrum of activity.

Penicillin Selection Technique. An aid to the isolation of chosen auxotrophic mutations in a bacterial population. It is based on the fact that penicillin kills growing bacterial cells, but does not affect nongrowing cells.

Pentitols: C_5-sugar alcohols. Naturally occurring P., are D-and L-arabitol, ribitol and xylitol.

Pentosans. Polysacchrides consisting of pentose residues, *e.g.* arabans and xylans. They are widely distributed in the plant kingdom, and they are important as cell wall and storage materials.

Pentose Phosphate Cycle, Hexose Monophosphate Shunt, Warburg-dickens-horecker Pathway, Phosphogluconate Pathway. An oxidative pathway of carbohydrate metabolism, in which glucose 6-phosphate (derived from glucose by phosphorylation; Kinases) is totally degraded to carbon dioxide, accompanied by the reduction of $NADP^+$ to $NADPH+H^+$. Overall equation:

$$C_6H_{12}O_6+7H_2O+12NADP^++ATP\rightarrow 6CO_2+12NADPH+12H^++ADP+P_i$$

The importance of pentose phosphate cycle lies in the production of reduced NADPH which is required for biosynthesis (*e g.* of fatty acids), and in the production of pentoses required for the synthesis of nucleosides, nucleotides and nucleic acids. By the action of a transhydrogenase system, or NADPH-cytochrome reductase, NADPH can be reoxidized to produce energy (36 molecules ATP per molecule glucose), but operation of P., for the sole purpose of energy production is unusual.

Pentoses. Aldoses containing five carbon a toms. They are an important group of monosaccharides. Naturally occurring pentoses include D-and L-arabinose, L-lyxose, D-xylose, D-ribose and 2-deoxy-D-ribose, and the ketopentoses (pentuloses) D-xylulose and D-ribulyse. They occur chiefly in the furnase form. They are not fermented by the usual yeasts. By distillation with dilute acids. They are converted into furfural, a reaction which serves for the detection of these, and their differentiation from hexoses.

Pepsin (EC 3.4.23.1). A protease in the stomach of all vertebrates with the exception of stomachless fish (*e.g.* carp). Purified pepsin shows maximal activity at pH 1—2, but in the stomach the optimal pH is 2—4. Above pH 6 it is inactivated by denaturation. It preferentially catalyses the hydrolysis of peptide bonds between two hydrophobic amino acids (Phe. Leu. Phe-Phe, Phe-Tyr). With the exception of protamines, keratin, mucin, ovomucoid and other carbohydrate-rich proteins, most proteins are attacked by pepsin.

Peptide Antibiotics. Oligopeptides with antibiotic activity. They are usually cyclic. In addition to L-amino acids they often contain nonprotein-D-amino acids and unusual amino acids, and other comonents like branched hydroxy-fatty acids. The cyclic structure is closed by the formation of amide bonds, or (depsipeptide antibiotics) by ester bonds. Owing to their ring structure and high content of nonprotein constituents peptide antibiotics are resistant to proteolytic enzymes.

Peptide Bond. The most important type of covalent bond between amino acid residues in peptides and proteins. Formally a peptide bond is an amide bond between the carboxyl group of one amino acid and the amino group of another. Shortening of the C—N bond by mesomerism of the P.b. has been confirmed by X-ray crystallographic analysis. Since free rotation around the bond axis is not possible, two planar conformation of the peptide bonds are possible *i.e.*, cis and trans (Fig.).

trans-peptide bond

cis-peptide bond

Peptide Hormones. A group of hormones with the chemical structure and physical properties of peptides. The smallest peptide harmonest thyreotropin releasing hormone, contains three amino acid residues (*i.e.* a tripeptide). Most peptide hormones are oligopeptides (up to 10 amino acid residues) Important peptide hormones are Oxytocin Vasopressin and the Releasing hormones.

Peptides. Organic compounds consisting of two or more amino acids joined covalently by peptide bonds (Peptide bond). Those containing ten or fewer amino acids are called aligopeptides while more than ten amino acid residues constitutes a polypeptide.

Peptidyltransferase (EC 2.3.2.12). An integral enzymatic activity of the large ribosomal subunit. It catalyses peptide bond formation between the NH_2 group of the amino acid of an aminoacyl tRNA and the COOH group of the terminal amino acid of the peptidyl-tRNA. The reaction appears to be sterically favored by the relative positions of the two reaction partners at the acceptor and donor sites of the ribosome. It is thought that it is located on the surface of the 50S subunit, where it makes contact with the 30S subunit.

Peptones. Mixtures of polypeptides produced by the partial degradation of protein. They are prepared by the acid or enzymatic (pepsin, trypsin) hydrolysis of dietary proteins. Autolysed and trypsin-treated yeast contains vitamins and other growth factors in addition to peptones. They are used in the formulation of many microbiological growth media.

Permeases. A term applied to carrier proteins involved in the transport of materials across membranes.

Peroxidase (EC 1.1.1.1.7). An oligomeric oxidase which utilizes hydrogen peroxide as an oxidant in the dehydrogenation of various substrates. The H_2O_2 is reduced to water :

$$AH_2+H_2O_2 \rightarrow A+2H_2O$$

In the cell, it is often accompanied by catalase and localized in the peroxisomes. The prosthetic group of plant peroxidase is ferriprotoporphyrin (a red hemin); animal peroxide contain an unidentified green hemin.

Pestalotin. A metabolic product from the culture filtrate of the fungus Pestalotia cryptomeriaecola Sawada. It has the properties of a synergist of gibberellins.

OCH₃ — O — O — OH

Pestalotin

Petite Mutants. Spontaneous mutants, chiefly yeasts, with chemical or physical defects in the respiratory chain They grow very slowly and form small colonies on nutrient agar ("petite" colonies). The same phenotype can be produced by a chromosomal mutation (segregational petite), or a mutation in the mitochondrial DNA (vegetative or neutral petite).

Petroselinic Acid. Cis-Δ^6-octadecenoic acid,

$$CH_3-(CH_2)_{10}-CH=CH-(CH_2)_4-COOH.$$

An unsaturated fatty acid. It is a glyceride component in the seed fats of many aromatic plants, *e.g.* parsley, celery, caraway.

Petunin. A blue anthocyanin present in the flowers of the blue garden petunia. The structures of more than 10 other naturally occurring glycosides of petunidin have been determined.

Phaeoplasts. Photosynthetic organelles in the brown algae and in diatoms. They are brown due to the presence of carotenoids like fucoxanthin and β-carotene.

Phage, Bacteriophase [Greek, phagein=to devour]. Viruses that attack bacteria. As a generalization, they all consist of a protein coat (the capsid or capsomer) surrounding the genetic material which is DNA or RNA. The nucleic acid is usually double stranded, but ΦX 174 (phi-chi, but usually called phi-ex) contains circular, single stranded DNA.

Length and M_r of circular DNA from some coliphages

Phage	Length of DNA (μm)	$M_r \times 10^6$
ΦX 174	1.77	1.7
λ	17	32
T2	56	130
T3	11.6	23
T4	50	125
T7	12.5	25

P, have various shapes and sizes, *e.g.*, ΦX 174 and Qβ (both coliphages) are spheroidal particles, and apical knob-like structures are also discernible on ΦX 174; coliphage fd is elongated, best described as a flexuous strand; PM2 of Pseudomonas is spheroidal with apical knobs and enclosed by an additional envelope. The most complicated P. structures are found among certain well studied coliphages (*i.e.* P. that attack

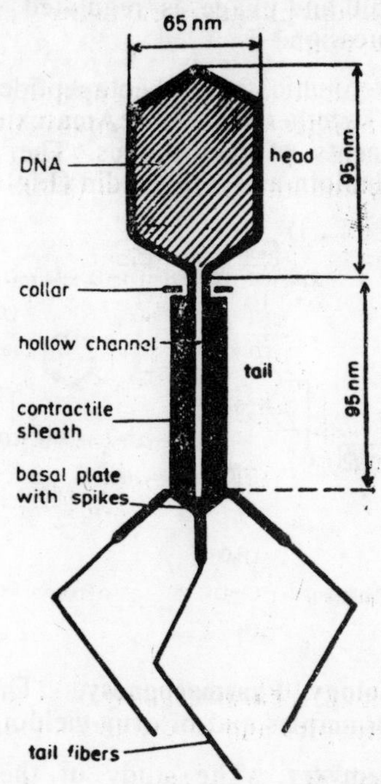

Schematic representation of coliphage T_2

Escherichia coli); these have been given arbitrary names based on letters and numbers, such as T_1, T_2, P_1, λ(lambda) etc. (T_2, T_4 and T_6 are called the T even P.)

The structure of coliphage T_2 (Fig.) may be taken as an example of the structure of T-even P.

T_1, T_3, T_5, T_7 and λ have similar structures to those of the T-even phages, but the tail is less rod-like (*i.e.* itis more flexuous), and it does not contract during the infection process.

Phage Development. Entry of phase DNA (or RNA) into the host cell initiates a series of processes leading to new phage progeny They can be divided into three phases :

1. Synthesis of early phage RNA and early phage proteins, and termination of the synthesis of all host nucleic acids and proteins;
2. Synthesis of late RNA and late proteins;
3. Morphogenesis of new phages. This complicated interaction of host cell and phage is regulated largely on the level of gene expression.

Phallotoxins. Heterodetic, cylic heptapeptides, present in Amanita phalloides. Together with the Amatoxins these are the main toxic components of this fungus. The cheif phallotoxins are phalloidin, phalloin and phallacidin (Fig).

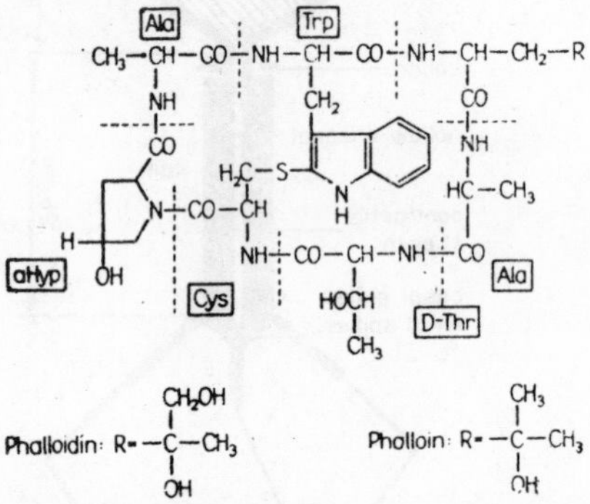

Phalloidin: $R = -C(CH_2OH)(OH)-CH_3$

Phalloin: $R = -C(CH_3)(OH)-CH_3$

Pharmaceutical Biology Pharmacognosy. The science of drugs and their active principles and of drug-yielding plants and animals.

Pharmaceutical Chemistry. The study of the chemical properties, development, preparation and analysis of pharmaceuticals.

Pharmacology. In the broad sense the study of all pharmaceutically active substances. In the narrow sense, the study of the action of substances foreign to the body, and the action of unphysiological concentrations of substances that occur naturally in the body.

Pharmacy. The study and practice of pharmaceutical preparation and analysis, involving a knowledge of dispensing techniques and the laws governing the use and sale of pharmaceuticals.

Phaseolin, Phaseollin. 7-hydroxy-3′, 4′-dimethylchromenylchromanocumaran, A Phytoalexin. It is produced by various species of bean in response to infection by phytopathogenic microorganisms. *e.g.* Phytophthera. The biosynthesis of phasedin (initiated by a polypeptide called monicolin A) starts from phenylalanine and proceeds via cinnamic acid and an intermediate isoflavone.

Phe. Abb. for L-phenylalanine.

Phenazines. Compounds based on the phenazine ring system. All known naturally occurring phenazines are produced only by bacteria, which excrete them into the growth medium. Both six membered carbon rings of phenazines are derived biosynthetically from the shikimate pathway of aromatic biosynthesis, via chorismic acid (not from anthranilate, as reported earlier).

L-phenylalanine, Abb. Phe: L-α-Amino-β-phenylpropionic acid. An aromatic proteogenic amino acid. Phe is essential in the animal diet, and it is both glucognic and ketogenic. The first stage in the metabolism of Phe is hydroxylation to L-tyrosine catalysed by Phe hydroxylase (a monooxygenase) (EC 1.14.16.1). L-Troside is then metabolized to furmarate and acetoacetate. Phenylketonuria is an hereditary defect in the synthesis of Phe hydroxylase (the enzyme may be absent or inactive).

L-phenylalanine Ammonia-lyase (EC 4.3.1.5). An enzyme responsible for the conversion of L-phenylalanine into cinnamic acid by nonoxidative deamination, present in plants and fungi, and important in the biosynthesis of Lignin. The products are trans-cinnamic acid and ammonia.

Pheromones. Predominantly low M_r substances produced by animals, especially insects (insect attractants), and secreted outside the body for purposes of communication (chemical biocommunication) with members of the same species. The main structural types are lower terpenes and higher unbranched fatty acids, and there are also ali-and heterocyclic representatives.

Depending on their mode of perception pharomones are classified as oral or olfactory.

Phlein. A high M_r reserve carbohydrate in plants. It is a straight chain polymer of fructofuranose units joined by 2, 6-glycosidic linkages. There is probably a D-glucose unit at the reducing end of the chain.

Phlorhizin, Phloridzin. A dihydrochalcome found in the root bark of bears, apples and other members of the Rosaceae. P. specificially blocks resorption of glucose by kidney tubules, thus inducing glucosuria. It therefore finds use in experimental physiology. Its activity may be due to inhibition of muta-rotase.

OH OH
HO—
O O-β Glc

Phlorhizin

Phlorin. 5, 22-dihydroporphyrin.

Phosphagens. Energy rich guanidimium or amidine phosphates, which function as storage depots for high energy phosphate in muscle. Excess energy rich phosphate (*i.e.* ATP) is transferred to P. from which ATP can be regenerated when required.

Phosphatases, Phosphoric Monoester Hydrolases (EC sub-sub-group 3 1.3). Esterases that catalyse the hydrolysis of monophosphate esters. They are widely distributed in living organisms, and they are mostly dimeric proteins with a catalytically important serine residue in the active center. They mucosa P. (M_r 140000; 2 chains, each of M_r 69000), placenta and bone P. (M_r 120000; 2 chains, each of M_r 60000) and the P. of Escherichia coli (M_r 85000; 2 chains, each of M_r 43000). The alkaline P. contain one or two essential zinc atoms per subunit, and also require Mg(II) ions for full activity; these cations have no effect on acidic P. In contrast to other esterases, P. are only slighty inhibited by diisopropylfluorophosphate.

Phosphatide Biosynthesis. Of particular importance is the biosynthesis of lecithins and cephalins, which are glycerophosphatides. The phosphatidic acid and 1,2-diacylglyceride precursors of glycerophosphatides are synthesized as described under Fat biosynthesis.

Phosphatide Degradation. The hydrolytic cleavage of phosphatides.

Phosphoadenosinephosphosulfate, 3′-phosphoadenosine-5′-phosphosulfate, Adenosine-3′-phosphate-5′-phosphosulfate. abb. **PAPS, Active Sulfate.** The product of sulfate activation. PAPS is a key intermediate in the reduction of sulfate and the formation of sulfate esters by sulfokinases. It is produced in a two stage reaction :

$$ATP+SO_4^{2-} \rightarrow APS+PP_i \text{(ATP) sulfurylase) (EC 2.7.7.4)}$$
$$APS+ATP \rightarrow PAPS+ADP \text{ (APS kinase) (EC 2.7.1.25)}$$
$$\text{Sum: } 2ATP+SO_4^{2-} \rightarrow PAPS+ADP+PP_i$$

Synthesis of PAPS is known as Sulfate activation. In the first stage, the terminal pyrophosphate of ATP is replaced by sulfate, forming adenosine phosphosulfate (APS); this reaction is catalysed by sulfate adenyltransferase (ATP sulfurylase, EC 2.7.7.4).

Phosphodiester. A phosphate ester in which two hydroxyl groups of the phosphoric acid are esterified with organic residues :

$$RO-PO_2H-OR'$$

For example, R and R′ may be nucleosides. All polynucleotides and nucleic acids are P. in which the 3′ and 5′ positions of neighboring pentose units are linked by esterification with a phosphate residue.

Phosphodiesterases. Enzymes that catalyse the hydrolytic cleavage of phosphodiesters, *e.g.* endonucleases, Ribonuclease and Deoxyribonuclease and the less specific exonucleases. The latter degrade both DNA and RNA stepwise in the 3′→5′ direction, producing 5′-mononucleotides (snake venom P., EC 3.1.4.1), or in the 5′→3′ direction, producing 3′-mononucleotides (spleen P. EC 3.1.16.1). P. have been used for the sequence determination of nucleic acids (especially RNA). 3′:5′-Cyclic nucleotide P. (EC 3.1.4.17) catalyses the hydrolysis of cyclic AMP.

Phosphofructokinase, 6-phosphofructokinase (EC 2.7.1.11). An oligomeric phosphotransferase, and a key control enzyme of glycolysis. It is induced by insulin, and its activity is subject to alosteric control; it is activated by AMP, fructose 6-phosphate, fructose 1, 6 bisphosphate, magnesium, potassium and ammonium ions, and inhibited by ATP and citrate. It catalyses the phosphorylation of fructose 6-phosphate by ATP in the presence of magnesium ions to form fructose 1, 6-bisphoshpate. The reaction is irreversible, and the conversion of fructose 1, 6-bisphosphate to fructose 6-phosphate is catalysed by the fluoride-sensitive enzyme, fructose bisphosphate, EC 3.1.3.11 (fructose diphosphatase).

Phosphoglyceric Acids. Monophosphate esters of Glyceric acid.

Phosphoketolase Pathway. A pathway of carbohydrate degradation found in various microorganisms, especially Lactobacillus, in which a ketopentose phosphate undergoes phosphorolylic cleavage to triose phosphate and acetyl phosphate. The key enzyme is phosphoketolase (EC 4.1.2.9), which catalyses the ATP dependent, irreversible cleavage of D-xylulose 5-phosphate to D-glyceraldehyde 3-phosphate and acetyl phosphate. The balance of pentose metabolism in Lactobacillus species is.

$$\text{Pentose}(C_5H_{10}O_5)+2\text{ADP}+2\text{P}'\rightarrow\text{Acetate}(C_2H_4O_2)+$$
$$\text{L-Lactate}(C_3H_6O_3)+2\text{ATP}$$

Phospholipases, Phosphatidases. A collective name for the carboxylic acid esterases, P.A_1, A_2 and B, and the phosphodiesterases, PC and D, which are specific for lecithins. Its activity is particularly high in liver and pancreas (P.A_1) in bee and snake venom (P.A_1 and P.A_2), in bacteria (P.C) and plants (P.D).

```
                          (A1)
                      H2C-O-C-R1
           O           |    ||
           ||          |    O
      R2-C--O-CH    (C) O  (D)
       (A2)     |       ||
              H2C-O-P-O-CH2-CH2-N⊕(CH3)3
                        |
                        O⊖
```

Sites of Attack by Lecithin-cleaving Enzymes. R_1 saturated fatty acyl residue. R_2 mono-or multi-unsaturated fatty acyl residue.

Phospholipids. Any lipids containing phosphoric acid as a mono-or diester. They are the basic constituents of Biomembranes and are especially abundant in the brain and myelin sheaths of the nerves.

```
R1—CO—O—¹CH2
          |
R2—CO—O—²CH     O
          |     ||
        H2³C—O—P—O—R3
                |
                OH
```

R_1=Alkyl
R_2=Alkenyl
$R_3=CH_2CH_2-\overset{\oplus}{N}(OH_3)_3$=α-Lecithins
$R_3=OH_2CH_2-NH_2$=Colamine (Ethanolamine) cephalins
$R_3=CH_2-\underset{NH_2}{\underset{|}{CH}}-COOH$=Serine cephalins

Phosphoglycerides

Phosphon D: 2, 4-dichlorobenzyltri-n-butyl-phosphonium chloride. A synthetic plant growth retardant. It inhibits, *e.g.* the growth of chrysanthemum stems and induces flowering.

Phosphoproteins. Conjugated proteins, containing phosphate esterified with the hydroxyl groups of serine or (less often) threonine residues. Well known examples are casein and ovalbumin.

5-Phosphoribosylamine, abb. **PRA**. An intermediate in Purine biosynthesis.

5-Phosphoribosyl I-pyrophosphate, 5-phosphoribose I-diphosphate, abb. **PRPP**. An energy-rich sugar phosphate formed by the transfer of a pyrophosphoryl residue from ATP to ribose 5-phosphate. It is concerned in various biosynthetic reactions, *e.g.* biosynthesis of purines, pyrimidines and histidine.

Phosphoroclastic Fission of Pyruvate. A special mechanism for the cleavage of pyruvate found only in saccharolytic Clostridia. It is responsible for the synthesis of ATP during nitrogen fixation. The first stage is the synthesis of acetyl phosphate :

$$CH_3-\overset{\overset{O}{\|}}{C}-COOH+P_i \rightarrow \text{Pyruvate}$$

$$CH_3-\overset{\overset{O}{\|}}{C}-O-PO_3H_2+CO_2+H_2$$

Acetyl phosphate

followed by synthesis of ATP from acetyl phosphate catalysed by acetokinase :

$$\text{Acetyl phosphate}+\text{ADP} \rightleftharpoons \text{ATP}+\text{Acetate}.$$

Phosphovitin. An egg yolk protein containing 10% phosphate. Serine constitutes 50% of the total amino acid content, and all the serine residues are phosphorylated. It is synthesized in the liver of the laying hen and transported in the blood to the developing egg.

Photocitral. A cyclization product of Citral.

Photolysis of Water. Cleavage of water by the light reaction of Photosynthesis. It is a property of photosystem II. It is not a simple photodissociation of water, but is the physiological counterpart of the Hill reaction. Electrons are withdrawn from water or OH^- ions, then-transported to $NADP^+$ via an electron transport chain; this results in the production of molecular oxygen with the formation of $NADPH+H^+$. Since Hill reagents oxidize water by withdrawal of electrons, $NADP^+$ is the natural Hill reagent.

Photophosphorylation. The synthesis of ATP in Photosynthesis. Its mechanism is similar to that of Oxidative phosphorylation by the respiratory chain; in both cases cytochromes are involved in electron transport. A distinction is drawn between cyclic and noncyclic photoplosphorylation. Both forms are found in green algae and higher plants. Cyclic photophosphorylation involves cyclic electron transport. Under the influence of light, electrons emitted from chlorophyl a return to chlorophyl a via an electron transport chain, thereby giving rise to ATP synthesis. Thus the positive holes left in the chlorophyll structure by the loss of electrons are refilled, and the electron excitation energy is transduced to the chemical energy of ATP. Cyclic photoplosphorylation involves cytochrome f, and the only product is ATP. In noncyclic photoplosphorylation ATP synthesis is linked to the transport of electrons from water (Photolysis of water) to $NADP^+$, thus producing both ATP and a reducing agent (NADPH). The production of molecular oxygen, the byproduct of water photolysis, is characteristic of photosynthesis in green plants and algae. Noncyclic photoplosphorylation can be considered as a Hill reaction coupled to the synthesis of ATP. Whereas cyclic P. requires only photosystem I, noncyclic P. depends upon the joint operation of both photosystems, which are connected in series; electrons are transported from OH^- ions of water to $NADP^+$ in an open-chain (noncyclic) system. The two kinds of P. are functionally and structurally separate in the chloroplast, but they are closely interrelated.

Photoproteins. Proteins responsible for luminescence in many light-emitting coelenterates. Light-emission by these proteins does not involve a luciferinluciferase system and the reaction proceeds in the absence of oxygen.

Photoreactivation. Repair of biological systems damaged by UV-irradiation, in a process promoted by light of a different wavelength. Pyrimidine dimers, which result from UV-irridation, can be monomerized by the action of UN light of shorter wavelength or light of longer wavelength, which promotes the action of repair enzymes. The enzyme binds only to UV-damaged DNA and converts pyrimidine dimers to monomers when irradiated with light of an appropriate wavelength.

Photorespiration. Light enhanced respiration in photosynthetic organisms. Illumination of C3-plants markedly increases the rate of oxygen utilization; this increase in respiration can be as high as 50% of the net photosynthetic rate. It thus results in a loss of yield in the photosynthesis of C3-plants. In C4-plants. It is either absent or extremely low. It is largely due to the oxygenase activity of Ribulosebisphosphate carboxylase which oxidatively cleaves rebulose 1, 5-bisphosphate into phosphoglycolate and 3-phosphoglycerate.

Photosynthesis

1. Any light-dependent synthesis.
2. The reductive synthesis of carbohydrate in green plants and Photosynthetic bacteria. It was formerly defined as the assimilation of carbon dioxide, but it is now recognized as primarily a process of energy transduction, in which light energy is converted into the chemical energy of oxidizible organic carbon compounds. P. in green plants (but not bacterial P.) can be represented by :

$$6CO_2 + 6H_2O \xrightarrow[\text{Chlorophyll}]{hv} C_6H_{12}O_6 + 6O_2$$

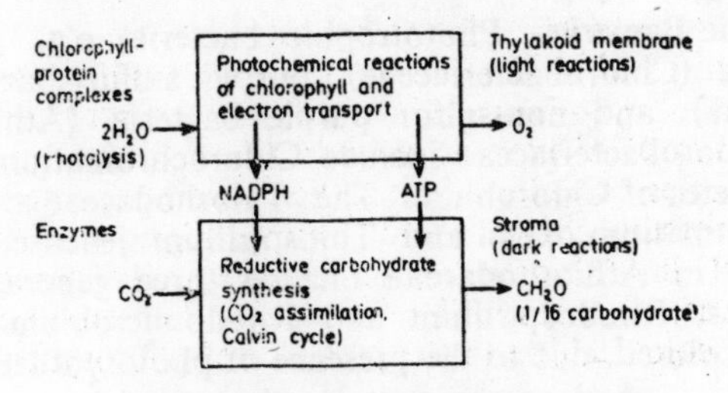

Relationship between light and dark reactions of photosynthesis

In principle, this general equation is the reverse of Respiration. The reaction is catalysed by chlorophyll a, which is structurally bound in the Thylakoids. In photosynthetic bacteria, chlorophyll a is replaced by bacteriochloropyll a is replaced by bacteria, chlorophyll a is replaced by bacteriochlorophyll a (Bacterial photosynthesis). Other Photosynthetic pigments serve as auxilliary pigments for light absorption and energy transfer, P. is the most important process for the production of organic material in the biosphere. All non-photosynthetic organisms are directly or indirectly dependent on P of phototrophic organisms. By comparison, Chemosynthesis plays quantitatively insignificant role in the carbon cycle of the biosphere.

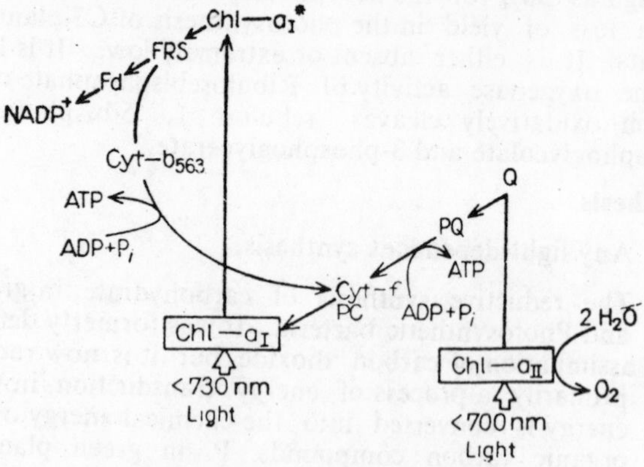

Coupling of the two light reactions in non-cyclic photophosphorylation. Chl chlorophyll, Cyt cytochrome, Fd ferredoxin, FRS ferredoxin reducing substance, Q quanching substance.

Photosynthetic Bacteria. Phototrophic bacteria, e.g., green sulfur bacteria (Chlorobacteriaceae), purple sulfur bacteria (Thiorhodaceae) and non-sulfur purple bacteria (Athiorhodaceae). The Chlorobacteriaceae include Chlorochromatium consortium and species of Chlorobium. The Thiorhodaceae are represented by Chromatium okenii and Thiospirillum jenense, which are of interest in Athiorhodaceae include three genera: Rhodopseudomonas, Rhodospirillum and Rhodomicrobium. All P.b. are deeply colored, due to the presence of photosynthetic pigments.

In place of chlorophyll a these bacteria contain bacteriochlorophyll a (2-devinyl-2-acetyl-3, 4- dihydrochlorophyll a). Green sulfur bacteria also contain bacteriochlorophylls c and d (formerly called Chlorobium chlorophyils), which absorb between 700 and 760 nm.

These bacteria neither produce nor consume molecular oxygen. They do not perform the photolysis of water, and they can live anaerobically, most of them being strict anaerobes. They contain one photosystem, analogous to the photosystem I of higher green plants algae and blue-green bacteria. Green and purple sulfur bacteria perform the photolysis of hydrogen sulfide :

$$2\,H_2S + CO_2 \xrightarrow{\text{light}} \underset{\text{Carbohydrate}}{[CH_2O]_n} + H_2O + 2S$$

Sufficient energy for this process is achieved in one photoevent (in contrast to the photolysis of water, which requires two photoevents coupled in series. Reducing power is trapped by ferredoxin, but some of the energized electrons are recycled, for the synthesis of ATP. Nonsulfur purple bacteria use various substances as donors of hydrogen and electrons, *e.g.* isopropanol :

$$\underset{\text{Isopropanol}}{2\,CH_3CHOHCH_3} + CO_2 \xrightarrow{\text{light}} \underset{\text{carbohydrate}}{[CH_2O]_n} + \underset{\text{Acetone}}{2\,CH_3COCH_3} + H_2O$$

Photosynthetic Carboxylation. The enzymatic fixation of carbon dioxide in photosynthesis. In C-3 plants, the photosynthetic carboxylation enzyme is ribulose bisphosphate carboxylase (EC 4.1.1.39). In C-4 plants it is phosphoenolpyruvate carboxylase (EC 4.1.1.31). Photosynthetic carboxylation is the first step of carbon dioxide assimilation in photosynthesis, and one of the dark reactions.

Photosynthetic Experimental Organisms and Systems. For technical reasons, certain systems are preferred for the investigation of photosynthesis, *e.g.* green algae such as Chlorella photosynthetic bacteria and isolated chloroplasts. Green algae and eugleoids (*e.g.* Euglena gracilis) can be cultured under defined conditions in an illuminated chemostat and they can also be grown in Synchronous culture. It is relatively easy to obtain chlorophyll-deficient mutants of these organism, which must grow heterophically Mutant technique). The production of plastids can be prevented by culture in the dark.

Photosynthetic Pigments. Pigments that take part in the trapping and utilization of light in Photosynthesis. Seed plants (Spermatophyta), ferms (Pteridophyta), mosses (Bryophyta), green algae (Chlorophyta), *e.g.* Chlorella) euglenoids (Euglenophyta, *e.g.*, Euglena) and brittleworts (Characeae) contain both chlorophylls a and b, and carotenoids, but no biliproteins. The latter are found in red algae Rhodoplasts) and Blue-green bacteria. Certain algae lack chlorophyll b (Chrysophyta Pyrrophyta and Cryptaphyta.

Photosystem, Pigment Systems. Structural functional units of the light reaction of Photosynthesis. The quantum efficiency of photosynthesis in chloroplasts falls sharply at wavelengths longer than 680 nm, although chlorophyll still absorbs light from 680 to 700 nm. This phenomenon is known as the "red drop". However, the quantum efficiency of light above 680 nm is increased by the simultaneous presence of presence of shorter wavelength light. This Emerson effect led to the proposal that photosynthesis depend upon the interaction of two light reactions (*i.e.*, two photosystems), both driven by light less than 680 nm, but only one by light of longer wavelengths.

Physostigmine, Eserine. An indole alkaloid from calabar beans, the ripe seeds of Physostigma venenosum (a woody vine indigenous to west coast of Africa). It contains a pyrrolidinoindole ring structure and a urethane group and exists in a stable form. It occurs with its N-oxide, geneserine m.p. 129°C, $[\alpha]_D -175°$ (acetone).

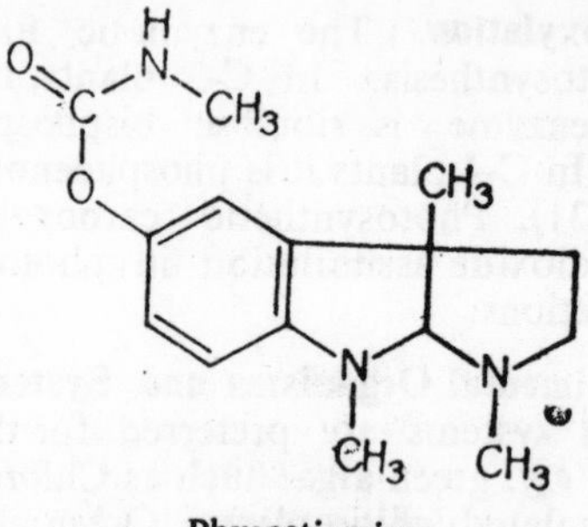

Physostigmine

It is used in opthalmic practice in the same way as pilocarpine, for pupil contraction and for the reduction of intraocular pressure.

Phytoalexins, Phytonicides Stress Compounds. Substances with antibiotic activity produced by plants in response to injury or stress, *e.g.*, infection with fungi, bacteria and viruses, mechanical wounding, UV-irradiation dehydration, cold, and treatment with phytotoxic chemicals (*e.g.*, heavy metals). They function as growth inhibitors of phytopathogenic organisms, cheifly fungi.

Phytochemistry. The chemistry of natural products from plants (Natural product chemistry). It is part of plant biochemistry, and it is concerned chiefly with secondary metabolites.

Phytoene. An aliphatic, colorless, hydrocarbon carotenoid. It is a polyisoprenoid. It contains six branch methyl groups, two terminal isopropylidene groups and nine double bonds, three of them conjugated. Only the Δ^{15} double bonds has cis configuration. Biosynthetically, it is derived from two molecules of geranylgeranyl pyrophosphate, and it serves as a C_{40}-starter molecular in the biosynthesis of other carotenoids; phytofluene -carotene. neurosporene and lycopene are formed by the stepwise dehydrogenation of P.

Phytofluene. An aliphatic, polyisoprenoid hydrocarbon carotenoid. It contains ten double bonds (live of them conjugated), six branch methyl groups and two terminal isopropylidene groups. It is an intermediate in lycopene.

Phytohormones, Plant Hormones. A group of natural (endogenous) plant growth regulators. They have multiple activities and low action specifities. The known stimulatory P. are Auxins, Gibberellins and Cytokinins; those with inhibitory activity are Abscissic acid, Flowering hormone and Fruit ripening hormone. They can be determined quantitatively by biological assay. The mode and site of action of P. at a molecular level are largely unknown.

Picrotoxin, Cocculin. A molecular compound of one molecule picrotoxinin and one molecule picrotin. It is a neurotoxin, which occurs in the seeds of Anamirta coculus, and is also found in Tinomiscium philippinense.

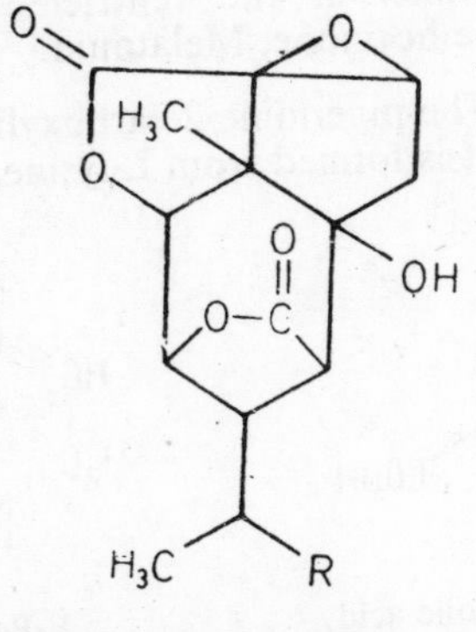

Pictrotoxinin, R=CH_2

R: $< \begin{matrix} OH \\ CH_3 \end{matrix}$

Phytopharmacology. The study of the action of synthetic and biogenic substances on plants. A knowledge of the action of biocides on plants is important for the application and development of herbicides, growth regulators and crop production agents.

Pigment 700 P_{700}. A Chlorophyll with an absorption maximum at 700 nm. P. 700 is a component of photosystem I and serves as an energy sink or trapping center in the light reaction of this system. It is important in the primary energy transformation of Photosynethesis.

PIH. Abbreviation for prolactin release inhibiting hormone.

Pilocarpine. An imidazole alkaloid, and the chief alkaloid from the leaves of Brazilian Pilocarpus species. It is used therapeutically as a diaphoretic, *i.e.*, to induce sweating, and especially in nephritis to relieve the kidneys and remove toxic metabolites. It is also used in opthalmology as an antagonist of atropine, and for regulating the intraocular pressure in glaucoma.

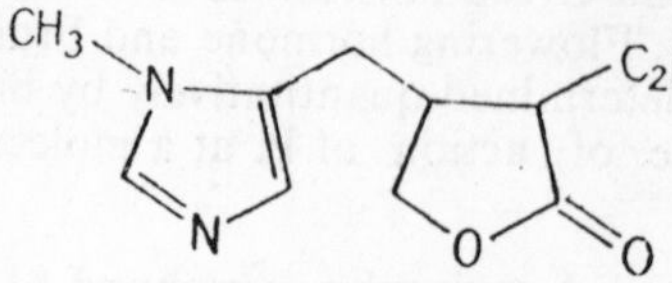

CH_3 N N C_2H_5 O O

Pilocarpine

Pineal Gland, Pineal Body, Epiphysis, Corpus Pineale. A small, cone-shaped, unpaired organ situated between the cerebral hemisperes on the roof the ventricle of the mammalian brain. It produces the hormone, Melatonin.

L-Pipecolic Acid. The piperidine-2-carboxylic acid, a nonproteogenic amino acid. It is formed from L-lysine, either by a α-deamina-

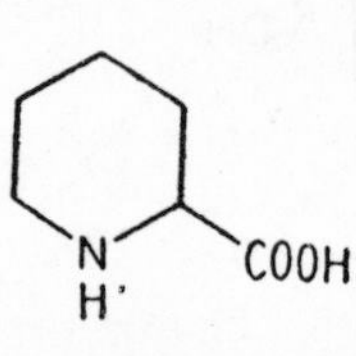

N H COOH

L-Pipecolic acid

H C HC CH_2 H_2C C N H COOH H

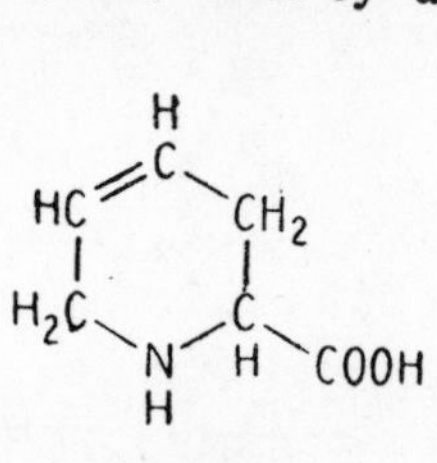

L-Baikiain

tion followed by cyclization and reduction, or as normal intermediate in the degradation of lysine to α-aminoadipic acid.

Piper Alkaloids. A group of alkaloids occurring in various species of Piper, especially black pepper (Piper nigrum). Structurally, these alkaloids consist of an aromatic carboxylic acid with an unsaturated side chain (*e g.*, piperic acid, sinapic acid) in amide linkage with a basic component, usually piperidine. The chief representative is Piperine.

Piperidine Alkaloids. A group of alkaloids containing the piperiding ring system. Simple piperidine alkaloids are the alkyl substituted piperidines which occur sporadically.

Piperine. Piperic acid piperidide, a Piper alkaloid the chief alkaloid of black paper (Piper nigrum) and responsibbe for its sharp taste.

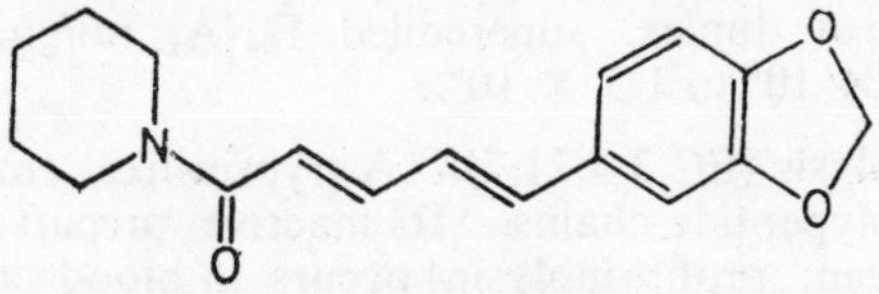

Piperine

Pisatin. A Phytoalexin. It is synthesized by pass (Pisum sativum) in response to infection to phytopathogenic microorganisms. It is biosynthesized from acetate and cinnamic acid.

PL. Abbreviation for placentalactogen.

Plant Mucilages. High Mr, complex, colloidal polysaccharides, which form gels and have adhesive properties. They are widely distributed in the plant kingdom, being found as secondary membrane thickening and as intercellular and interacellular material. They occur in root, bark, cortex, leaves, stalks, flowers, endosperm and seed coat. Some bulbs contain special muciage cells.

Plaque. A transparent area in a lawn of bacteria on the surface of a solidified growth medium. It is caused by lysis of bacteria in that area by bacteriophage. Under controlled conditions, each plaque represents a center of infection initiated by one infective bacteriophage particle. The number of plaques produced after evenly spreading a known volume of phase suspension over the surface of the bacterial culture is used as a simple assay of the number of infective phage particles.

Plasmakinins. Physiological, highly active oligopeptides, with hormone-like properties. They act upon the smooth muscle of blood vessels, gastrointestinal tract, uterus and bronchi. Important representatives are Bradykinin, kallidin and methionyl-lysyl-bradykinin.

Plasmalemma. The cell membrane.

Plasma Protein. A complex of predominantly conjugated proteins present in the blood plasma of vertebrates. The number of plasma proteins is estimated to be more t an 100. In mammalian plasma the concentration of P.p' is 6–8%. Serum proteins lack Fibrinogen and Prothrombin but are otherwise essentially the same as plasma proteins 60 P.p. have been isolated and characterized.

Plasmid. Extrachromosomal DNA in the bacterial cell. They carry genetic information and replicate independently of the bacterial chromosome. A plasmid which can become reversibly integrated into the host DNA is called as episome. Plasmids are circular molecules of duplex, supercoiled DNA, ranging in size M_r about 1.5×10^6 to 1.5×10^8.

Plasmin, Fibrinolysis (EC 3 4.21.7). A trypsin-like enzyme, containing two polypeptide chains. Its inactive precursor or zymogen (plasminogen, profibrinolysin) occurs in blood at a concentration of 50–100 mg/100 ml serum. It is responsible for fibrinolysis, *i.e.* dissolution of blood clots by the proteolytic degradation of fibrin to solute peptides.

Plasmochromic Pigments. The plant pigments contained in plastids, *e.g.* Chlorophyll and carotenoids.

Plasmon. The total extrachromosomal hereditary complement of a eukaryotic cell. Chondrome; Plastome; Cytoplasmon.

Plastids. Organelles in the cells of eukaryotic plants. These contain DNA, and are self replicating. Division of P. and replication of P. DNA are not synchronized with nuclear division and replication of nuclear DNA. They are usually elipsoid, 1–10 μm long.

Plastome, Plastidom. The total genetic information contained in the DNA of the plastids of a eukaryotic plant cell.

Plastoquinone. A polysioprenoid quinone. It is structurally similar to ubiquinone, but P, contains a methyl group (not a methoxy group) in the aromatic ring. It can be isolated from chloroplasts; it acts as a reversible redox component in photosynthetic electron transport (Photosynthesis).

Platelet Activating Factor, PAF. A phospholipid released by IgE-sensitized leucocytes in the presence of antigen, and possibly endogenous to platelets as well. It is present during anaphylactic shock, and appears to mediate inflammation and allergic responses.

Plumierid. An Iridoid.

Poisons, Toxins. Compounds which damage an organism in relatively small amounts, and which may kill it at high enough doses. Plant poisons are found in all parts of certain plants they are not produced by specialized cells. They tend to affect the heart and circulation.

Pollinastanol. 4, 4-demethylcycloartenol. A plant constituent related structurally to the sterols and to cycloartenol.

Poly A. Abb. for Polyadenylic acid.

Polyadenylic Acid, abb. **Poly A.** A Homopolymer consisting entirely of residues of adenylic acid.

Poly A sequences of varying lengths are found at the 3′ end of many eukaryotic mRNA molecules. They are present in the nucleoplasmic, giant messenger-like RNA, and in the polysomal cytoplasmic mRNA. The poly A sequence is not synthesized by transcription, but by enzymatic addition of adenylic acid residues to the newly synthesized nuclear RNA.

Polyamino Acids. Naturally occurring or synthetic polymers, consisting of identical amino acids linked by peptide bonds. These in the capsular substance of anthrax bacteria contain residues of D-glutamic acid linked by γ-peptide bonds. Poly-γ-D-glutamic acid precipitates antibodies against anthrax, a property not shared by poly-γ-L-glutamic acid.

Poly A Polymerase. An enzyme which specifically catalyses the synthesis of poly A sequences, a process which does n t require DNA.

Polyketides, Acetogenins. Natural products containing several recurring two-carbon units, formally equivalent to the condensation products of several molecules of acetate, Biosynthesis of these alkaloids occur on a multienzyme complex.

Examples of P are Tetracyclines, Griesofulvin, Macrolide antibiotics, Cycloheximide, and various fungal products such as orsellinic acid, 6-methylsalicyclic acid and cyclopaldic acid.

Polymerases. A collective term for enzymes that catalyse the formation of macromolecules from simple components, *e.g.* separate entries for each of the following : DNA-polymerase, RNA-polymerase, RNA-synthetase, RNA-dependent, DNA-polymerase, Polynucleotide phosphorylase, Poly A-polymerase.

Polymorphism. A genetically determined Heterogeneity of proteins, especially enzymes. It occurs when the frequency of a genetic variant in a population is greater than 1%. Frequencies of this order develop by positive selection or by the effect of incidental genetic drift on rare mutations that have a heterozygotic advantage.

Polymyxins. Heteromeric, homodetic, cyclic, branched peptides, produced by Bacillus polymyxa, possessing antibiotic activity against Gram-negative bacteria polymixins are used to treat infections of the intestinal and urinary tracts, sepsis and endocarditis.

Polynucleotide Ligase, DNA-ligase. An enzyme that joins two DNA fragments by catalysing the formation of an internucleotide ester bond between phosphate and deoxyribose. These are active in the repair of damaged DNA, and in the linkage of Okazaki fragments during DNA replication.

Polynucleotide-methyltransferases, Polynucleotide Methylases. Specific enzymes that cataiyse the methylation of purine and pyrimidine bases, or sugars in the intact polynucleotide chain by transfer of methyl groups from S-adenosyl-L-methionine. Methylated bases and/or sugars are found in tRNA, mRNA, rRNA and DNA.

Polynucleotide Phosphorylase. An enzyme that catalyses the synthesis in vitro of polyribonucleotides. 5′-Nucleoside diphosphates are added to oligonucleotide starter molecules (primers), with the release of phosphate. The resulting sequenee depends upon the availability of components for the reaction.

Polynucleotides. Linear sequences of about 20 nucleotides, in which the 3′-position of each monomeric unit is linked to the 5′ of the neighboring unit via a phosphate group The valency angles between phosphate and sugar residues are such that the sugar-phosphate backbone forms a helix, with the bases projecting sideways. The sugar may be D-ribose (in ribopolynucleotides) or 2-deoxy-D-ribose (in deoxyribopolynucleotides).

Polynucleotide Thioltransferases. Thiolases that catalyse the specific thiolation of purine and pyrimidine bases in the synthesis of Rare nucleic acid components.

Polyporenic Acid, Polyporenic Acid A. A tetracyclic triterpene carboxylic acid. It is structurally related to 5 α-lanostane. It occurs in the fungus Polyporus spp. growing on birch trees.

Polyprenols, Polyprenyl Alcohols. Acyclic, polyisoprenoid alcohols. P: occur free or esterified with higher fatty acids in microorganisms, plants and animals.

Polyribosomes, Polysomes, Ergosomes (obsolete). The structural unit of Protein biosynthesis consisting of several to many Ribosomes attached along the length of a strand of mRNA.

Polysaccheride Sulfate Esters, Polysaccharide Sulfates. Sulfate esters of polysaccharides, synthesized by the action of sulfokinase. There are two possible mechanisms of biosynthesis ;

1. Sulfation of the polysaccharide chain by phosphoadenosinephosphosulfate (PAPS);
2. Polymerization of sugar sulfates, which occur as sulfated sugar nucleotides.

Polyterpenes, Polyisoprenes, Polyisoprenoids. Acyclic, unsaturated, terpene hydrocarbons or alcohols, of the general $(C_5H_8)_n$, and consisting of a large number of isoprene units. They are classified as shown in the table.

TABLE

Classification of Polyterpenes

Compound	No. of isoprene units	Double bonds
Polyprenols	6—24	cis/trans
Gutta, Balata	about 100	trans
Natural caoutchouc	10000	cis

Poly U. Abb. for polyuridylic acid.

Polyuridylic Acid, abb. **Poly U.** A Homopolymer of uridylic acid, containing an indeterminate number of nucleotide units. In a cell-free, ribosomal protein biosynthesis, poly U acts as synthetic mRNA, and its translation product is polyphenylalanine. It is often used to determine the synthetic capacity of such systems.

Polyuronides. Macromolecular compounds found in plants, consisting of units of uronic acids in the pyranose form. The chief components are D-glucuronic acid, D-galacturonic acid and D-mannuronic acid. Examples are pectins, alginic acid and plant mucilages.

Ponasterone A, 2β, 3β, 14α, 20 ι (R), 22β (R)-pentahydroxy-5β-cholest-7-en-6-one, a phytoecdysone.

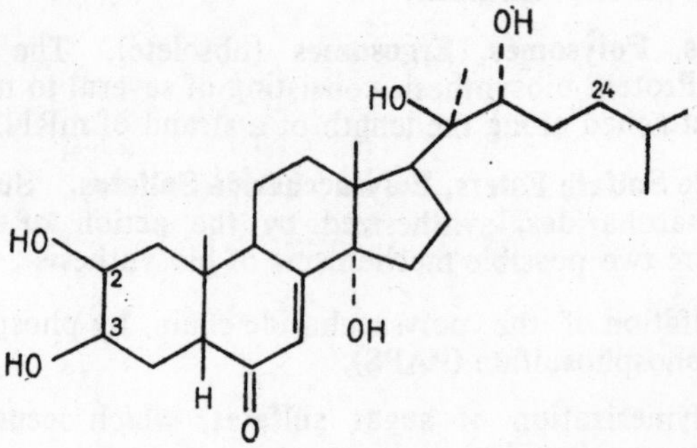

Ponasterone A

Poriferstero1. (24 S)-5α-stigmasta-5,22-dien-3 β-ol, a marine zoosterol. It is a characteristic sterol of sponges, and has been isolated from. *e.g.* Haliclona variabilis, Cliona celata and Spongia lacustris If differs from Stigmasterol by an altered configuration at C24.

Porphin. The parent tetrapyrrole of the Porphyrins. It is synonymous with porphyrin

Porphodimethene. 5, 10, 15, 22-tetrahydroporphyrin.

Porphomethene. 5, 15-dihydroporphyrin.

Porphyrinogen. 5, 10, 15, 20, 22, 24-hexahydroporphyrin.

Porphyrins. Cyclic ,tetrapyrroles, which can be considered as derivatives of the parent tetrapyrrole, porphin. The eight β-hydrogens of the parent porphin are completely or partly substituted by side chains, *e.g.* alkyl, hydroxyalkyl, vinyl, carbonyl or carboxylic acid groups. The different porphyrins are classified on the basis of these side chains, *e g.* protoporphyrin, coproporphyrin, etioporphyrin, mesoporphyrin, uroporphyrin.

Positive Control. Activation of transcription, or the activity of enzymes by activators or metabolites.

Post Translational Modification of Proteins. Processing In a general sense, any difference between a functional protein and the linear polypeptide sequence encoded between the initiation and termination codons of its structural gene can be regarded as a post translational modification of proteins. Thus, the folding of the polypeptide chain, stabilized by weak, noncovalent interactions, and the association of the subunits of an oligomeric protein both represent a modification that occurs after translation.

Post Transcriptional Modification of RNA, Maturation of RNA, Processing of RNA. Most RNA is modified after transcription by irreversible cleavage by endonucleases, producing smaller functional molecules. The superfluous, nonfunctional fragments are completely degraded. Almost 50% of newly synthesized eukaryotic 45S RNA is degraded after transcription (Ribosomes). During the maturation of high M_r nuclear RNA to Messenger RNA, over 90% of the molecule is degraded. The precursors of Transfer RNA lose about one third of their nucleotides, whereas precursors of prokaryotic ribosomal.

PP_i. Inorganic pyrophosphate; the anion of pyrophosphoric acid, $P_2O_7^{4-}$.

PP. inorganic pyrophosphate.

PP-factor. Pellagra preventative factor; biologically equivalent to nicotinic acid and nicotinamide.

PPQ. Abb. for pyrroloquinoline quinone.

PPR. Abb. for phosphoribosylamine.

Progress Curve. In enzyme kinetics, a plot of the concentration of a reactant, or several reactants (*e.g.* substrates, products, enzyme-substrate complexes) or an enzymatic reaction as a function of the time for which the reaction has progressed.

Precipitation. Formation of an insoluble, inactive antigen-antibody complex from a soluble antigen and a bivalent specific antibody (the precipitin). Monovalent or incomplete antibodies do not precipitate the soluble antigens; the resulting complexes cannot form cross-linkages and therefore remain soluble. In the quantitative determination of the precipitation reaction (due to Heidelberger), increasing quantities of antigen are added to a constant amount of antibody. At the Equivalence point the supernatant above the precipitate contains neither antigen nor antibody.

Precipitation Curve, Heidelberger Curve. A plot of the quantity of precipitate formed during the titration of antibody with an antigen, or vice versa. The antibody must be at least bivalent. The precipitation curve reaches a maximum at the Equivalence point, then decreases as the concentration of one component exceeds its optimum (Fig.).

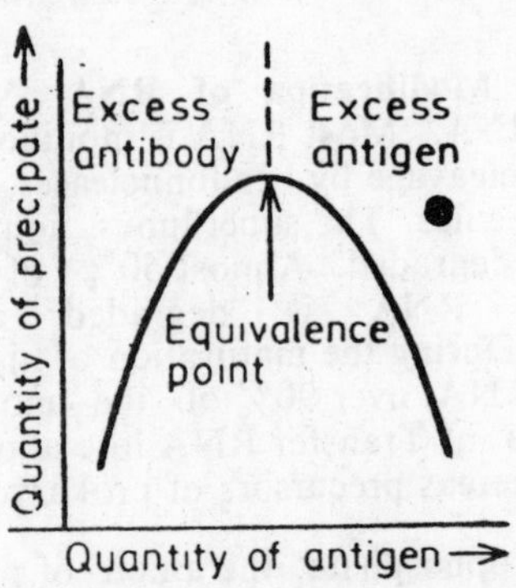

Precipitation, Curve. An increasing quantity of antigen is added to a constant amount of antibody. The resulting precipitate is removed by centrifugation, and antibody and antigen are determined in the supernatant.

Precursor. A starting compound in the biosynthesis of a Metabolite.

Prednisolone: 1, 2-dehydrocortisol, 11β, 17α, 21-trihydroxy-pregna-1, 4-diene-3, 20-dione. A synthetic steroid prepared by chemical or microbiological dehydrogenation of cortisol. It has powerful antiinflammatory and antiallergic activity, very similar to that of prednisone (1, 2-dehydrocortisone, the corresponding dehydro derivative of cortisone): both compounds are used in the treatment of disorders such as asthma, arthritis and eczema.

Prednisolone

(DOBC—17)

5β-Pregnane-3α, 20α-diol, Pregnanediol. A biological degradation product of progesterone. It occurs as the glucuronide, particularly in pregnancy urine.

5β-Pregnane-3α, 20α-diol-glucuronide

Pregnant Mare Serum Gonadotropin, abb. **PMS.** A hormone produced by the uterine endometrium. It is poorlyexcreted by the kidneys, and it accumulates in the blood. Its action is similar to that of Follicle stimulating hormone.

Pregnenolone, 3β-hydroxy-pregn-5-en-20-one. A steroid ketoalcohol formally derived from the parent hydrocarbon pregnane (Steroids). In animals and plants, it is produced from cholesterol via 20, 22-dihydroxy-cholesterol, and it is the biosynthetic precursor of progesterone.

Pre-proprotein. A precursor in the biosynthesis of a secretory protein which possesses an N-terminal signal peptide sequence in addition to the proprotein sequence, *e.g.* pre-proinsulin, pre-protrypsin, pre-proparathormone, pre-promellitin, etc.

Preprotein. The biosynthetic precursor of a secretory protein containing a very short-lived signal peptide sequence (Signal hypothesis), *e.g.* preovomucoid and prelysozyme.

PRH. Abb. for Prolactin Releasing Hormone (Releasing hormones).

Pribnow Box. A sequence of seven nucleotide pairs, which is the same or very similar in all promoters. It is located five to seven nucleotide pairs from the initiation point of RNA transcription.

The structure of the pribnow box is : 5′-TATPuATG-3′
3′-ATAPyTAC-5′

where Pu is a purine and Py a pyrimidine.

Primary Metabolism. Those metabolic processes that are basically similar in all living cells, and are necessary for maintenance and survival. It includes the fundamental processes of growth (synthesis of biopolymers and their constituents, synthesis of macromolecular superstructures of cells and organelles), energy production and transformation, and the Turnover of body and cell constituents.

Primer

1. A small polymer that is required as a starter for the synthesis of a larger biopolymer. In nucleic acid synthesis for example, an oligonucleotide serves as a primer, and it is extended by the enzyme-catalysed addition of further nucleotide units from nucleoside triphosphates.

2. A Pheromone that causes a long term physiological change.

Primobolane 1-methyl-17β-acetoxy-5α-androst-1-en-3-one. A synthetic Anabolic steroid. It is a less potent (one tenth) androgen, but a more potent (five times) anabolic agent than testosterone propionate and it is used therapeutically.

Pro. Abb. for L-proline.

Proazulenes, Azulenogens, Hydroazulenes. A group of natural cyclic sesquiterpenes, which can be thermally dehydrogenated or dehydrated to Azulenes. They are chiefly compounds of the guaiane type, *e.g.* guaiol.

Process Control. A term used in industrial biochemistry, particularly with reference to control of production of microbial fermentation products.

Processing. Modification of protein molecules after translation or modification of RNA after transcription.

Prochirality. A molecule (or atom) is prochiral if it becomes chiral (asymmetrical or dissymmetrical) by the replacement of one point ligand by a new point ligand. A prochiral carbon atom possesses two identical (a, a) and two different (b, c) substituents.

Prodigiosin. A red pigment and secondary metabolite of the bacterium, Serratia marcescens. In its biosynthesis L-proline enters intact, and contributes a greater number of carbon atoms than any other amino acid.

Production Strains, Industrial Microorganisms: Microorganisms used for the synthesis of industrial products, or for certain conversion stages in industrial syntheses.

Production strains showing very high conversion rates are called high performance strains. P.s. can be patented.

Progesterone, Corpus Luteum Hormone, Luteo-hormone: The pregn ·4-ene-3,20-dione; Δ^4-pregnane-3,20-dione. It is structurally related to the parent hydrocarbon, pregnane. It is the natural progestin, and an antagonist of the Estrogens. It promotes proliferation of the uterine mucosa. It promotes implantation and further development of the fertilized ovum in the uterine mucosa (secretion phase). During pregnancy, it prevents further maturation of follicles, and stimulates development of the lactatory function of the mammary gland.

Progesterone

Prolactin, Lactotropin, Lactogeuic Hormone, Mommotropin, Luteo-mammotropic Hormone, Luteotropic Hormone, abb. **L.T.H. Luteotropin:** A gonado·tropin. Phylogenetically, It is one of the oldest adenohypophysial hoamones. It acts primarily on the mammary gland by promoting lactation in the postpartal phase, and in rodents it also acts on the ovary. Its activity in males is not clear.

Prolamines. A group of simple (unconjugated) proteins. They occur in cereals, and contain up to 15% proline and 30—45% glutamic acid, but they have only low contents of essential amino acids. The chief representatives are gliadin (wheat and rye), zein (maize; contains no tryptophan or lysine) and hordein (barley; contains no lysine). Olats and rice do not contain P.

l-Proline, abb. **Pro**. pyrrolidine-2-carboxylic acid, a proteogenic amino acid. It is biosynthesized chiefly from L-glutamate via glutamic-γ-semialdehyde. Some may also be formed from exogenous ornithine via pyrroline carboxylic acid.

Pronase. A mixture of at least 4 proteolytic enzymes from Streptomyces griseus. Two peptide esterase components resembling chymotrypsin and trypsin have been separated and further characterized. Both are inhibited by chicken ovoinhibitor.

Propionic Acid: $CH_3{\cdot}CH_2{\cdot}COOH$. A simple fatty acid. It occurs as its salts (propionates) and esters in many plants. It is especially important in the metabolism of propionic bacteria.

Propionyl-coenzyme A, Propionyl-CoA. Activated propionic acid, formed by attachment of coenzyme A by a thioester linkage. Propionyl-CoA is important in fatty acid biosynthesis and in fatty acid degradation.

Proplastids. Rounded (0.2—1 μm diam), colorless and largely structureless organelles in the meristematic tissues of higher plants, or in unicellular algae cultured in the dark. These are the biogenetic precursors of Plastids.

Proproteins. Inactive protein precursors, which are activated by the removal (a highly specific reaction) of a peptide sequence.

Prostaglandins. A group of animal hormones, biosynthesized from C20 unsaturated fatty acids.

It exhibit a wide variety of pharmacological properties. Of particular importance are: bronchospasmolytic activity (treatment of acute asthma); control of gastric secretion (possible ulcer therapy); antagonist of the hypotensive and diuretic action of angiotensin (treatment of essential hypertension and cardiovascular disorders); initiation of ovulation, *e.g.* in cows, pigs and sheep (used to synchronize mating in large animal herds); relief of pain during parturition; and finally, very small amounts of PG cause abortion.

Prosthetic Group. Nonproteogenic. In an enzyme, this group is a catalytically active group attached to the enzyme protein (aboenzyme).

Protamines. A group of strongly basic, simple proteins associated with DNA in the cell nucleus. They replace the somatic Histones in sperm, at least during spermogenesis.

Proteases. All enzymes that catalyse the exergonic hydrolysis of peptides. They are divided into two groups, depending on their site of attack on the polypeptide chain ;

$$
\begin{array}{c}
\begin{array}{ccccccc}
 & R_1 & & & & R_2 & \\
 & | & & & & | & \\
-N- & C- & C- & - & N- & C- & C- \\
| & | & \| & & | & | & \| \\
H & H & O & & H & H & O
\end{array} \\
\downarrow +H_2O \\
\begin{array}{ccccc}
 & R_1 & & R_2 & \\
 & | & & | & \\
-N- & C- & C-OH+H_2N- & C- & C- \\
| & | & \| & | & \| \\
H & H & O & H & O
\end{array}
\end{array}
$$

1. Endopeptidases (proteinases) catalyse the hydrolysis of bonds within the peptide chain, forming variously sized cleavage peptides.
2. Exopeptidases catalyse the hydrolytic removal of only terminal amino acids from the polypeptide chain.

Proteid. An obsolete name for a conjugated protein.

Proteinases. A group of Proteases.

Protein Biosynthesis. A cyclic, energy-requiring, multistage process, in which free amino acids are polymerized in a genetically determined sequence to form polypeptides. It represents the translation of the genetic information carried by mRNA.

Protein Deficiency Diseases. The result of long-term nutritional deficiency in the form of too little complete protein. In small children it is called kwashiorkor. The symptoms are swelling of the legs face and liver (starvation edema), liver damage, gastrointestinal malfunction, anemia and osteoporosis. It is often fatal.

Proteins. Naturally occurring polymers of high M_r, consisting predominantly of amino acids linked by peptide bonds.

Proteinoids. Heteropolyamino acids; artificially prepared polypeptides ($M_r > 1000$) formed in 20–40% yield by heating a mixture of several amino acids for 16 h at 170°C (thermal) condensation.

Proteohormones, Protein Hormones. The proteins (often glycoproteins) with hormonal function. Like other proteins, they are synthesized by the translation of appropriate mRNA, and degraded by proteolysis.

Proteolysis, Protein Degradation. Hydrolysis of proteins by the action of proteolytic enzymes, or nonenzymatically by acids (*e.g.* 6 M HCl at 110°C for 24 h or longer) or alkalis. Ultimate products of P. are amino acids. Dietary proteins are hydrolysed to L-amino acids by firoteolysis in the intestine. After absorption, these amino acids are used in the synthesis of new proteins specific to the organism.

Prothrombin Factor II. An enzymatically inactive, calcium-binding, single chain α_2-glycoprotein of blood plasma. M_r 72000; carbohydrate content 14.7% (bovine), or 11.8% (human). Synthesis of prothromycin occurs in vertebrate liver and requires vitamin K.

Protoheme, Heme, Ferroheme, Ferroprotoporphyrin, Protoheme IX. 17, 12-diethenyl-3, 8, 13, 17-tetramethyl-21 H, 23 H-porphine-2, 18-dipropanoate (2-)-N^{21}, N^{22}, N^{23}, N^{24}]-iron; or 1, 3, 5, 8-tetramethyl-2, 4-divinylporphine-6, 7-dipropionic acid ferrous complex. Protoheme crystallizes as fine brown needles with a violet sheen.

It is the prosthetic group of a number of hemoproteins, *e.g.* hemoglobins, erythrocruorins, myoglobins, some peroxidases, catalase and cytochromes b. The four coordinate bonds of the iron lie in the plane of the nearly planar porphyrin ring structure, while the two unoccupied sites of the iron are perpendicular to it.

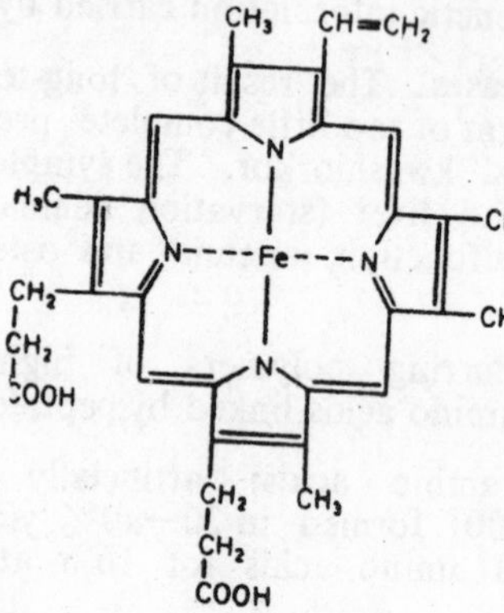

Protoheme

Protopectins (s). A ground substance in plant cell walls. P. consists of insoluble Pectins and are probably not pure homoglycans. They are present in the cell wall as salts of calcium and magnesium. The constituent polygalacturonic acid chains of protopections are linked to one another by salt linkages, phosphate bonds and esterification with arabinose.

Prototrophism. The property of being able to grow at the expense of usual or common nutrients, with no special requirement for Growth factors. It is shown by prototrophic organisms.

Provitamins. Inactive precursors of Vitamins. These are mostly of vegetable origin, and are converted into active vitamins after absorption from the diet.

PRPP. Abb. for 5-phosphoribosyl 1-pyrophosphate.

Pseudoalkaloids. A group of alkaloids earlier assigned to other groups (*e.g.* some were grouped with the terpenes) with which they show a close structural relationship. At the time, their nitrogen content seemed incidental.

Pseudo-isoenzymes. Multiple forms of an enzyme, which catalyse the same reaction. They have similar properties to isoenzymes, but do not have genetically determined differences of primary structure. Their multiplicity is the result of enzymatic or non-enzymatic modification of one original primary sequence, either in vivo or in vitro (*i.e.* during isolation).

Pseudopelletierine, ψ-pelletierine, Pseudopunicine, 9-methyl-3-granatanone. 9-methyl-9-azabicyclo-3, 3, 1-nonan-3-one. The most important representative of the Punica alkaloids, present in the root bark of Punica granatum. Its structure is based on the meso form of granatane (9-azabicyclo-3, 3, 1-nonane).

Pseudouridine, 5-β-D-ribofuranosyluracil, 5-ribosyluracil, ψ, ψ rd. A structural analog of uridine containing a C—C bond between C-5 of uracil and C—1 of ribose. ψ is a Rare nucleic acid component found in tRNA.

Psi (ψ) factor. A protein responsible for the specific initiation of the RNA polymerase reaction at the promotor sites of the genes for rRNA in bacteria.

Psilocybin. 4-phosphoryloxy-N, N-dimethyltryptamine. This and the related compound psilocin (4-hydrovy-N, N-dimethyltryptamine, are jointly responsible for the psychotropic action of the fruiting body of the Mexican hallucinogenic fungus Teonanacatl (Psilocybe mexicana).

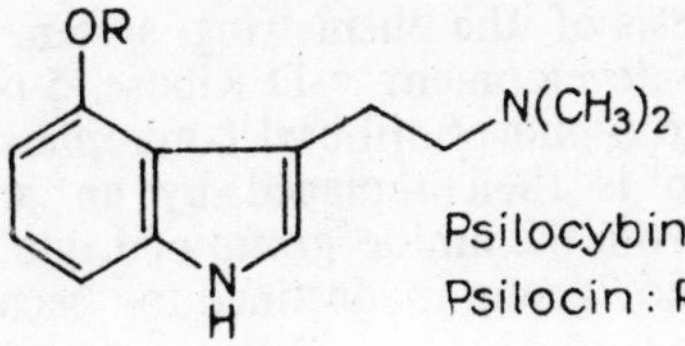

OR

$N(CH_3)_2$

N
H

Psilocybin : R = H_2PO_3
Psilocin : R = H

Psychotropic Agents. The chemical compounds that influence the human psyche. They are used in psychiatry. They include Narcotics and Hallucinogens and are almost exclusively of vegetable origin.

Pteridines. A group of compounds containing the pteridine ring system (pyrimidino- [4, 5b]-pyrazine, or pyrazine-[2, 3d]-pyrimidine). Naturally occurring pteridines are chemically related to pterine (2-amino-4-oxodihydropteridine). 2, 4-Dioxotetrahydropteridine is known as lumazine.

Punica Alkaloids. A group of piperidine alkaloids, originally isolated from the bark of the pomegranate tree (Punica granatum L., official drug Cortex granati), and subsequently isolated from other plant families. Decoctions of the drug, or the isolated alkaloids, have some use as vermifuges.

Purine. A heterocyclic compound with a condensed pyrimidine-imidazole ring system. Purin ring system is found in combination with ribose in the nucleoside antibiotic, Nebularine.

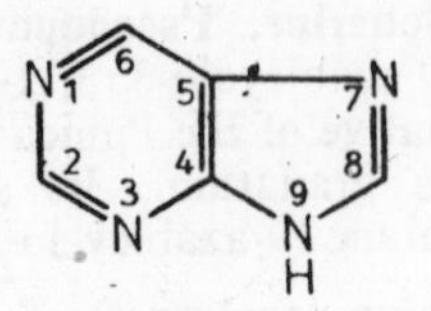

Purine

Various substituted and oxidized purine derivatives occur naturally, and are of considerable biological importance. The purine derivatives, Adenine and Guanine are present in DNA and RNA, and they are commonly referred to as purine bases.

Purine Antibiotics. Purine derivatives with antibiotic activity. They occur as nucleosides polypeptides or free bases (8-Azaguanidine).

Purine Biosynthesis, De Novo Purine Biosynthesis. A common pathway for the biosynthesis of the purine ring system found at all levels of evolutionary development α-D-Ribose 5-phosphate is pyrophosphorylated to 5-phosphoribosyl 1-pyrophosphate. The pyrophosphate group is then replaced by an amino group, which is transferred from the amide group of L-glutamine. The nitrogen of this amino group is destined to become N-9 of purine ring system.

Purine Degradation, Purine Catabolism. A series of reactions in which purines are degraded by cleavage of the purine ring. It is usually aerobic, but anaerobic purine degradation occurs in certain microorganisms.

Puromycin. A nucleoside antibiotic from Streptomyces alboniger. It inhibits protein biosynthesis on 70S and 80S ribosomes. It is a structural analog of the 3′-terminal end of aminoacyl-tRNA (Fig.).

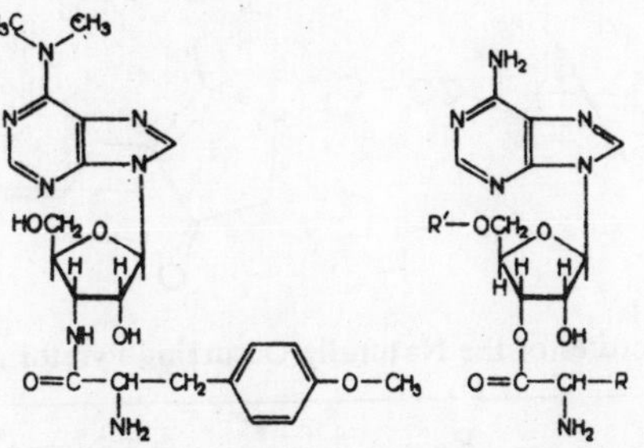

Comparison between the Structure of Puromycin (left) and the 3'-terminal end of an Aminoacyl-tRNA (Right)

Purpurin. 1, 2, 4-trihydroxyanthraquinone. A red anthraquinone pigment, m.p. 263 °C. Its glycoside occurs in madder root (Rubia tinctorum) (accompanied by alizarin), and in other members of the Rubiaceae. It is formed from its glycoside during storage, and there is no appreciable quantity of this in the fresh root. It is used as a reagent for the detection of boron, for the histological detection of insoluble calcium salts, and as a nuclear stain. It forms colored lakes with various metal salts, and is used as a fast cye in cotton printing.

Putrescine. Tetramethylenediamine, $H_2N\text{-}(CH_2)_4\text{-}NC_2$. A biogenic amine present in ribosomes and bacteria. It is formed from arginine during the bacterial degradation of proteins. Increased protein decomposition (*e.g.* in cholera) leads to the appearance of putrecine in urine and feces.

Pyr. Abb. for pyroglutamic acid.

Pyrethrins. Diterpene insecticides present in the flowers of Chrysanthemum cinerariaefolium (syn. Pyrethrum cineariaefolium). The dried flowers, known as "pyrethrum", also have insectidal activity and serve as starting material for the preparation of pyrethrins.

Pyridine Alkaloids. A group of alkaloids containing the pyridine ring system, which occur in various unrelated plants, and as metabolic products of microorganisms. Important examples are Nicotiana alkaloids, Areca alkaloids, Gentiana alkaloids, and Valeriana alkaloids. They are biosynthesized either from nicotinic acid, or as products of terpene synthesis.

Pyridine Nucleotide Coenzymes. Nicotinamide·adenine·dinucleotide, and Nicotinamide-adenine-dinucleotide phosphate.

CO–O R2 R1 O

Structures of the Naturally Occurring Pyrethrins

R_1	R_2	Name
$-CH_3$	$-CH_3$	Cinerin I
$-CH_3$	$-C_2H_5$	Jasmolin I
$-CH_3$	$-CH=CH_2$	Pyrethrin I
$-CO_2CH_3$	$-CH_3$	Cinerin II
$-CO_2CH_3$	$-C_2H_5$	Jasmolin II
$-CO_2CH_3$	$-CH=CH_2$	Pyrethrin II

Pyridine Nucleotide Cycle. A cycle of reactions in which nicotinamide, produced in the degradation of NAD^+, is reutilized for the synthesis of NAD^+. The pathway is therefore a salvage loop, and has a sparing effect on the dietary and/or biosynthetic requirement for nicotinamide.

Pyridoxal Phosphate PalP. The coenzyme form of Vitamin B_6. It is stable in aqueous solution when kept refrigerated and protected from light. It is particularly sensitive to photodecomposition in the solid state and in alkaline solution. It plays an important central role in amino acid metabolism, acting as the coenzyme in many different metabolic conversions of amino acids. It is formed from pyridoxal by a kinase reaction :

$$\text{pyridoxal} + \text{ATP} \xrightarrow{Mg^{2+}} \text{pyridoxal 5-phosphate} + \text{ADP}$$

With amines and amino acids, Palp forms Schiffs bases (azomethines) The substrate of a pyridoxal phosphate enzyme is the

Schiff's base of the amino acid with PalP; the action specificity, *i.e.*, transamination, decarboxylation, racemization etc. is determined by the apoenzyme.

Pyridoxal Phosphate Enzymes. Enzymes that contain the coenzyme Pyridoxal phosphate.

Pyrimidine 1-3-diazine, A hecterocyclic compound, consisting of a six-membered ring with 2 nitrogen atoms. Pyrimidine ring system is present in many natural compounds, *e.g.* antibiotics (nucleoside antibiotics), pterins purines and vitamins; it is especially important in the pyrimidine bases, Cytosine, Uracil and Thymine which are constituents of nucleic acids Pyrimidine itself does not occur naturally. Pyrimidine analogs can also become incorporated into nucleic acids.

Pyrimidine Analogs, Antipyrimidines. Pyrimidines and pyrimidine nucleosides structurally related to, but different from the natural compounds. They therefore act as antimetabolites and and selectively inhibit certain biochemical pathways, especially nucleic acid synthesis. Most pyrimidine analogs are modified bases or their nucleosides, but there are also pyrimidine hucleoside analoges with modified sugar components. The m os common chemical modifications are the introduction of su bstituents (*i.e.*, halogens on C5 of uracil and cytosine), replace ment of an OH with an SH group (*e.g.*, 2-thiourabil) and re-replacement of a ring carbon with nitrogen (5-azauracil). Arabinonucleosides (steric inversion of the OH at C2 of the ribose) and Xylosylnucleosides) (inversion at C3) of natural pyrimidine bases are also active pyrimidine analogs.

Pyrimidine Antibiotics. Structurally modified Pyrimidine derivatives with antibiotic activity. They occur as nucleosides polypeptides (*e.g.* albomycin and grisein), or free bases (*e.g.* bacimethrin). The P. a. Toxoflavin and Fervenulin, are biosynthesized from purines.

Pyrimidine Biosynthesis, de Novo Pyrimidine Biosynthesis. Total synthesis of the pyrimidine ring of uracil, thymine, cytidine and their derivatives from cells. (The pyrimidine ring of thiamine [vitamin B_1] has a different biosynthetic origin;

Pyrimidine Degradation, Pyrimidine Catabolism Reductive or (in special cases) oxidative reactions leading to the cleavage of the heterocyclic ring of natural pyrimidines.

1. *Reductive P.d.* To a certain extent, this process represents a reversal of Pyrimidine biosynthesis. The pyrimidine ring is partially hydrogenated, and the resulting dihydrocompound is

cleaved hydrolytically. Cytosine is converted to uracil by deamination, and uracil is degraded to β-alanine. Thymine is degraded to β-aminoisobutyratd. These endproducts are transaminated and metabolized to common metabolic intermediates.

2. *Oxidative P. d.* In Corynebacterium and Mycobacterium uracil is oxidized to barbituric acid, which is cleaved hydrolytically to urea and malonic acid. Thymine is oxidized to 5-methylbarbituric acid, followed by hydrolysis to urea and methylmalonic acid.

Pyroglutamic Acid, Pyrrolidine Carboxylic Acid, Abb. **Pyr or >Glu, 5-oxoproline.** An internal cyclic lactam of Glutamic acid representing a condensation of the α-amino with the γ-carboxyl group. N-Terminal residues of Pyr are found in certain peptide hormones, *e.g.*, thyrotopin releasing hormone.

Pyrroles Pyrrole Derivatives. The compound containing the pyrrole ring. They are subdivided into mono, di, tri and tetrapyrroles. The tetrapyrroles may be noncyclic or cyclic. Bile pigments and the chromophores of Biliproteins are linear, while Porphyrins and Corrinoids are cyclic tetrapyrroles.

Pyrrolidine Alkaloides. A group of Alkaloids with simple structures. They are either derivatives of proline (*e.g.*, stachydrin and its diastereoisomer, betonicine), are they are drived from a N-methyl-2-alkylpyrrolidine (*e.g.* hygrin and cuskhydrin). The latter occur together with the tropane alkaloids, with which they share the same biogenetic precursors, ornithine and acetate.

Pyrrolizidine Alkaloids, Senecio Alkaloids. A group of ester alkaloids, in which amino alcohols (necines) are esterified with necic acids The necines are derivatives of the pyrrolizidine ring system (also knowh as 1-azabicyclo [0,3,3] octane) (Alkaloids, Table)

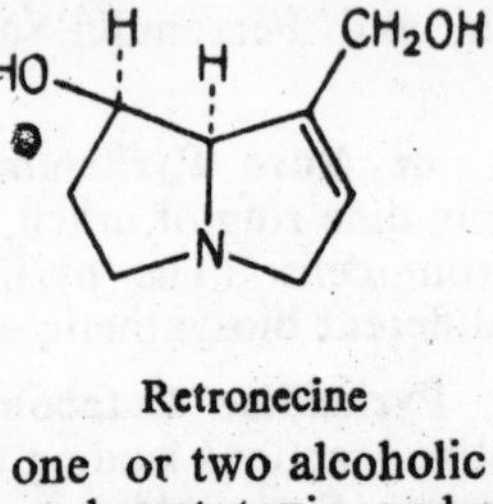

Retronecine

and they possess one or two alcoholic hydroxyls, *e.g.*, retronecine (Fig.). They are hepatotoxic, and can cause liver cirrhosis in grazing animals.

Pyrrolo Quinoline Quinone, PPQ. 2, 7, 9,-tricarboxy-1 H-pyrrolo [2, 3-f] quinoline-4, 5-dione. The cofactor of the enzyme

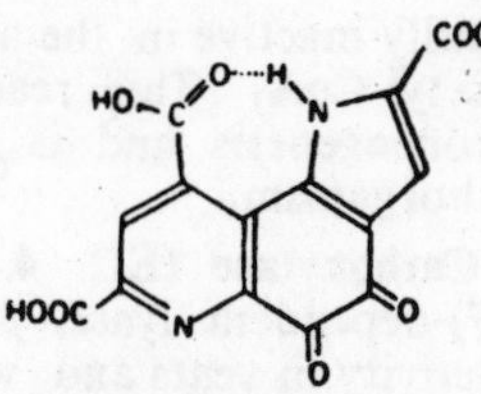

Pyrrolo quinoline quinone

methanol dehydrogenase (EC 1.1.99.8) from Hyphomicrobium X and Methylophilus methylotrophus, and of glucose dehydrogenase (EC 1.1.99.—) from Acinebacter calcoaceticus.

Pyruvate. The anion of pyruvic acid. It is an important metabolic intermediate in aerobic and anaerobic metabolism.

Pyruvate is synthesized from phosphoenolpyruvate in Glycolysis. Phosphoenolpyruvate is an enol ester and an energy rich compound with a free energy of hydrolysis of 50.24 kJ (12 kcal) per mol. During catalysis by pyruvate kinase, this free energy is exloited for the transfer of the phosphate group to ADP, resulting in the synthesis of ATp and pyruvate.

Pyruvate is also produced in the metabolism of certain amino acids, in particular transamination of alanine, dehydration of serine, and desulfhydration of cysteine.

Pyruvate is reduced to lactate in anaerobic glycolysis.

Pyruvate is converted to ethanol in anaerobic Alcoholic fermentation.

By the action of the pyruvate dehydrogenase complex. under aerobic conditions, pyruvate is oxidatively decarboxylated to acetyl coenzyme. A. The latter is an important metabolite in various other biosynthetic and biodegradative processes. Equation for oxidative decarboxylation of P.

$$\underset{\text{(Pyruvate)}}{CH_3COCOO^-} + \underset{\text{(Coenzyme A)}}{HSCoA} + NAD^+ \rightarrow$$

$$\underset{\text{(Acetyl coenzyme A)}}{CH_3CO\text{-}SCoA} + CO_2 + NADH + H^+$$

Pyruvate Carboxylase (EC 6.4.1.1). A biotin-dependent ligase, in animals and plants, which catalyses the addition of CO_2 to pyruvate :

$$\text{Pyruvate} + CO_2 + ATP + H_2O \xrightleftharpoons{Mn^{2+}} \text{Oxaloacetate} + ADP + P_i$$

The enzyme is practically inactive in the absence of its positive allocsteric effector, acetyl-CoA. This reaction is an important early stage of Gluconeogenesis and is an example of CO_2 fixation in the animal organism.

Pyruvate Decarboxylase, Carboxylase (EC 4.1.1.1). A thiamine pyrophosphate (TPP)-dependent lyase, absent from animals, and present in high activity in yeats and wheat seedlings. It is a specific enzyme of alcoholic fermentation, which catalyses the cleavage of pyruvate (via actiue acetaldehyde) into acetaldehyde and CO_2. The cofactors are Thiamine pyrophosphate and magnesium ions.

Pyruvate Kinase, Phosphopyruvate Kinase (EC 2.7.1.40). A widely distributed, metal ion-dependent phosphotransferase, present in yeast, muscle, liver, erythrocytes and other organs and cells It catalyses the last reaction of glycolysis :

$$\text{Phosphoenolpyruvate (PEP)} + \text{ADP} \rightleftharpoons \text{Pyruvate} + \text{ATP}$$

(Substrate level phosphorytation).

Each subunit of P.k forms an intermediate, cyclic, ternary metal bridge complex

P k.—Mn—ADP
P EP

in which the PEP and ADP are bound to the enzyme via a manganese (II) ion. Tetrameric pyruvate kinase from muscle and erythrocytes shows Michaelis-Menten type kinetics (plot of of fnitial velocity against substrate concentration is a reactangular hyperbola), whereas the yeast enzyme, is an allosteric enzyme, showing sigmoid kinetics.

Pyruvic Acid. $CH_3-CO-COOH$, the slmplest and most important α-ketoacid (2-oxoacid).

Pythocholic Acid. 3α, 7α, 16α-trihydroxy -5β-cholan-24-oic acid. A bile acid possessing three hydroxyl groups. It is a characteristic component of the bile of many snakes, and has been isolated from the bile of the tiger snake, python and boaconstrictor, among others.

Q

Q. Abbreviation for coenzyme Q.

Quantasome. The smallest structural unit of photosynthesis; small elementary units of the thylakoid, measuring 18×15×10 nm, containing 230 chlorophyll molecules, cytochromes, copper and iron. They may also be observed as granular units in the chloroplast lamella. The functional status of quantasomes is not clearly defined; they may be involved in both electron transport and photosynthetic ATP synthesis and therefore analogous to the electron transport particles of the respiratory chain.

Quantum Requirement. The number of light quanta required for the formation of one molecule of O_2 in Photosynthesis. Two quanta are required per electron. The theoretical value of quantum requirement is eight, sincc the production of one molecule of O_2 proceeds according to the following equation, with transfer of four electrons from water to $NADP^+$.

$$2\,H_2O \rightarrow O_2+4H^++4e^-$$

It is influenced by the physiological state of the experimental system.

Queen Substance. Originally a term for the entire mandibular gland secretion of the queen bee, which contains about 30 substances. It is now the trivial name for 9-oxo-trans-2-decenoic acid. This compound, together with 9-hydroxy-trans-2-decenoic acid, is very important as a pheromone for the maintenance of the division of labor within the hive. In the course of caring for the young, the worker bees lick the pheromone mixture off the

R=O Queen substance
R=H, OH 9-Hydroxy-trons-2-decenoic acid

queen. This causes their ovaries to shrink, and they are inhibited from building queen cells. Larvae in queen cells are not fed honey, but royal jelly, a mixture of pollen and secretions.

Royal jelly does not contain Q. It is recommended as a health product, but its effectiveness is disputed.

Quinazoline Alkaloids. A group of about 30 alkaloids which, occur in higher plants (in families which are taxonomically very distant from one another), animals and bacteria. They are derived biosynthetically from anthranilic acid.

Quinine. The most important of the cinchona alkaloids. In quinine a quinoline ring system is connected via a secondary hydroxyl on C4 to a quinuclidine structure. It occurs in nature in association with its stereoinsomers quinidine, the C-9 empimer and epiquinine and epiquinidine, the C 8′ epimers. It is used therapeutically as a drug against malaria and bacterial influenza. By reducing the rate of tissue respiration, it has an antipyretic effect.

Quinoline Alkaloids. A group of alkaloids based on the quinoline skeleton. The are found both in microorganisms and in higher plants.

Quinolizidine Alkaloids. A group of alkaloids based on the quinolizidine (norlupinane) skeleton. The most important of these are the lupine alkaloids, which are synthesized from lysine via cadaverine.

Quinones. Aromatic dioxo compound derived from benzene or multiple-ring hydrocarbons such as naphthalene, anthracene, etc. They are classified as Benzoquinones, Naphthoquinones Anthraquinones etc. on the basis of the ring system. The C=O groups are generally ortho or para, and form a conjugated system with at least two C=C double bonds: hence the compounds are colored, yellow, orange or red. This type of chromophore is found in many natural and synthetic pigments.

Quinones are biosynthesized from acetate/malonate via shikimic acid. A few quinones are used as laxatives and worming agents, and others are used as pigments in cosmetics histology and aquarell paints.

R

Raffinose, Melitose. A nonreducing trisaccharide. M.p. 120°C, $[\alpha]_D^{20}$ + 123°C (water). It contains units of D-galactose, D-glucose and D-fructose. The galactose and glucose are linked by an α-1, 6-glucosidic bond, and the fructose is linked to the glucose by an α, β-1, 2-glycosidic bond. It is easily fermented by yeasts.

Ramachandran Plots, Conformational Maps. The plots of rotation about the αC-carbonyl-C bond (ψ) in a peptide linkage against rotation about the αC-amino-N bond (Φ). A general Ramachandran plot is constructed with the aid of models and computers.

Raphanatin. 7-glucosylzeatin. It is formed from the cytokinin, zeatin, and it has no cytokinin activity. It is a storage form of zeatin, present in ranish seedlings. Glucosylation at position 7 of the purine ring probably serves to protect zeatin from enzymatic degradation.

Rare Nucleic Acid Components, Unusual Nucleic Acid Components, Minor Nucleic Acid Components. Nucleic acid components of relatively infrequent occurrence, formed by the enzyme-catalysed modification of either the base of sugar of the usual nucleic acid constituents, *i.e.* modification of adenine, guanine, cytosine, uracil, thymine or ribose.

Rate Equation. In enzyme kinetics, an equation expressing the rate of a reaction in terms of rate and the concentrations of enzyme species, substrate the product. When it is assumed that steady state conditions obtain, the Michaelis-Menten equation is a suitable approximation.

Rauwolfia Alkaloids. A group of about 50 structurally related indole alkaloids from the roots and rhizomes of various species of Rauwolfia, Aspidosperma and Corynanthe. All Rauwolfia alkaloids contain a β-carbolene skeleton; they are classified into 3 types :

1. yohimbine (corynanthine),
2. ajmaline,

3. serpentine. The large number of R.a. is due to the existence of stereoisomers. Thus Rauwolfia contains seven stereoisomers of yohimbine.

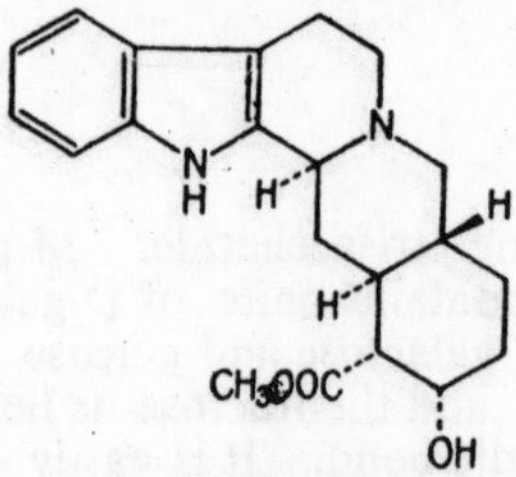

Yohimbine 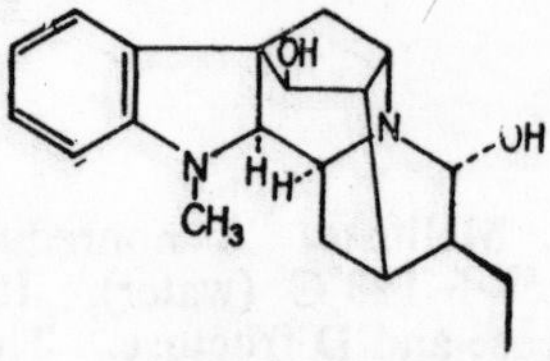 Ajmaline

Receptor Proteins. Mostly membrane-bound, but also soluble proteins with high specific affinity for hormones, antibodies, enzymes and other biologically active compounds. Binding of proteohormones to membrane-associated R.p. represents the first stage in the expression of hormonal activity; this is followed by activation of the membrane-bound adenylate cyclase, which catalyses formation of the second messenger, cyclic AMP.

Redoxin. An electron-transferring protein. Redoxins that contain iron, bound as a functional group to S, N or O ligands of the protein, are known as Ferredoxins.

Reduction. The addition of electrons. It is the converse of Oxidation. In biochemical systems it may involve electron transfer only (*e.g.* R. of cytochromes and ferredoxin), but the majority of biochemical reduction involve the addition of hydrogen (hydrogenation). The pyridine nucleotide coenzymes, NAD^+ and $NADP^+$, play an important part in reduction.

Redundancy

1. The occurrence (frequent in eukaryotes, only in isolated instances in prokaryotes) of linearly arranged, largely identical, repeated sequences of DNA.

2. Terminal redundancy. The existence of identical genetic information at each end of a viral chromosome.

Regulation. A group of structural genes, whose gene products (enzymes) are involved in the same reaction pathway, and which are regulated together. The individual genes are in different regions of the chromosome, *i.e.* they do not lie on neighboring sequences of DNA, as in an operon.

Relaxin. A female sex hormone, formed in mammals during pregnancy. It is a heterodetic, cyclic polypeptide, the A-chain contains 22, the B-chain 26 amino acid residues.

Release Factors, Termination Factors. Catalytically active proteins necessary for the termination step of RNA synthesis and protein biosynthesis.

Releasing Hormones, Releasing Factors, Liberins, Statins. A group of peptide neurohormones, synthesized in various distinct nuclei of the hypothalamus. They are released into the capillaries of the portal vessels in the median eminence of the hypothalamus, then carried to the anterior pituitary (adenohypophysis) where they regulate the production and secretion of tropic hormones (*e.g.* thyreotropin, somatotropin, gonadotropins).

Renaturation. Conversion of a denatured protein or nucleic acid into its native configuration. Nucleic acids; proteins.

Renin (EC 3.4 99.19). An endopeptidase, formed in the juxtaglomerula cells (cells next to the glomerula) of the kidney. In the plasma, it releases antiotensin I (a decapeptide) from the N-terminal sequence of angiotensinogen (an α_2-plasma globulin).

Rennet Enzyme, Renin, Chymosin. A pepsin-like proteinase (Proteases) which is formed as inactive prorennin in rennet bags, and probably in the stomach of all nursing mammals. It is converted to active rennin by pepsin or autocatalytically.

Repair Enzymes. Enzymes that catalyse stages in the repair of DNA. They include exo- and endonucleases, DNA polymerases and ligases. Deoxyribonucleic acid (Repair of DNA).

Repeatability. The repeatability refers to measurements in one laboratory over a short time period, whereas reproducibility refers to measurements over longer periods of time and/or in different laboratories.

Replication. DNA replication.

Replication Site. The site of DNA replication in vivo. In bacteria, the replication site is on the cell membrane. The circular DNA is bound to the cell membrane at its initiation region, together with DNA-polymerase and initiation proteins. Deoxyribonucleic acid.

Replicon. A term proposed by Jacob and Brenner for the replication unit. In prokaryotes or viruses, it is the complete circular or linear DNA (or RNA in RNA viruses), representing the total chromosome of the organism.

Repressor. An allosteric protein, which regulates the transcription of structural genes, and is encoded by a regulator gene. It binds to the operator region of DNA, thereby preventing synthesis of the mRNA for the structural genes of an operon or regulation (Enzyme repression).

Reproducibility. This is not the same as Repeatability.

Reserpine. Therapeutically the most important Rauwolfia alkaloids. Hydrolysis of reserpine with alcoholic KOH produces reserpinic acid (structurally similar to yohimbine), trihydroxybenzoic acid and methanol. It is found widely in the genus Rauwolfia, and is responsible for the sedative properties of these plants.

Resilin. A structural protein from the exoskeleton of arthropods, especially insects. It has a high glycine content, and no cystine. It is located between the chitin lamellae, and endows the arthropod exoskeleton with a certain elasticity. A notable component of R. is trityrosine.

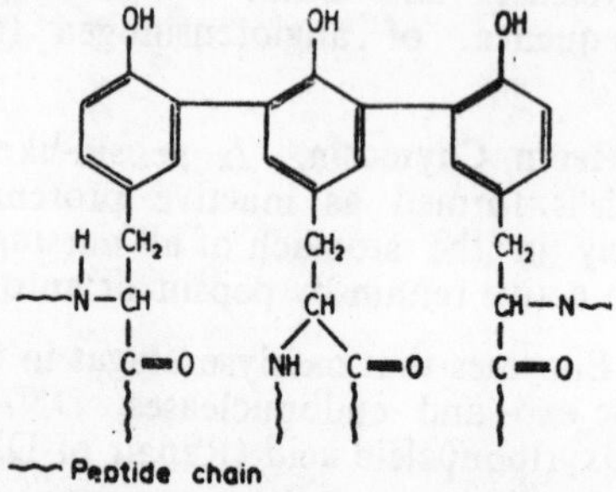

Trityrosine Residue in Resilin

Resin Acids, Resinic Acids. Hydroaromatic diterpenes, the acid components of resins. The most important representatives are abietic acid, neoabietic acid, dextro-pimaric acid and neopimaric acid.

Resinates. Salts and esters of resin acids.

Resins. Largely amorphous, solid or half-solid, transparent, odorless and tasteless organic substances, usually of vegetable origin.

Crude resins and refined resin components are widely used in the production of paints, varnishes, textile conditioners, cosmetics and pharmaceuticals.

Respiration, Oxidative Metabolism. A process by which cells derive energy in the form of ATP from the controlled reaction of hydrogen with oxygen to form water. The hydrogen is derived from the degradation of organic substrates by oxidases of dehydrogenase, which in turn release it to a system of redox catalysts located in the inner mitochondrial membrane, or in the cell membrane of bacteria.

Respiratory Chain, Electron Transport Chain. A series of redox catalysts which transport electrons from respiratory substrates to oxygen. The energy of this electron flow may be used in the synthesis of ATP. The couping of ATP synthesis with electron transport in the respiratory chain is known as Oxidative phosphorylation.

Respiratory Inhibitor. A compound which interferes in some way with the respiratory chain. There are three types :

1. Uncouplers which prevent the synthesis of ATP without stopping the flow of electrons.
2. inhibitors of oxidative phosphorylation in the narrower sense, and
3. inhibitors of electron transport along the respiratory chain.

Respiratory Poison. Respiratory inhibitor.

Restriction Endonucleases (EC 3.1.23.1 to EC 3.1.23.45). Enzymes present in a wide variety of prokaryotic organisms, where they serve to cleave foreign DNA molecules (*e.g.* phage DNA). These recognize specific palindromic sequences (Palindrome) in double stranded DNA. Many endonucleases are known, all with different specificities (Table). These represent a powerful set of tools, that are used for the analysis of chromosome structure.

Reserved Electron Transport. Reversal of Oxidative phosphorylation in which NAD^+ is reduced by an ATP-dependent reverse transport of electrons. It occurs in organisms that oxidize hydrogen donors whose redox potential (Oxidation) is more positive than that of the pyridine nucleotide coenzymes, and it operates in the oxidation of substrates not specific for NAD

$$e.g.\ \text{Succinate} + \text{NAD} + \xrightarrow{\text{ATP}}$$

Fumarate + NADH + H^+. The redox system succinate/fumarate (E_o 0.00 V) is 320 m V more positive than the redox system $NAD^+/NADH + H^+$ (E_o —0.32 V); electrons are passed from succinate to flavoprotein in the respiratory chain, then via NADH-dehydrogenase to NAD^+.

l-Rhamnose. 6-deoxy-L-mannose. A deoxyhexose. It is a component of many glycosides, *e.g.* anthocyanins, and of plant mucilages. It is biosynthesized from glucose.

Rhein. 1, 7-dihydroxy-3-carboxyanthraquinone. A yellow anthraquinone, present free or as a glycoside in the roots of many higher plants, particularly rhubarb. It has a purgative action, and various R-containing drugs (Radix Rhei, Folia Sennae) are used therapeutically.

Rhesus Factor, abb. **Rh-factor**. Several closely related, blood group-specific erythrocyte antigens, present in 85% of europeans (Rh-positive). The antigen is absent from the remaining 15%, who are Rh-negative. The natural antibody does not normally occur in humans, but is formed, in rabbits or guinea pigs after immunization with rhesus monkey erythrocytes. If Rh-negative individuals come into contact with Rh-antigen, *e g.* by blood transfusion, or from an Rh-positive fetus (the antigen crosses the placenta), Rh-antibodies are formed. Repeated transfusion may then lead to hemolysis, or an Rh-positive fetus may suffer hemolytic damage (erythroblastosis of the newborn).

Rh-factor. Rhesus factor.

Rhizobia. Bacteria of the genus Rhizobium, which can live free in the soil, or enter into a symbiotic relationship with leguminous plants. As leguminous symbionts. These are responsible for the formation of root nodules and the fixation of atmospheric nitrogen (Nitrogen fixation). In the nodule of the host plant, these are present as bacteroids.

Rhodoplasts. Photosynthetic organelles of the red algae (Rhodophyta). These are red or red-violet, due to the presence of important light-trapping photosynthetic pigments, called Biliproteins. These are responsible for the characteristic color of these marine algae.

Rhodopsin. Vitamins (Vitamin A).

Rhodoxanthin. 3, 3′-dioxo-β-carotene, a xanthophyll. It is as a pigment in brown-red ("copper") leaves, in the needles of various conifers (*e.g.* yew) and in bird feathers.

Ribitol. An optically inactive, C_5-sugar alcohol, biosynthesized by reduction of ribose. It is a component of the flavin molecule, *e.g.* riboflavin.

Ribonuclease (EC 3.1.27.5). A pancreatic phosphodiesterase specific for RNA. It catalyses hydrolysis of the phosphate ester bond between pyrimidine nucleoside 3-phosphate residues and the 5-hydroxyl group of the neighboring ribose residues. The cleavage products are 3′-ribonucleoside monophosphates and oligonucleotides possessing a terminal pyrimidine nucleoside 3′-phosphate.

Ribonucleic Acid, Abb. RNA, Obsolete Pentose Nucleic Acid. A biopolymer of ribonucleotide units, present in all living cells and some viruses. Structure. The mononucleotides of RNA consist of ribose phosphorylated at C3, and linked by an N-glycosidic bond to one of four bases; adenine, guanine, cytosine or uracil. Many other bases (chiefly methylated bases) also occur, but are less common (Rare nucleic acid components). The mononucleotides form a linear chain via 3′, 5′-phosphodiester bonds (Nucleic acids). The base composition of RNA shows wide variations, but these do not lend themselves to mathematical interpretation and prediction, as for DNA. The complete base sequences of certain small species of RNA (tRNA,5S-RNA) have been determined. RNA does not form a double stranded a α-helix. The single chains show partial folding into α-helical regions, probably by hydrogen bonding

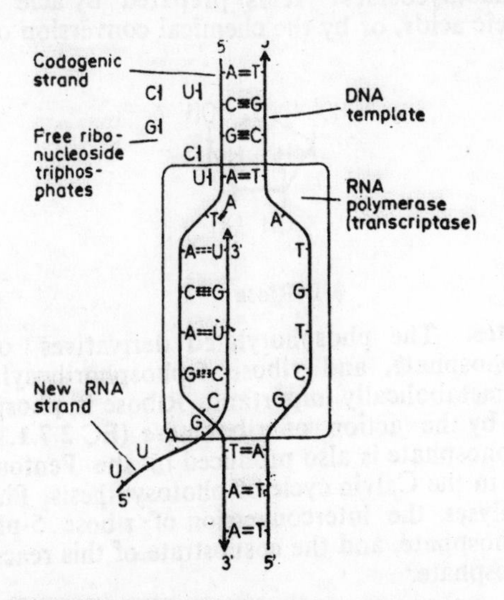

Schematic representation of transcription on a DNA template. A adenine, C cytosine, G guanine, U uracil, T thymine

between complementary bases. These α-helical regions are separated by regions of single stranded, unordered RNA.

There are three main types of RNA, classified on the basis of their function : Messeger RNA (mRNA), ribosomal RNA (rRNA) and Transfer RNA-tRNA (they also have different secondary and tertiary structures. Viral RNA is structurally and functionally very similar to mRNA.

RNA Degradation. RNA is continually degraded in the cell. It is cleaved by various ribonucleases, polynucleotide phosphoryleses and phosphodiesterases. In strong acid RNA is hydrolysed completely to bases phosphate and ribose; alkaline hydrolysis produces 2′-and 3-nucleoside monophosphates.

Ribonucleotide Reductase (EC 1.17.4.1 **or** 1.17.4.2). An enzyme system that catalyses reduction of ribonucleotides to 2-deoxyribonucleotides. This is a stage in the biosynthesis of DNA precursors, and is the only metabolic route for the reduction of ribose to deoxyribose.

D-Ribose. A monosaccharide pentose. It is not fermented by yeasts and the dry solid occurs as the pyranose form. It is a component of RNA, some coenzymes, vitamin B_{12}, ribose phosphates and various glycosides. It is prepared by acid hydrolysis of yeast nucleic acids, or by the chemical conversion of arabinose.

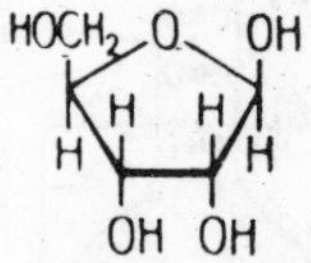

β-D-Riose

Ribose Phosphates. The phosphorylated derivatives of ribose of Ribose 1-phosphate, and ribose 5-phosphoribosyl 1-pyrophosphate are metabolically important. Ribose is phosphorylated in position 5 by the action of ribokinase (EC 2.7.1.15) and ATP. Ribose 5-phosphate is also produced in the Pentose phosphate cycle, and in the Calvin cycle of photosynthesis. Phosphoribomutase catalyses the interconversion of ribose 5-phosphate and ribose 1-phosphate, and the cosubstrate of this reaction is ribose 1, 5-bisphosphate.

Ribosomal Proteins. Integral proteins of ribosomes. Prokaryotic ribosomes contain 35—40% protein, and eukayotic ribosomes contain 48-52% protein. The most extensively studied ribosomal proteins are those of Escherichia coli. The 50S-subunit

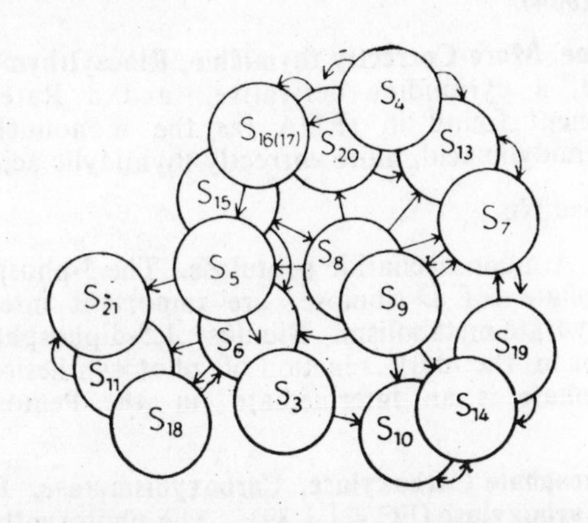

Three dimensional arrangement of the 21 S-proteins of the 30S subunit of the Exc herichia coli ribosome. Arrows indicate the nature and intensity of the interactions between individual subunits. 16S RNA provides the framework for the assembly process. For further details concerning the order and nature of the interactions, Normura, M. (1973) Science, 179, 869.

contains 34 different L-proteins (L=large sub-unit), and the 30S subunit contains 21 different S-proteins (S=small subunit).

Ribosomes, Monosomes. The sities of Protein biosynthesis in the cell. These resemble giant multienzyme complexes. They are spherical to ellipsoid, highly hydrated cell organelles, 15 to 30nm

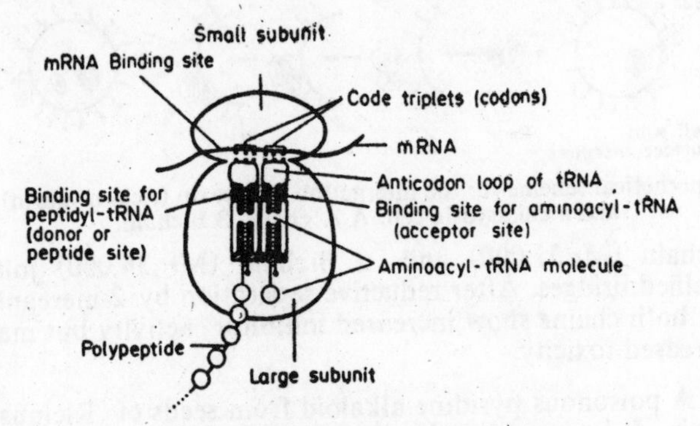

Schematic representation of a ribosome

in diameter, normally present in the cytoplasm as Polysomes. They were first described in 1953 by Palade (Nobel prize 1974).

Ribothymidine, More Correctly thymidine, Ribosylthymine. 5-methyluridine, a pyrimidine derivative, and a Rare nucleic acid component found in tRNA (as the mononucleotide unit of ribothymidylic acid, more correctly thymidylic acid).

Ribotides. See Nu.

D-Ribolose. A monosaccharide pentulose. The 5-phosphate and 1.5-diphosphate of D-ribulose are important intermediates of carbohydrate metabolism. Ribulose 1,5-diphosphate is the CO_4 acceptor in the dark reaction of photosynthesis, and ribulose 5-phosphate is an intermediate in the Pentose phosphate cycle.

Ribulosebisphosphate Carboxylase, Carboxydismutase, Pentose Phosphate Carboxylase (EC 4.1.1.39). The photosynthetic carboxylation enzyme of green plants and purple and green bacteria. It is structurally bound in the thylakoids, but is easily solubilized during isolation. The enzyme consists of two types of subunit, the larger subunit is encoded in the nuclear genome of the plant, the smaller subunit in the plastid DNA. A high proportion of total leaf protein is accounted for by ribulosebisphosphate; in spinach it makes up 16% of leaf protein.

Ricin. A toxalbumin phytotoxin from Ricinus seeds. 493 amino acid residues. Itinhibits protein biosynthesis (causes dissociation of polysomes) and has antitumor properties. It consists of an

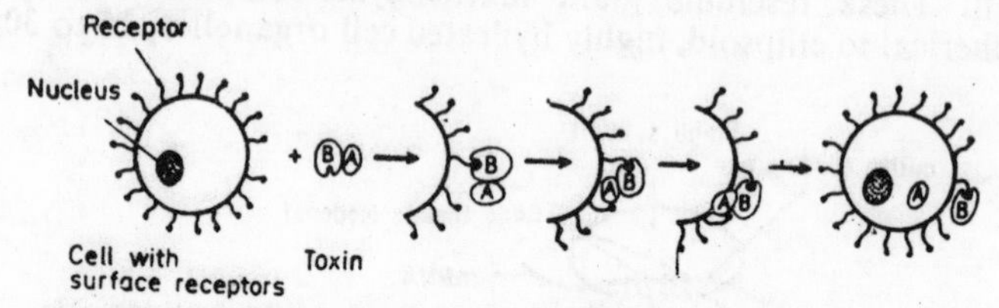

Hypothetical scheme for the interaction of a toxin (ricin or abrin) and a eukaryotic cell. A A chain, B B.chain.

A-chain (Mr 32,000) and a B-chain (Mr 34,000) joined by disulfied bridges. After reductive separation by 2-mercaptoethanol, both chains show increased inhibitor activity but markedly decreased toxicity.

Ricinin. A poisonous pyridine alkaloid from seeds of Ricinus communis. It is an exceptional alkalold, in that it occurs in only one type of plant and is not accompanied by other alkaloids. It is biosynthesized from nicotinic acid Biosynthetic precursors of nicatinic acid are aspartic acid and a 3-carbon compound (probably hydroxyacetone phosphate).

```
      OCH3
       |   CN
      / \ /
     ||   |
      \  / \
       N    O
       |
      CH3
```

Ricinin

Ricinoleic Acid. 12-hydroxyoleic acid,

$$CH_3-(CH_2)_5-CHOH-CH_2-CH=CH-(CH_2)_7-COOH$$

a fatty acid, is present in the glycerides of castor oil (Ricinus oil) where it accounts for 80-85% of the total fatty acids. It is also present in maize oil, wheat oil and various other vegetable oils.

Rickets. Vitamin D deficiency disease.

Rieske Protein. A very electropositive iron-sulfur protein, containing a 2 Fe/S center, and accompanying the b cytochromes and cytochrome c_1 in the respiratory chain complex III. The R p. is in rapid equilibrium with cytochrome c_1.

Rifamycins. A group of antibiotics produced by Streptomyces mediterranei. They contain a naphthalene ring system bridged between position 2 and 5 by an aliphatic chain. Rifamycin SV and rifampicin inhibit DNA-dependent RNA synthesis in prokaryotes, chloroplasts and mitochondria, but not in the nuclei of eukaryotes.

Rishitin. A sesquiterpene Phytoalexin, isolated from Phytophthera-infected potatoes (Solanum tuberosum and Solanum demissum).

RNA. Abbreviation for Ribonucleic acid.

RNA-dependent DNA-polymerase, Reverse Trans Scriptase. An enzyme present in oncogenic (cancer-causing) viruses, *e,g.*, avian myeloma virus and various leukemia viruses. These are RNA viruses, and the enzyme catalyses the synthesis of the provirus DNA, using the viral RNA as a template. The resulting DNA is then incorporated into the genome of the infected cell.

RNA-polymerase, DNA-dependent RNA-polymerase, Nucleoside Triphosphate : RNA-nucleotidyl Transferase, Transcriptate (EC 2. 7.7.6.) An enzyme that catalyses the synthesis of RNAnucleoside triphosphates, using a template of DNA; the base sequence of the resulting RNA is complementary to that of the DNA template. Ribonucleic acid).

Several RNA·P have been isolated from eukaryotic cells. differing in their properties and cellular location RNA-P. I is found in the nucleolus and preferentially catalyses the synthesis of RNA.

RNA-Synthetase, RNA-replicase, RNA-dependent RNA-polymerase. An enzyme that appears in bacterial, animal and plant cells, following infection with RNA viruses. Using a template of single stranded (viral) RNA, the enzyme catalyses the synthesis of a complementry strand of RNA.

Rocellic Acid. (S)-2-dodecyl-3-methylbutanedio acidic d-2 dodecyl-3-methylsuccinic acid. A branched chain dicarboxylic acid found in lichens. It is biosynthesized from itaconic acid.

rRNA. Ribosomal RNA.

Rubixanthin. 3(R)-β,ψ-carotene-3-01; 3(R)-hydroxy-γ-carotene. A xanthophyll. It is copper-red pigment of various rose species and some other higher plants. The 5′-cis-isomer, gazaniaxanthin, is the pigment of various Gazania species.

Rubrosterone. 2β, 3β, 14α-trihydroxy-5β-androst-7 one 6,17-dione. A plant steroid. It was isolated, together with ecdysterone, from the roots of Amarantha obtusifolia and A, rubrofusca, and is considered to be a biogenetic degradation product of ecdysone.

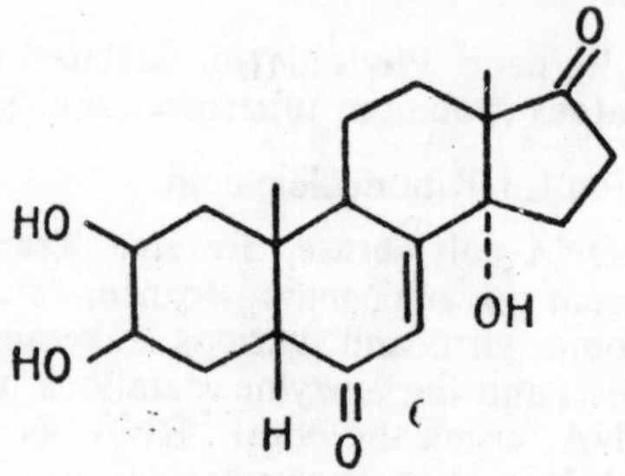

HO HO H O OH O

Rubrosterone

Rubredoxin. A redoxin, functionally similar to Ferredoxin. It was isolated from Clostridium pasteurianum; synthesis of R, appears to be promoted by a relative deficiency of iron. It contains less iron than ferredoxin, usually containing one F atom/molecule of protein. Under acid conditions it is more stable than ferredoxin, and it has more positive redox potential (E_o—0 057 V.)

Rutacease Alkaloids. Alkaloids from common rue (Ruta graveolens L.) and other members of the Rutaceae. They are biosynthesized from anthranilic acid. They furanoquinolines have spasmolytic activity, and the drugs are sometimes used therapeutically.

S

Sabininic Acid: 12-hydroxylauric acid, $HOCH_2$-$(CH_2)_{10}$-COOH. A fatty acid, present as a typical wax acid in the wax of many pines.

Safynol: trans, trans-3, 11-tridecadiene-5, 7, 9-trine-1, 2-diol, a Phytoalexin. Safynol and Δ^8-dehydrosafynol are formed by safflower (Carthamus tinctorius), following infection by Phytophthera. ED_{50} (Dose) is 12 μg/ml for S. and 1.7 μg/ml for the dehydroderivative.

Salamander Alkaloids. Toxic steroid alkaloids secreted by the skin glands of salamanders, *e.g.* Salamandra maculosa (european) fire salamander) and Salamandra atra (alpine salamander.

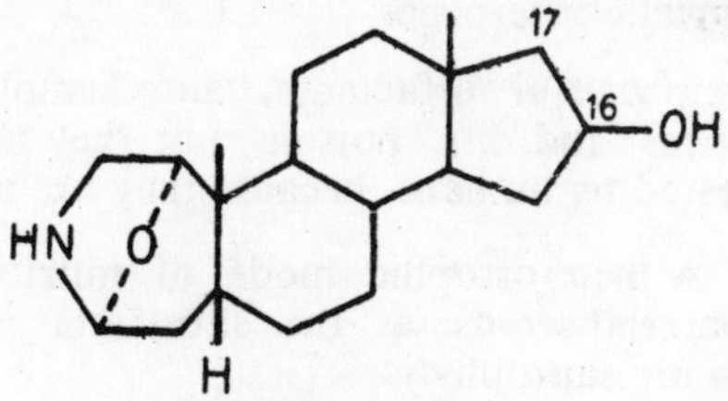

17
16
OH
HN
O
H

Samandarin

These are modified steroids, in which the A-ring is extended to a seven membered ring by a nitrogen between C2 and C3 (A-azahomosteroids). They excite the central nervous system and cause paralysis.

Salamander Toxins. Toxins secreted by the skin glands of Salamandra maculosa (european fire salamander) and Salamandra atra (alpine salamander).

Salsola Alkaloids. A group of simple isoquinoline alkaloids, occurring in Salsola spp. Chief representative is salsoline (1-methyl-6-hydroxy-7-methoxy-1, 2, 3, 4-tetrahydroisoquinoline), which occurs in the D-form, m.p. 215—216°C, $[\alpha]_D + 40°$ (Water) and in the DL-form.

Salvage Pathway. Utilization of performed purine and pyrimidine bases for nucleotide synthesis. In addition to de novo synthesis, this pathway represents an alternative pathway for formation of purine and pyrimidine nucleotides.

SAMP. Abb. for adenylosuccinate.

Sangivamycin: 4-amino-5-carboxamide-7-(D-ribofuranosyl)-pyrrolo-(2, 3-d)-pyrimidine. A deazaadenine-type antibiotic from Streptomyces spec. Its antibiotic activity spectrum is similar to that of Toyocamycin.

Saponins. A large and widely distributed group of plant substances, named from their ability to form strongly foaming, soap-like solutions with water. They are all glycosides, and are classified according to the nature of the aglycon (also called a genin). *i.e.* steroid, triterpene and steroid-alkaloid saponins; the latter are also known as glycoalkaloids. Aglycons of the steroid saponins are also called sapogenins; these are spirostane-type, C_{27}-steroids. They all possess a 3β-hydroxyl group (spirostanols), which forms a glycosidic linkage with the sugar, *e.g.* D-glucose, D-galactose or L-arabinose. The triterpene saponins are less well studied; their aglycons are chiefly tetra and pentacyclic triterpenes.

S. are powerful surfactants, cause hemolysis and are potent plasma toxins and fish poisons, but they have no toxic effects when ingested by humans, because they are not absorbed.

Saprophytism. A heterotrophic mode of nutrition, in which dead, organic material serves as the substrate. Many bacteria and most fungi are saprophytic.

Sarcine. An obsolete name for hypoxanthine.

Sarcosine. abb. **Sar, N-methylglycine:** CH_3-NH-CH_2-COOH. An intermediate in the metabolism of choline (choline→betaine aldehyde→betaine→dimethylglycine→monomethylglycine (Sar) →glycine). Methyl groups lost in this pathway are transferred to tetrahydrofolic acid.

Sargasterol: (20S)-stigmasta-5, 24(28)-diene-3β-ol. A phytosterol in brown algae, *e.g.* Sargassum ringolianum. It differs from Fucosterol by the opposite configuration at C20.

Satellite DNA. DNA fractions that can be separated from the main DNA by CsCl-gradient centrifugation. It has been demonstrated in the nuclei of many eukaryotes, accounting for about 1% of nuclear DNA in man, and about 10% in the mouse. It consists of a linear double helix, and differs markedly from the rest of the nuclear DNA with respect to base composition and density.

Saxitoxin. A neuromuscular blocking agent, which prevents nerve transmission by blocking sodium pores in postsynaptic membranes. It is produced by dinoflagellates of the genus Gonyaulax, found in "red tides". S. accumulates in shellfish that ingest the dinoflagellates, hence cases of poisoning from eating the Californian sea mussel (Mytilus californianus), the Alaskan butterclam (Saxidomus giganteus) and the scallop.

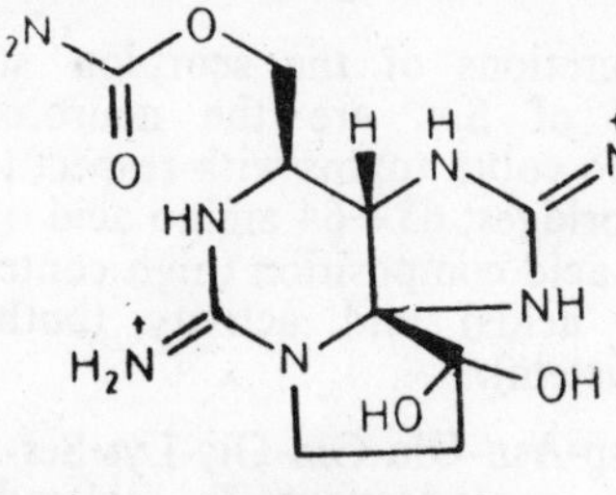

Saxitoxin

Schottenol: 5α-stigma-7-ene-3β-ol. A widely distributed phytosterol isolated, *e.g.* from the cactus Lophocereus schottii. It is an essential dietary constituent for the insect Drosophila pachea, which lives on this plant.

Scillaren A, Glucoproscillaridin A, Transvaalin. A bufadienolide cardiac glycoside. The carbohydrate residue is the disaccharide, scillabiose (6-deoxy-4-O-β-D-glucopyranosyl-L-mannose; 4-O-D-glucopyranosyl-L-rhamnose), attached glycosidically at C3 of the aglycon. S.A. is the chief active component of Scilla maritima (squill), used since antiquity as a diuretic, cardiac stimulant and mouse poison.

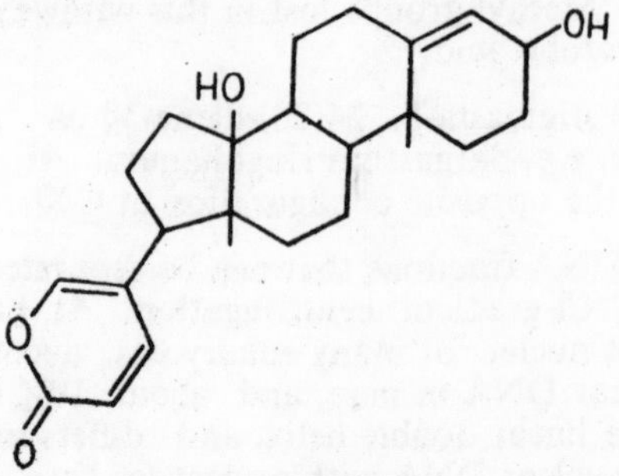

Scillarenin

Scopolamine: α-(hydroxymethyl) benzeneaceticacid 9-methyl-3-oxa-9-azatricyclo [3.3.1.0^{2,4}] non-7-yl ester; 6β, 7β-epoxy-3α-tropanyl S-(-)-tropate, a tropane alkaloid, from Solanaceae, spp., especially Datura metel L. and Scopola carniolica Jacq. DL-Scopolamine (atroscine), forms an effluorescent hydrate, m.p. 55—57°C (also reported as 82—83°C). It has similar pharmacological activity to that of hyoscamine, but has comparatively less activity on the peripheral nervous system. It is a highly toxic, anticholinergic agent.

Scorpion Venoms. Secretions of the scorpion stinging apparatus. Active principles of S.v. are the neurotoxic scorpamines. They are similar to cobra toxins with respect to M_r (M_r 6800—7200; 4 disulfide bridges; 63—64 amino acid residues of known sequence), amino acid composition (high contents of basic and aromatic amino acids) and activity (both peripheral and central nervous system).

Scotophobin: Ser-Asp-Asn-Gln-Gln-Gly-Lys-Ser-Ala-Gln-Gln-Gly-Gly-Tyr-NH_2. A pentadecapeptide isolated from the brains of rats trained to avoid the dark. It induces dark avoidance in untrained animals.

Scurvy. Vitamin C deficiency disease.

Secondary Metabolites. Substances such as pigments, alkaloids, antibiotics, terpenes, etc. which occur only in certain organisms, organs, tissues or cells, and are the products of secondary metabolism.

Secretin. A polypeptide hormone, M_r 3050, containing 27 amino acid residues. S. shows considerable sequence homology with

	5	10
Secretin:	His- Ser- Asp-Gly- Thr- Phe- Thr- Ser- Glu- Leu- Ser- Arg- Leu- Arg-	
Glucagon:	His- Ser- Gln- Gly- Thr- Phe- Thr- Ser- Asp- Tyr- Thr- Lys- Tyr- Leu-	
VIP:	His- Ser- Asp- Ala- Val- Phe- Thr- Asp- Asn- Tyr- Thr- Arg- Leu- Arg-	
GIP:	Tyr- Ala- Glu- Gly- Thr- Phe- Ile- Ser- Asp- Tyr- Ser- Ile- Ala- Met--	

15	20	25
Asp- Ser Ala- Arg- Leu- Gln- Arg- Leu- Leu- Gln- Gly- Leu- Val- NH_2		
Asp- Ser- Arg- Arg- Ala- Gln- Asp- Phe- Val- Gln- Trp- Leu- Met- Asp- Thr		
Lys- Gln- Met- Ala- Val- Lys- Lys- Tyr- Leu- Asn- Ser- Ile- Leu- Asn- NH_2		
Asp- Lys- Ile- Arg- Gln- Gln- Asp- Phe- Val- Asn- Trp- Leu- Leu- Ala- Gln- Gln		

Amino acid sequences of secretin, glucagon, VIP and GIP. Amide groups are present at the C-termini of secretin and VIP. The sequence of GIP is shown incomplete; it continues:-Lys-Gly-Lys-Lys-Ser-Asp-Trp-Lys-His-Asn-Ile-Thr-Gln (total 43 residues).

Glucagon. Vasoactive intestinal peptide (VIP) and Gastric inhibitory peptide (GIP), and it is thought that these four hormones evolved from a common ancestral protein by a process of gene multiplication.

Secretory Proteins. The proteins synthesized intracellularly, often in specialized secretory organs (*e.g.* digestive glands), then secreted. These proteins that are also enzymes are called secretory enzymes. In cells actively engaged in synthesizing these proteins the rough endoplasmic reticulum (RER) is highly developed.

Sedimentation Coefficient, Sedimentation Constant. A measure of the rate of sedimentation, used in the determination of M_r of macromolecules by ultracentrifugation. The sedimentation coefficient (s) is equal to the rate of sedimentation of a marcomolecule per unit centrifugal field; specifically,

$s = \frac{dx/dt}{\omega^2 . x}$, where s is the sedimentation coefficient, ω is the angular velocity of the centrifuge rotor (radians/sec), x is the distance from the center of rotation, dx/dt is the velocity of sedimentation.

d-Sedoheptulose, D-altro-2-heptulose. A monosaccharide from Sedum (stonecrop.) The 7-phosphate is an intermediate of carbohydrate metabolism aldolase, reactions of sedoheptulose 7-phosphate give rise to D-erythrose 4-phosphate, which is a precursor of Aromatic biosynthesis.

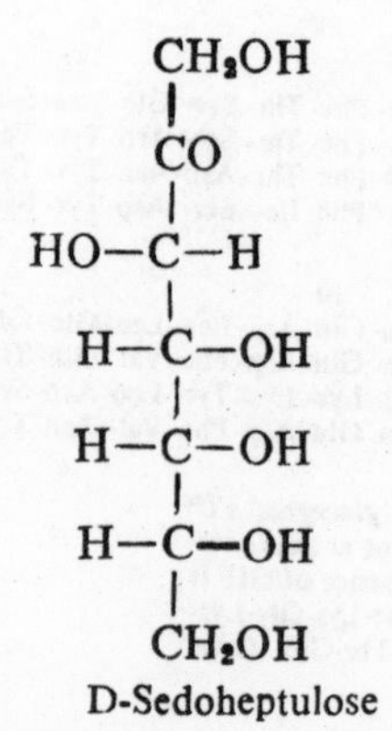

D-Sedoheptulose

Sedum Alkaloids. A group of piperidine alkaloids from Sedum spp. They are 2- or 2, 6-substituted piperidine derivatives, similar in structure and biosynthesis to the Punica and Lobelia alkaloids.

Selenium Se. An element toxic in large quantities, but an essential micronutrient for mammals, birds, many bacteria, probably fish and other animals. It is an essential component of the enzyme glutathione peroxidase, which is important in the protection of red cell membranes and other tissues from damage by peroxides :

$$2GSH + 2H_2O_2(\text{or R-OOH}) \rightarrow GSSG + 2H_2O \quad (\text{or } H_2O + \text{R-OA})$$

Selenoamino Acids. Amino acids containlng selenium (Se) in place of sulfur (S), *e.g.* Se-methyl-selenocysteine. They are formed in plants growing on Se-rich soils.

Sephadex. A trade name for a series of polydextrans used in gel filtration chromatography.

Sequence Polymers. Synthetic amino acid polymers, consisting of multiple repeats of one short sequence. In contrast to Polyamino acids.

Ser. Abb. for L-serine.

L-serine, abb. **Ser:** L-α-amino-β-hydroxypropionic acid, HO-CH_2-$CH(NH_2)$-COOH. It is a proteogenic, glucogenic amino acid. It is a major component of silk fibroin. In phosphoproteins, phosphate is esterified chiefly with the hydroxyl groups of Ser residues During acid hydrolysis of proteins, a large proportion of Ser is destroyed. It is converted quantitatively into formaldehyde by periodate oxidation.

Serine Hydrolases. Hydrolases which have a catalytically active serine residue in their active center, *e.g.* trypsin, chymotrypsin A, B and C, thrombin and B-type carboxylic acid esterases.

Serine Proteases. A group of well studied animal and bacterial endopeptidases which have a similar action mechanism, and a catalytically active serine residue in their active centers (serine residue 195 in chymotrypsin). In all serine proteases catalysis involves formation of an ester between the hydroxyl group of the catalytically active serine and the carboxyl group of the cleaved peptide bond (acyl-enzyme intermediate); this is hydrolysed in the deacylation stage of the reaction, restoring the free hydroxyl group of the serine and releasing the cleavage peptide.

Serotonin: 5-hydroxytryptamine. A biogenic amine, occurring as a hormone in plants and animals, produced by hydroxylation of L-tryptophan to 5-hydroxytryptophan, followed by decarboxylation. Synthesis occurs in the central nervous system, lung, spleen and argentaffine "light" cells of the intestinal mucosa.

Serratomolide. A cyclic depsipeptide produced by Serratia marcescens. Chemically, it is the cyclic dimer of serrataminic acid (D-β-hydroxydecanoyl-L-serine).

Sesquiterpenes. Aliphatic, mono, di or tricyclic terpenes, formed from three isoprene units ($C_{15}H_{24}$). About 100 structural types are known and about 1000 natural representatives, forming the largest class of terpenes. Most are found in the volatile oils of plants.

TABLE

Some sesquiterpenes and their function

Function	Sesquiterpenes
Juvenile hormones	Juvabione, farnesyl derivatives
Phytohormones	Abscisic acid
Plant sex hormones	Sirenin
Pheromones	Farnesol
Antibiotics	Trichothecin
Proazulenes	Guaiol
Alkaloids	Nupharidine
Scents	Santalols, cedrenes
Bitter principles	Cnicin
Phytoalexins	Ipomeamarone

Sesterterpenes. Terpenes formed from five isoprene units ($C_{25}H_{40}$). They have a tricyclic skeleton, and have been isolated from insect secretions and lower fungi. Biosynthetic studies show that cochliobolin B (zizanin B) (m.p. 175°C, $[\alpha]_D+300°$), for example, is formed via geranylfarne-sylpyrophosphate by head-to-tail linkage of five isoprene units Fig.).

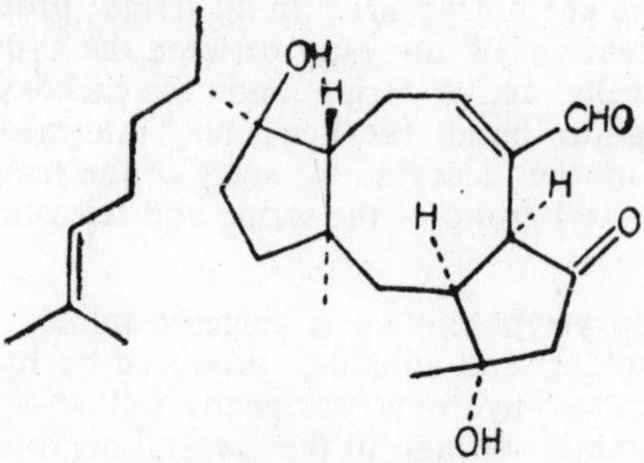

Cochliobolin B

Sexual Attractants. Natural products involved in sexual interaction.

They are usually produced by the sexually mature female in order to attract and predispose the male to copulation.

Plant sexual attractants are called gamones, or plant sex hormones.

SF. Abb. for Sulfation factor.

Showdomycin: 2-β-D-ribofuranosylmaleinimide. A C-substituted Nucleoside antibiotic from Streptomyces showdoenis, structurally related to uridine and pseudourine. It selectively inhibits enzymes of uridine and orotic acid metabolism; the

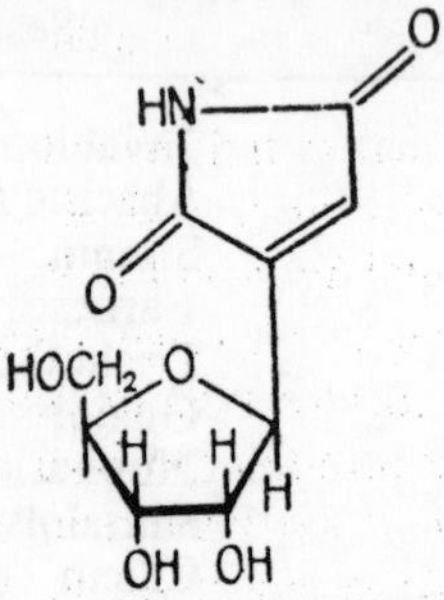

Showdomycin

maleinimide moiety reacts with sulfhydryl groups of the affected enzymes. It is especially active against Streptococcus haemolyticus.

Sickle Cell Hemoglobin. Abb. HbS one of the most frequently occurring abnormal hemoglobins, especially in negroids. As a result of a point mutation, the glutamic acid residue at position 6 in the β-chain (normal hemoglobin) is replaced by valine. The α-chain is normal ($\alpha_2\beta_2$ 6Glu→Val).

Siderochromes. Iron-containing, red-brown, water soluble secondary metabolites produced by microorganisms. They include a series of antibiotics, the sideromycins (albomycin, ferrimycin danomycin, etc.) and a class of compounds with growth factor properties for certain microorganisms, the sideramines (ferrichrome, coprogen, ferrioxamine, ferrichrysin, ferrirubin, etc.). The sideramines competitively inhibit the antibiotic activity of the sideromycins.

Siderophilins. Nonheme, iron-binding, single chain animal glycoproteins, carbohydrate content about 6%. On the basis of their occurrence, they are classified as transferrin (vertebrate blood), lactoferrin (mammalian milk and other body secretions) and conalbumin or ovotransferrin (avain blood and avain egg white). They differ in their physical, chemical and immunological properties, but each possesses two iron-binding sites for iron (III).

SIF. Abb. for Somatotropin release inhibiting factor.

Signal Hypothesis. A mechanism proposed by Blobel for the segregation of secretory proteins during their biosynthesis. The mRNA for secretory proteins is thought to contain a sequence of signal codons directly after the initiation codon. The signal peptide sequence, which contains between 15 and 30 amino acid residues, is therefore translated at the beginning of protein biosynthesis. It causes specific receptor proteins in the membrane of the rough endoplasmic reticulum to aggregate, and a complex is formed between mRNA, ribosome and membrane receptor protein.

SIH. Abb. for Somatotropin release inhibiting hormone.

Silicon, Si. An essential trace element in human nutrition. Si is a cross-linking agent in connective tissue. High levels of Si (as SiO_2) are present in plants and diatoms.

Single-strand Break. A break in a double-stranded DNA molecule which involves only one of the two strands, so that the molecule remains together.

Single-substrate Enzymes. Enzymes which catalyse reactions involving only one substrate. They are usually isomerases or hydrolytic enzymes; in the latter case, the water involved in the reaction is regarded as a constant, and there is frequently no special enzyme-water complex formed.

Sirenin. The first plant sexual attractant or gamone to be structurally elucidated. It is a sesquiterpene. It occurs naturally in the L-form. It is produced by the female gametes of the fungus Allomyces, which lives in damp soils. Gametes swim from the mycelium, and it acts as a chemotactic stimulus to attract the male to the female gametes. It is active at a concentration of 10^{-10} M.

Siroheme. The heme prosthetic group found in sulfite reductase of Escherichia coli and nitrite reductase of green plants.

Sitosterols. A group of phytosterols structurally related to the parent hydrocarbon stigmastane.

Slow Reacting Substance A. Abb. for slow reacting substance of anapylaxis. It is produced by sensitized cells as part of the immune response to antigens. Leukotrienes.

Slow Reacting Substances. Substances that cause smooth muscle to contract slowly in vitro.

Snake Venomes. A mixture of toxins produced in the venom glands (parotid gland, or salivary gland of the upper jaw) of venomous snakes (asps or hooded snakes, *e.g*, the cobra; sea snakes; vipers, *e.g.* puff adder, rattlesnake). They consist of highly toxic, antigenic polypeptides and proteins (which cause paralysis and death of the prey), and enzymes (which facilitate the spread of the toxins, and initiate digestion of undivided swallowed prey).

α-Solamarine. A Solanum glycoalkaloid, first isolated from woody nightshade (Solanum dulcamara). The aglycon is tomatidenol (22S : 25S)-spirosol-5-ene-3 β-ol, M_r 413.67, m.p. 239°C; $[\alpha]_D$ −37.8 (chloroform), and the carbohydrate residue is a trisaccharide of one molecule of D-glucose and two molecules of L-rhamnose.

α-Solanine, Solanine. A solanum alkaloid, and the chief toxic alkaloid of the potato (Solanum tuberosum), also present in other Solanum spp. It is a glycoalkaloid containing the aglycon solanidine (solanide-5-3ene-β-ol), and the trisaccharide β-solatriose.

Solanum Alkaloids. Steroid alkaloids that occur in plants of the nightshade family (Solanaceae), belonging to the genera Solanum. Lycopersicon, Cyphomandra and Cestrum. These are structurally related to the parent hydrocarbon, cholestane.

α-Solasonine, Solasonine. A solanum steroid alkaloid occurring in many Solanum spp., *e.g.* S. sodomeum, S. aviculare, S. laciniatum and S. nigrum. It is a glycoalkaloid containing the aglycon, solasodine (22R : 25R)-spirosol-5-ene-3 β-0 1], and a branched trisaccharide.

Somatomedins. A collective term for a group of peptides comprising somatomedins A, B and C. It particular, it causes an increased incorporation of sulfur into cartilage, and is therefore called sulfation factor. All these have insulin-line activity.

Somatotropin, Somatotropic Hormone, abb. **STH, Growth Hormone,** abb. **GH.** A fundamentally important hormone, which in conjunction with other hormones (insulin, thyroxin, etc.) controls growth, differentiation and the continual renewal of body substances. GH is a single chain polypeptide containing 190 amino acid residues and two disulfide bridges. It is synthesized in the anterior pituitary in response ro SRF and SIH.

d-Sorbitol. A C_6-sugar alcohol found widely in plants. It can be prepared by catalytic or electrochemical reduction of the configurationally related D-glucose. D-fructose or L-sorbose. It is used in the food industry as a preservative and as a softening agent in sweets. Since it is well tolerated, it is used as a sweetener in diabetic diets.

```
      CH2OH
        |
    H—C—OH
        |
  HO—C—H
        |
    H—C—OH
        |
    H—C—OH
        |
      CH2OH
```

Sorbitol

l-Sorbose. A monosaccharide hexulose, present in certain plant juices, *e.g.* rowan berries, and biosynthesized from D-sorbitol. It is an intermediate in the commercial synthesis of ascorbic acid.

Soybean Trypsin Inhibitor, abb. **STI.** The best known plant trypsin inhibitor. With bovine trypsin, at pH 8.3, it forms a stoichiometric, enzymatically inactive, stable complex with an association constant of 5.10^9 per mol STI. It also inhibits other vertebrate and invertebrate trypsins and plasmin.

Specificity Constant, Physiological Effectivity. A measure of the turnover of a substrate. It is the ratio of the catalytic constant and the Michaelis constant : k_{cat}/K_m. It is equal to the rate constant of a reaction for the rate equation :

$$v = k_{cat}\, E_o S/(K_m + S),$$

where E_o is the total enzyme concentration, and S is the substrate concentration, when $S \gg K_m$.

Spectrin. A protein which makes up about 75% of the "skeleton" of the erythrocyte membrane. It is not found in any other type of cell. S. is a heterodimer or tetramer of two polypeptides, band 1 with a M_r of 240000 and band 2 with a M_r of 222000.

Spermaceti. A solid animal wax, obtained from the head of the sperm whale, Physeter macrocephalus. It consists chiefly of cetyl palmitate, accompanied by smaller quantities of cetyl laurate and myristate. It is used in the pharmaceutical and cosmetic industries as a basis for creams.

Spermine. $H_2N\text{-}(CH_2)_3\text{-}NH(CH_2)_4\text{-}NH(CH_2)_3\text{-}NH_2$. A biogenic, aliphatic tetramine. It occurs in high concentrations, together with the triamine, spermidine (mono-(γ-aminopropyl)-putrescine), in human sperm. It is also present in ribosomes and some viruses.

Sphingosine. 2-amino-4-octadecene-1, 3-diol. a long-chain amino alcohol which a component of sphingomyelins (Phospholipids) and Glycolipids.

Spider Toxins. Toxic substances produced in the venom glands of many spiders. They serve to paralyse and kill prey, and are dangerous to humans only in rare cases. They contain hyaluronidase and proteolytic activity, but phospholipases and hemolytic or blood clotting activities are absent..

Spinasterols A group of very similar phytosterols found in higher plants. Chief representative is α-spinasterol (5α-stigmasta-7-22-diene-3 β-ol; Fig.).

Spinochromes. Derivatives of 1, 4-naphthoquinone responsible for the red or orange color of sea urchin shells.

Spirostane. The oxygen-containing parent structure of the steroid saponins. This system is formally derived from the parent hydrocarbon cholestane.

Spongosine. 9-β-D-ribofuranosyl-2-methoxy-adenine. A nucleoside with a modified base, isolated from sponges.

Spongosterol. (24R)-5α-ergost-22-end-3 β-ol. A marine zoosterol occurring as a typical sterol of sponges (Spongia) and isolated, *e.g.* from Suberitis domuncula and Suberitis compacta.

Sporidesmolides. Cyclic depsipeptides from the fungus Pithomyces chartarum. Sporidesmolide I is cyclo-(-Hyv-D-Val-D-Leu-Hyv-Val-MeLeu-); sporidesmolide II contains D-allo isoleucine in place of D-valine, white sporidesmolide III contains L leucine in place of L-N-methylleucine.

Sporopollenin. The material of the outermost cell wall layer (exine) of pollen grains and spores of pteridophytes, and also present in small amounts in fungal zygospore walls (*e.g.* zygospore wall of Mucor mucedo contains 1% S).

S-protein. A cleavage product of ribonuclease, representing amino acid residues 21—124 of the ribonuclease primary sequence. It is produced together with S-peptide (positions 1—20), by the action of subtilisin on ribonuclease.

Squalene. Biochemically the most important aliphatic triterpene. It is the intermediate in the biosnythesis of all cyclic triterpenes.

SRH. Abb. for somatotropin releasing hormone. Releasing hormones.

SRNA. Abb. for the obsolete name, soluble RNA, now known as Transfer RNA or tRNA.

SRS. Abb. for slow reacting substance. Leukotrienes.

Stachyose. A nonreducing tetrasaccharide found in plants. The four sugar residues are linked in the order D-galactose-D-galactose-D-glucose-D-fructose.

Starch. A high M_r polysaccharide, formula $(C_6H_{10}O_5)_n$; the chief storage carbohydrate in most higher plants, consisting of about 80% water-insoluble Amolopectin and 20% water-soluble Amylose. In plant metabolism starch first appears as an assimilation product in the chloroplasts. It is then degraded, the products of degradation are translocated, and starch is resynthesized as storage starch.

Starch is biosynthesized from adenosine-diphosphate-glucose.

Steady State. A chain of chemical reactions is in a steady state when the concentration of all intermediates remains constant, despite a net flow of material through the system, *i.e.* the concentration of intermediates remains constant, while a product is formed at the expense of a substrate.

Stearic Acid. N-octadecanoic acid, $CH_3s(CH_2)_{16}$-COOH. A fatty acid Together with palmitic acid, it is one of the most plentiful and most widely distributed fatty acids, occurring in practically all animal and plant oils and fats.

Sterane. An earlier name for gonane.

Steroid Alkaloids. Nitrogen-containing steriods present in plants. These are especially common in the plant families, Solanaceae (nightshades), Liliaceae (lilies), Apocyanaceae (periwinkles) and Buxaceae (boxwoods), where they often occur as glycoalkaloids, or esterified as ester alkaloids.

Steriod Hormones. Groups of steriods that function as hormones. They comprise Adernal corticosteriods sex hormones, hormones Ecdysone and related moulting hormones, and the plant sex hormone, Antheridiol. They are biosynthesized from cholesterol. For synthesis, storage and secretion of S.h., Hormones.

Steriods. A large group of terpenoids, including many important biological compounds, *e.g.* Sterols, Steroid hormones, Bile acids, Cardiac glycosides, Steroid alkaloids and steroid saponins Synthetic S., *e.g.* Ovulation inhibitors and Anabolic steroids, and structurally modified steroid hormones are pharmacologically important. Structurally, S are derivatives of the hydrocarbon cyclopentanoperhydrophenanthrene.

Sterols. A group of naturally occurring steroids, possessing a 3 α-hydroxyl group and a 17 α-aliphatic side chain. These are structural derivatives of the parent hydrocarbons, cholestane, ergostane and stigmastane. They occur in animal and plant cells as free sterols, and as glycosides and esters. According to their origin, they are classified as zoosterols (animals), phytosterols (plants), mycosterols (fungi) and marine. Sterols (marine animals and plants).

Steviol. The aglycon of Stevioside.

Stevioside. A tetracyclic diterpene, from the bush, Stevia rebaudiana, which is native to Paraguay. It is both a glycoside and a Glucose ester. Enzymatic hydrolysis releases the aglycon, steviol

($R_1 = R_2 = H$). Acid hydrolysis causes a Wagner-Meerwein rearrangement of rings C and D to produce isosteviol. It is 300 times sweeter than sucrose and would be an ideal sweetening agent; commercial exploitation is difficult on account of its low concentration (about 6g S. per kg dried leaves). It possesses a gibberellin-like growth stimulating activity.

R_1 = Glucose-Glucose
R_2 = Glucose

Stevioside

STH. Abb. for somatotropic hormone.

STI. Abb. for Soybean trypsin inhibitor.

Stilbenes. Polyketides formed from one molecule of cinnamic acid and three molecules of manonyl-CoA. The immediate precursors of S: are the corresponding stibenecarboxylic] acids,[which have a carboxyl group at C 2 of ring A (Fig.).

Cinnamic acid precursor of ring B	R_1	R_2	Stilbene
Cinnamic acid	H	H	Pinosylvin
p-Coumaric acid	H	OH	Resveratrol
Caffeic acid	OH	OH	Piceatannol
Isoferulic acid	OH	OCH_3	Rhapontigenin

Stigmasterol, 5 α-stigmasta-5, 22-diene-3 α-ol, a widely distributed phytosterol, first isolated from calabar beans, and later from many other sources, *e.g.* soybeans, carrots, coconut oil and sugar cane wax. It is used as a starting material in the technical synthesis of steroid hormones.

Stilbestrol, Stilboestrol, Trans-stilbestrol. A synthetic: nonsteroid Esterogen, used in estrogen, therapy.

Stoichiometric Model, Stoichiometric Reaction Scheme. In enzyme kinetics, a chemical reaction equation in which molecules involved in the reaction, including the enzyme are represented as latters. The simplest example of a S.M. in enzyme kinetics is :

$$E + S \rightleftharpoons EL \rightarrow E + P$$

More complicated reactions are more conveniently expressed by enzyme graphs.

Strand Polarity, Antiparallel Conformation. The polarity of nucleotide chains, with reference to the sequence of 3′, 5′-phosphodiester bonds. Pollynucleotide chains have a 3′-end (the terminal sugar residue is linked to the preceding residue via its 5′-hydroxyl, and the 3′-hydroxyl is free or phosphorylated) and a 5′-end (5′-hydroxyl is free or phosphorylated).

Strand Selection. The ability of DNA-dependent nucleic acid polymerases to choose the codogenic strand of the double-stranded DNA.

Streptogenin Peptides. Natural products (*e.g.*, liver extracts, peptones, partial hydrolysates of proteins) or synthetic peptides, which stimulate the growth of microorganisms, especially lactic acid bacteria. The unknown growth stimulant is called streptogenin.

Streptomycin. An antibiotic from Streptomyces griseus. It is an aminoglucoside, in which streptidine is linked glycosidically to the disaccharide, streptobiosamine. It inhibits protein biosynthesis on 70S ribosomes. It becomes bound to the 23S core protein of the 30S ribosomal subunit. This protein appears to be responsible for the binding of mRNA, which is prevented by S.

l-Streptose. 5-deoxy-3-formyl-L-lyxose, A monosaccharide, with a branched carbon chain. In combination with 2-deoxy-2-methylamino-L-glucose, it forms streptobiosamine, the disaccharide component of Streptomycin. It is biosynthesized by the rearrangement of an unbranched hexose.

Stroma Matrix. Colorless, homogeneous (by light microscopy) ground substance of cell organelles, like Chloroplasts and Mitochondria.

Strophanthins, Stropanthosides. Cardenolide cardiac glycosides from Strophanthus spp., *e.g.*, g-strophanthin (ouabain) from Strophathus gratus, and k-strophanthin from Strophanthus kombe. The aglycon of k-strophanthin is strophanthidin. Strophanthindin is the aglycon of other cardiac glycosides, *e.g.*, convallatoxin of Convallaria majalis (lily of the valley).

Structural Colors, Schemochromes. Colors created by optical effects, due to the physical nature of surfaces, *e g.*, interference, refraction and diffraction on very thin layers.

Structural Proteins, Skeletal Proteins, Scleroproteins, Fibrous Proteins. A group of simple animal proteins with structure and support functions. They are generally insoluble in water and salt solutions. The best known structure proteins are the cystine-rich Keratins. Others are Collagen, Elastin Crystallins, silk fibroin, chondrin, spongin, etc.

Strychnine. A Strychnos alkaloid. It is the chief alkaloid in the seeds of tropical Strychnos spp., from which it is obtained commercially, the total synthesis by Woodward is uneconomic and only of scientific interest. It has many therapeutic properties, but its high toxicity precludes its extensive use in medicine.

Strychnos Alkaloids. A group of indole alkaloids from the tropical plant genus Strychnos. The highly toxic chief alkaloids, Strychnine and Brucine, contain a heptacyclic ring structure. These are biosynthesized from tryptophan and a trepenoid C_{10}-unit.

Suberic Acid. Octanedioic acid, $HOOC—(CH_2)_6—COOH$. A higher saturated dicarboxylic acid. It is formed by the oxidation of cork or ricinus oil with nitric acid, along with its homologs azelaic and sebacic acids. It is obtained in higher yields from cyclooctene (technical synthesis).

Substance P. Arg-Pro-Lys-Pro-Gin-Gln-Phe-Phe-Gly-Leu-Met-NH_2. A linear peptide present in the digestive tract of various mammals, and in the brain of humans, mammals, birds, reptiles, fishes, etc. It is a powerful promoter of muscular contraction in the intestinal tract, and it decreases blood and salivary pressures.

Substrate Level Phoshhorylation. Adenosine triphosphate (ATP) synthesis not involving photosynthetic phosphorylation or axidative phosphorylation in the respiratory chain. It occurs in Glycolysis and in the oxidative decarboxytation of 2-oxoglutarate in the Tricarboxylic acid cycle.

Substrate Specificity. The ability of an enzyme to recognize and specifically bind its subtrate. It is a function of protein structure.

Subtilisin (EC 3.4 21.4). An extracellular, single chain, alkaline serine protease from Bacillus subtilis and related species.

Subunit

1. In protein chemistry the smallest protein or polypeptide unit that can be separated from an oligomeric protein without breaking covalent bonds. In allosteric enzymes, *e.g.*, aspartate transcarbamylase, nonidentical subunits can be further classified into regulatory and catalytic subunits.
2. The term is also applied to the two subunits of Ribosomes.

Succinate Dehydrogenase (EC 1.3.99.1) An oligomeric flavoenzyme of the TCA-cycle, which catalyses the oxidation of succinate to fumarate. It consists of two nonidentical, iron-containing subunits (M_r 70000 and 30000).

Succinate-glycine Cycle, Glycine-succinate Cycle, Shemin Cycle. A bypass of the TCA-cycle of particular importance in the

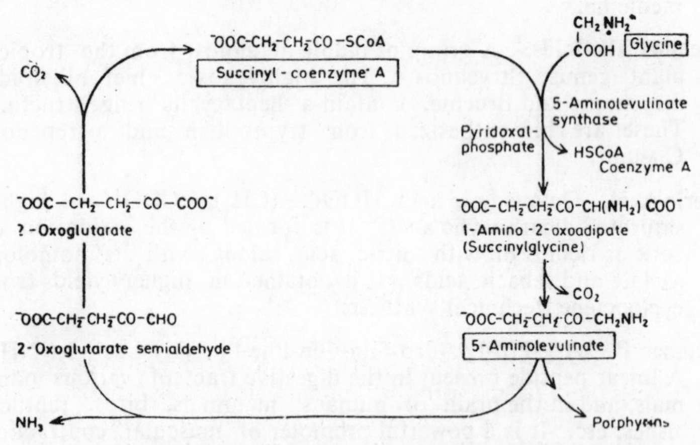

Synthesis of 5-aminolevulinate via the succinate-glycine cycle

metabolism of red blood cells. It converts succinyl-CoA and glycine into 5-aminolevulinate, which is the biosynthetic precursor of the Prophyrins.

Succinic Acid, Ethane Dicarboxylic Acid. HOOC-CH_2-CH_2-COOH. Succinate is an important intermediary metabolite generated both in the Tricarboxylic acid cycle and in the Glyoxylate cycle. It can be used via the reactions of the tricarboxylic acid cycle for the synthesis of amino acids and carbohydrates. Succinyl-coenzyme A is the starting point for the synthesis of porphyrins from intermediates of the Succinate-glycine cycle. Succinyl-coenzyme A is also formed by carboxylation of propionyl-coenzyme A in the degradation of valine and isoleocine, and from 2-methylmalonyl-coenzyme A in the degradation of fatty acids with an odd number of carbon atoms.

Succinic Acid 2, 2-dimethylamide, Alar 85, B9 :

$$HOOC-(CH_2)_2-CO-N(CH_3)_2$$

A synthetic growth regulator. It is used to stimulate blossoming in apples and to accelerate the development of fruit color and fruit loosening in cherries.

Succinic Acid 2, 2-dimethylhydrazide, Dimethazide, B-995 :

$$HOOC-(CH_2)_2-CO-NH-N(CH_3)_2.$$

A synthetic growth regulator. It acts as a gibberellin antagonist and retards growth especially actively in dicortyledonus plants. It increases the number of blossoms on fruit trees and prevents the loss of immature fruit.

n-Succinyladenylate. A denylosuccinate, Purine biosynthesis.

Sucrose, Cane Sugar, Beet Sugar. α-D-glucopyranosyl-β-D-fructofuranoside. A nonreducing disaccharide, biosynthesized from fructose 6-phosphate and UDP-glucose. It is a trehalose-type disaccharide and therefore does not give typical sugar reactions, *i.e.*, does not form an osazone or oxime, or show mutarotation. Hydrolysis with dilute acid or enzymes, *e.g.*, α-glucosidase (maltase) (EC 3 2.1.20), or β-D-fructafuranosidase (invertase, EC 3.2.1.26) cleaves sucrose into equal parts of D-glucose and D-fructose.

Sugar. In the narrow sense a general name for commercially available Sucrose. In the wider sense a term used for Carbohydrates in particular mono and oligosaccharides.

Sugar Alkohol. The polyhydric alcohols, which occur widely as metabolic reduction products of monosaccharides. They are biosynthesized by reduction of the correspanding monosaccharide with NADH or NADPH. Important natural representatives are D-glycerol, erythritol, ribitol, xylitol, D-sorbitol, D-mannitol, dulcitol.

Sugar Anhydrides. Internal acetals formed by the intramolecular removal of water from a sugar molecule. They are chemically similar to glycosides, and can be hydrolysed to the corresponding sugars by water or dilute acids.

Sugar Esters. Esters of mono- or oligosaccharides with organic or inorganic ocids. The phosphate esters, *e.g.*, glucose 6-phosphate are fundamentally in intermediary carbohydrate metabolism. Condroitin and mucoltin sulfates, and heparin are typical animal sulfate esters.

Sulfate Activation. Formation of the active sulfates, Phosphoade nosinephosphosulfate (PAPS) and adenosine 5′-phosphosulfate (APS). PAPS is the subsirate of sulfotransferases in the synthesis of sulfate esters. APS is the substrate of sulfate assimilation and sulfate respiration.

Sulfate Assimilation, Assimilatory Sulfate Reduction. The reductive assimilation of oxidized inorganic sulfur leading to the biosynthesis of L-cystein. It is a property of plants and bacteria, whereas protozoa and animals can only perform the first step of S.a., *i.e.*, sulfate activation. The sulfate ion SO_4^{2-} is reduced to the oxidoreduction level of the sulfide ion, S^{2-}, or the mercapto group of L-cysteine.

Sulfate Esters. Products formed by the transfer of a sulfuryl group to the oxygen function or organic compounds. Donor for the sulfuryl transfer is Phosphoadenosinephosphosulfate (PAPS). Naturally occurring sulphate esters are Polysaccaride sulfate esters, and sulfate esters of phenols, steroids, choline, cerebrosides and flavonoids. Synthesis of phenol sulfate esters in mammalian liver was the first biological sulphate ester synthesis to be recognized, and the first to be performed in vitro.

Sulfate Reduction. Reduction of the sulfate ion SO_4^{2-} to the sulfide ion S^{2-}, in which the hexavalent, positive sulfur of the sulfate is converted to the divalent, negative form.

The enzymology of sulfate reduction has been studied in particular in enzyme preparations from baker's yeast (Saccharomyces cerevisiae) and the anaerobic bacterium, Desulfovibrio desulfuricans.

Sulfate Respiration, Dissimilatory Sulfate Reduction. A form of respiration in which the sulfate ion replaces oxygen as the terminal electron acceptor. The sulfate ion must first be activated. It is an anaerobic process in which sulfate is reduced to hydrogen sulfide, which is excreted, Ecologically, it contributes to desulfurication, and is important for the sulfur cycle of the biosphere.

Sulfhydryl Reagents, SH-reagents. Substances that react with thiol groups (syn. sulfhydryl or SH-groups) of proteins. In vivo they cause metabolic changes or alterations of function and they are generally toxic. In vitro they are used to detect, titrate and characterize SH-groups.

Superoxide Dismutase, SOD. Hyperoxide Dismutase, Superoxide : Superoxide Oxidoreductase, (EC 1.15.1.1) : When it was found that copper and zinc-containing proteins, hemocuprein (bovine erythrocytes), hepatocuprein (bovine liver) and cerebrocuprein (human brain) have a common identity, they were given the single name, cytocuprein.

SOD catalyses the dismutation (disproportionation) of superoxide : $O_2^- + O_2^- + 2H^+ \rightleftharpoons O_2 + H_2O_2$. There are two main types of SOD : 1. cyanide-sensitive, Cu and Zn-containing, eukaryotic enzymes, cyanide-insehsitive, Fe or Mn-containing, prokaryotic enzymes, SOD from liver mitochondria contains Mn and has M_r 80000 (4 subunits, M_r 20000).

Suppressor. With reference to Amber mutants and Ochre mutants, a S. mutation is a secondary mutation in a tRNA cistron, which restores the ability of the tRNA to recognize the nonsense codon. The resulting new tRNA is called suppressor tRNA. The nonsense codon then becomes a sense codon and codes for a specific proteogenic amino acid during translation.

Symbiosis. Close spatial coexistence of different species of advantage to all partners. It is therefore also called mutualistic S. Biochemically relevant symbiotic systems are not nodule bacteria—Leguminous plants; algafungus in lichens; rumen flora—ruminants; fungus—higher plant. Partners showing large differences in size are called macrosymbionts and microsymbionts.

Synchronous Culture. A synchronized cell population, in which all cells divide and pass through subsequent phases of the cell cycle at the same time. Synchronization can be achieved in various ways, *e.g.* by nutrient limitation, light stimulation, temperature change, treatment with antimetabolities of nucleic acid metabolism.

Synergists. Substances or factors that increase the biological activity of another substance or factor, but are inactive alone in the same quantity.

Synzymes, Enzyme Analogs. Synthetic macromolecules with enzymatic activity. They may be prepared by polymerization of amino acids or their derivatives, or by the attachment of catalytic groups to nonprotein materials.

T

T. Abb. for thymine.

Taka-amylase. A bacterial α-amylase (EC 3.2.1.1). isolated and crystallized from Aspergillus oryzae Taka-diastase preparations. It is a calcium-containing, single chain protein with N-terminal alanine and C-terminal serine.

Taraxasterol, α-lactucerol, α-anthesterol. A simple, unsaturated, pentacyclic triterpene alcohol, structurally a derivative of the parent hydrocarbon, 5α-taraxastane. It occurs as the acetate in the latex of the dandelion (Taraxacum officinale) and other members of the Compositae.

Taraxerol, Alnulin, Skimmiol. A simple, unsaturated, pentacyclic triterpene alcohol, structurally a derivative of the parent hydrocarbon, 5α-taraxerane.

Tarichatoxin. The main toxin of North American salamanders (Taricha torosa, T. rivularis). It is identical with Tetrodotoxin.

Tectoquinone. 2-methylanthraquinone, a yellow anthraquinone, m.p. 179° C, present in teak wood. It is one of the few nonhydroxylated, naturally occurring anthraquinones.

Teichmann's Crystals, Chlorhemin Crystals. Rhombic erystals formed by heating hemoglobin with sodium chloride and glacial acetic acid. T.c. are used for the microscopic detection of blood.

Teichoic Acids. The polymers present in the cell walls of Gram-postive bacteria. They consists of chains of glycerol or ribitol residues joined by phospate groups; in addition sugars are linked to the glycerol or ribitol, and some of the hydroxyl groups are esterified with residues of D-alanine. *E.g.* T.a. from Staphylococcus aureus H consists of eight ribitol units joined 1 → 5 by phospodiester linkages; the sugar, N-acetyl-glucosamine, is attached to position S of the ribitol chiefly by β-linkages with some α-linkages.

Template. A micromolecule that determines the structure of another micromolecule In protein translation, 3 nucleotides in the mRNA determine the nature of the amino acid at each position in the polypeptide (Genetic code).

Terminal. Adjective for the chain end component of a biopolymer, *e.g.* N and C-terminal amino acids.

Serminal Oxidase. The terminal enzyme of the respiratory chain. In most organisms it is Cytochrome oxidase but in various plant systems other T.o. are present or have been proposed. In aerobic nitrate respiration, it is nitrate reductase.

Termination. The final phase in the biosynthesis of bipolymers.

Termination Codon, Stop Codon Punctuation Codon. A sequence of three nucleotides in mRNA, which signals the end of polypeptide synthesis and release of the polypeptide in the process of Protein biosynthesis 5'-UAA, S'-UAG (Amber condon) and UAG are rermination codons.

Terminus. The chain end of a biopolymer, *e.g.* N-or C-terminus of a protein, meaning the N or C-terminal amino acid.

Terpene Alkaloids Isoprenoid Alkaloids. Alkaloids containing a terpene structure, with 10-30 carbon atoms. They are conveniently classified according to the plant general in which they occur.

TABLE

Classification of Terpene Alkaloids

Terpene type	Name
Monoterpene	Gentiana alkaloids Valeriana alkaloids
Sesquiterpene	Nuphara alkaloids Dendrobium alkaloids
Diterpene	Aconitum alkaloids Erythrophleum alkaloids
Triterpene (Steroid)	Solanum alkaloids Veratrum alkaloids Funtumia alkaloids Holarrhena alkaloids Buxus alkaloids Salamander alkaloids.

Terpenes, Terpenoids, Isoprenes, Isoprenoids. An extensive group of natural products whose structures are compo› of isoprene units. The number of carbon atoms is usually a multiple of 5. These are biosynthesized from the active 5-carbon unit, isopentenyl pyrophosphate.

Testane. Earlier name for 5α-androstane.

Testosterone. 17β-hydroxy-androst-4-en-3-one. An important member of the androgens. It is synthesized in the interstitial cells of testicular tissue, and stimulates growth of the prostate and seminal vesicles, and promotes sperem maturation and development of male secondary sexual characteristics. Apart from mammalian testes, it also occurs in blood and urine.

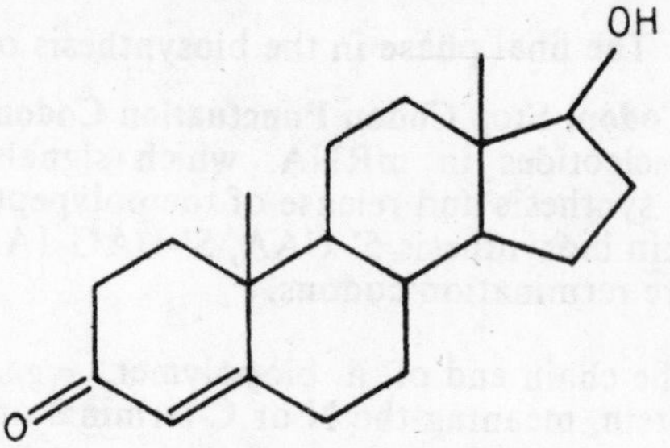

Testortene

19-nor-Testosterone. A synthetic anabolic steroid. It differs from Testosterone by the absence of the C 19 methyl group. It has a higher anabolic but lower androgenic activity than testosterone.

Tetracyclins. A group of antibiotics from various Streptomyces spp. T. contain four linearly fused six-membered rings; individual T. differ according to the nature of substituents (Fig. and Table).

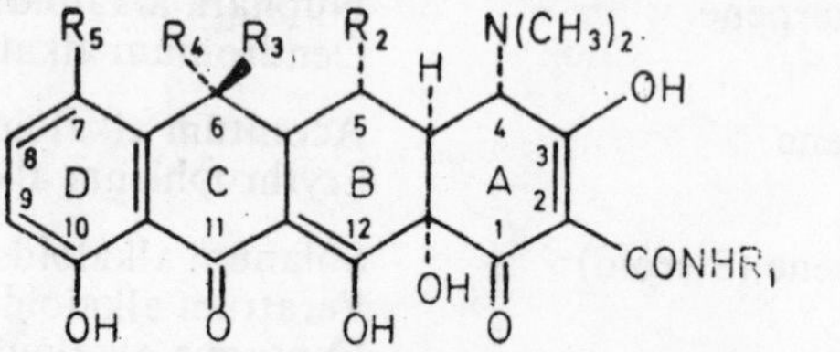

TABLE

The Structures of some Tetracyclins

Name	R_1	R_2	R_3	R_4	R_5
Chlortetracyclin (aureomycin)	H	H	OH	CH_3	Cl
Oxytetracyclin (terramycin)	H	OH	OH	CH_3	H

1	2	3	4	5	6
Tetracyclin	H	H	OH	CH_3	H
Methacyclin (rondomycin)	H	OH	CH_2=		H
Doxycyclin (vibramycin)	H	OH	H	CH_3	H

They inhibit protein biosynthesis by preventing the binding of aminoacyl-tRNA to ribosomes. Next to the pencillins. They were one of the most widely used antibotics, particularly in the treatment of bronchits, pneumonia, bile duct and urinary infections, plague and cholera.

Tetrahydrofolic Acid. abb. THF, folate-H_4 coenzyme F. 5, 6, 7, 8-tetrahydropteroylglutamic acid. The coenzyme responsible for the binding, activation and transfer of all active one carbon units, with the exception of carbon dioxide (the F in coenzyme F stands for formylation).

Tetraterpenses. Terpenes comprising eight isoprene units ($C_{40}H_{64}$). Naturally occurring tetraterpenes are almost all carotenoids, and the group contains no polycyclic compounds.

Tail-to-tail condensation of two molecules of geranylgeranylpyrophosphate gives phytoene, which undergoes stepwise dehydrogenation to produce the all-transconfiguration of the true caretenoids.

Tetrodotoxin, Tetrodontoxin; Spheroidine, Tarichatoxin, fugu Poison. Octahydro-12-(hydroxymethyl)-2-imino-5, 9 :7, 10a-dimethano-10a H-[1, 3] dioxocino [6, 7-d) pyrimidine-4, 7, 10, 11, 12-pentol, a guanidine derivative that exists in 2 tautomeric forms. It is an extremely potent toxin from the ovaries, liver and skin (but not present in the blood) of many species of Tetrodontidae, especially the globe fish, Spheroides rubripes. It acts on the membranes of nerve fibres, and is an autagonist of Batrachotoxin.

Tetrose. A monosaccharide containing four carbon atoms, *e.g.* threose, erythrulose. T. occur as intermediates in carbohydrate metabolisms, usually as their phosphates.

Thaumatin. A sweet tasting, strongly basic, histidine and carbohydrate-free, single chain protein. It is 750-1600 times (weight basis) or 30000-100000 times (molar basis) sweeter tasting than sucrose. It occurs in the fruits of Thaumatococcus daniellii (a monocotyledon of the arrowroot or Maranta family). It shows considerable sequence homology with the B-chain of another sweet protein, Monellin. These two proteins are immunologically related, and it is thought that a tripeptide sequence (-Glu-Tyr-Gly-) near the surface of each molecule is a common antigenic determinant.

THC. Abb. for Δ^1-tetrahydrocannabinol.

Thd. Abb. for Ribothymidine.

Thermolysin. (EC 3.4.24.4) A heat-stable, zinc and calcium-containing neutral protease, from Bacillus thermoprotelyticus, with a substrate specificity similar to that of Subtilisin. After one hour at 80°C, it still has 50% original activity. This high

$$(H_3C)_2\overset{\oplus}{S}-CH_2-COO^{\ominus}$$

Dimethylthetir

heat stability of thermolysin is attributed to the large number of hydrophobic regions and the presence of four bound calcium ions, which serve in place of disulfide bridges to maintain the compact shape of the molecule. It is neither a thiol nor a serine enzyme.

Thermostable Enzymes, Heat Stable Enzymes. A small number of enzymes, mostly hydrolases, whose temperature optima lie between 60 and 80°C. They usually have a compact structure, stabilized by many disulfide bonds and/or extensive hydrophobic regions, and a low α-helix content.

Thetins. Sulfonium compounds, *e g.* dimethyl-thetin, which can be used as a methylating agent.

Thiamine Pyrophosphate, abb. TPP, Aneurin Prophosphate, abb. APP, Cocarboxylase. The pyrophosphoric ester of thiamine (aneurin, vitamin B, and the prosthetic group (or coenzyme) of various thiamine pyrophosphate enzymes, *e.g.* Pyruvate decarboxylase, pyruvate dehydrogenase and 2-oxoglutrate dehydrogenase, Transketolase glyoxylate carboligase and oxalyl-CoA decarboxylase.

In its role as a coenzyme, TPP reacts with the substrate of TPP-enzymes to form active aldehydes.

Thin-layer Chromatography, TLC. A form of chromatography in which the solid carrier material is spread in a thin layer on a glass or plastic plate. The advantages of the method are the short distances required for good separation and the correspondingly short development times, high sensitivity, separation of very small amounts of substances, and, if inorganic carrier materials are used, the possibility of using caustic detection reagents.

TLC can be used preparatively as an "open column", Depending on the size of the plate, upto 100 g of material can be separated by preparative TLC while analytical TLC can be used for amounts between 10 mg and 10 mg.

Thioester, Acylmercaptan. A compound of the general formula RS~CO-R_1. The thioester (acylmercaptan) bond is energy-rich. All fatty-acyl coenzyme A derivatives (activated fatty acids, *e.g.* acetyl-CoA) are thioesters. During substrate phosphorylation on glyceraldehyde 3-phosphate dehydrogenase a thiol group of the enzyme forms an energy-rich intermediate T.

Thiol Enzyme SH-enzyme. An enzyme whose activity depends upon the presence of a certain number of free thiol groups. Total enzyme are found amongst the hydrolases, oxidoreductases and transferases.

Thiol Group, Sulfhydrl Group, Mercapta Group. SH. The functional group of thiols (mercaptans), RSH, where R is the remainder of the molecule. The functional form of lipoic acid and thioredoxin is a dithiol.

2-Thiomethyl-N[6]. Isopentenyladenosine 2-methylmercapto-6-isopentenyladenosine, an adenosine derivative and one of the rate nucleic acid components found in tRNA from wheat. It is an active cytokinin. The hydroxylated derivative, 2-methylmercapto-6-(4-hydroxy-3-methyl-cis-2-enylamino)-purine has also been found in some species of tRNA.

Thioredoxin. A heat-stable, acidic, metal-free redoxin, Mr 12000. It is a component of deoxyribose synthase, in which thioredoxin and thioredoxin reductase form a hydrogen transfer system linked to the reduction of ribose or ribonucleoside phosphates by NADPH+H^+.

Thr. Abb. for L-threonine.

L-Threonine. abb. **Thr.** L-threo-α-amino-β-hydroxybutyric acid, H_3C-CH(OH)-CH (NH_2)-COOH. A proteogenic, essential amino acid with two asymmetric C-atoms. A useful reaction for the determination of L-T is oxidation with periodate to acetaldehyde, glyoxylate and ammonia. Enzymatic hydrolysis of peptide bonds involving L-T. appears to be particularly difficult, which may be relevant to the nutritional physiology of this amino acid.

Thrombin. A blood coagulation enzyme, responsible for the conversion of fibrinogen to fibrin. It is a glycoprotein (5% carbohydrate), Ar produced - by activation of Prothrombin. Degradation of bovine Prothrombin (582 amino acid residues) by factor X_a and by thrombin itself produces the two chains of α-T. of known primary structure (A-chain, Mr 5700, N-terminal theonine, 49 amino acids; B-chain N-terminal isoleucine, 259 amino acids, with carbohydrate attached), which are linked by one disulfide bridge.

Thy. Abb. for thymine.

Thylakoids. Internal membrane structures of the chloroplast. Under the electron microscope, these appear as disc-shaped, flattened vesicles, about 600 nm diameter. These are arranged in stacks, which are the grana observable under the light miscroscope.

Thymidine, more correctly deoxythymidine, Abb. dThd, thymine deoxyriboside. A Nucleoside of thymine and D-2-deoxyribose. It should not be confused with Ribothymidine, which is also (and more correctly) called thymidine.

Thymidine Phosphates, more correctly deoxythymidine phosphates. Nucleotides of thymine; phosphate esters of deoxythymidine. Although T.p. contain deoxyribose, the prefix deoxy is usually omitted, because the corresponding ribose derivatives hardly ever occur naturally. Thymidine 5'-monophospaate abb. TMP, thymidylic acid (more correctly deoxythymidine 5'-monophosphote, abb. dTMP, deoxythymidylic acid) a component of DNA, and an intermediate in the synthesis of TPP. Stepwise phosphorylation of TMP leads to thymidine 5' diphosphate, abb. TDP (more correctly deoxythymidine 5'-diphosphate, abb. dTDP), which serves as the activating group in certain Nuclaoside diphosphate sugars and to thymidine 5'-triphosphate, abb. TTP (more correcrly deoxythymidine 5'-triphosphate, abb. (dTTP), Mr 482.18, a sub-state of DNA synthasis.

Thymine, abb. T or Thy 2, 6-dihydroxy-5-methylpyrimidine, 5-methyluracil. A pyrimidine base present in DNA. It is formed by the degradation of thymidine 5'-monophosphate; the methyl group of T. does not arise from methionine, but from an active one carbon unit.

Thymonucleic Acid, Thymus Nucleic Acid. Nucleic acid from the thymus gland; effectively an obsolete term for DNA.

Thymopoetin, Thymin, Thymosin. A polypeptide hormone from the thymus (49 amino acid residues of known sequence). It is required for the general differentiation of thymobytes, but has no influence on the aciquisition of the immunological repertoire.

Thyroid Gland. Glandula thyreoidea, A well vasculated gland at the front of the neck. It is paired in amphibians and birds, and unpaired in elasmobranch fish and mammals, synthesizes, stores (in the thyroid follicles) and secretes Thyroxin and triiodothyronin, under the influence of the anterior pituitary hormone thyrotropin.

Thyrotropin, Thyroid Simulating Hormone, abb. **TSH.** A glycoprotein hormone, containing 23% carbohydrate. Primary structure of some TSH molecules is known. It consists of an α-and a β-chain, and the α-chain is structurally similar to Luteinizing hormone (see). Synthesis occurs in the basophilic cells of the anterior pituitary. Both synthesis and secretion are stimulated by thyrotropin releasing hormone from the hypothalmus, and inhibited by thyroxin.

Thyroxin, 3, 5, 3, 5-tetraiodothyronine, abb. **T_4.** A hormone produced by the thyroid gland and absolutely essential for growth and development. T_4 and the second thyroid hormone, 3-5-3′-triiodothyronine are synthesized from L-tyrosine residues in thyroglobulin, a diameric glycoprotein that constitutes the bulk of the thyrpid follicle. Tyrosine residues in thyroglobulin become iodinated, so that the protein contains several mono- and diiodotyrosine residues.

Tin, Sn. A metal occurring in many tissues and dietary components. The redox potential of $Sn^{2+} \rightleftharpoons Sn^{4+}$ is 0.13 volt, near to the redox potential of the flavin enzymes, suggesting a possible biological role. It has been reported that Sn is essential for the growth of rats.

Tissue Hormones. Hormones produced in specialized, single cells scattered through a tissue rather than clumped in a gland. They fall into three groups :

1. Secretin, Gastrin and Cholecystokinin from the gastrointestinal tract:
2. Angiotensin and Bradykinin which occur as inactive precursors in the blood; and
3. Biogenic amines such as Histamine, Serotenin, Tryamine and Melatonin.

*T*m. value **Melting Point.** The temperature, in °C, at which a double standard nucleic acid becomes 50% denatured to the single stranded form. A DNA solution is heated and its absorbance at 260 nm is plotted against temperature. Transition from double to single stranded DNA occurs over a relatively narrow temperature range, and is characterized by an increase in absorbance at 260 nm. Tm is taken as the temperature at the mid point (half the final increase of absorbance at 260 nm) of the S-shaped curve.

TN. Abb. for troponin.

Toad Toxins. The poisons found in the secretions of the skin glands of toads (Bufonidae). These are classified as :

1. bufadienolides (bufogenins) with digitalis-like effects on the heart *e.g.* bufotoxin. They strengthen and slow the heartbeat. The bufadienolides are present in toad blood at a dilution of 1 ; 5000 to 1 : 20000, and they are necessary for normal heat activity.

2. Alkaline toxins which are alkaloids derived from tryptamine or indole, *e.g.* bufotenine, dehydrobufotenine and O-methylbufotenine.

α-Tomatine, Tomatine. A Solanum alkaloid, and the chief alkaloid of the tomato (Lycopersicon esculentum). also occurring in other Lycopersicon and Solanum spp. It is a glycoalkaloid of the aglycon tomatidine [(22S : 25S)-5α-spirosolane-3β-ol, and the tetrasaccharide tβ-lycotetraose. α-tomatine, imparts a bitter taste and protects the tomato plant from attack by the Colorado potato beetle. It also has antibiotic activity against the causative agents of tomato wilt and other pathogenic fungi.

Toxicology. The study of toxins, their physiology, biochemistry and the molecular basis of their activity.

Toxic Proteins. Mostly low Mr single chain, nonenzymic proteins, produced especially by snakes and invertebrate animals, but also by some plants (phytotoxins) and virulent strains of bacteria. With the exception of bacterial enterotoxins and Botulinus toxins these show practically no oral activity, and are only toxic when injected, *i.e.* when the digestive tract it bypassed.

Toxoflavin. 3, 8-dimethyl-2-4-dihydroxypyrimido (5, 4-e)-as-triazine. An antibiotic from Pseudomonas coccovenans, with high antibacterial activity, but no activity against fungi. In its biosynthesis C8 of a purine precursor is removed, and the as-triazine ring is formed by introduction of the aminomethyl group of glycine.

Toyocamycin. 4-Amino -cyano-7-(D-ribofuranosyl)-pyrrolo-(2, 3-d)-pyrimidine, 6-amino-7-cyano-9-β-D-ribofuranosyl-7 deazapurine a 7-deazaadenine antibiotic from Streptomyces toyocaensis and S.rimosus. Biosynthesis is analogous to that of Tubericidin *i e.* the carbon atoms of the pyrrole ring are derived from 5-phosphoribosyl l-pyrophosphate. It is particularly active against Candida albicans, Saccharomyces cerevisiae and Mycobacterium tuberculosis.

TPN. Abb. for triphosphopyridine nucleotide.

TPP. Abb. for Thiamine pyrophosphate.

Trace Elements, Microelements. Elements required in very small quantities by living oranisms. They act catalytically, or are components of catalytic systems and other mineral nutrients is not always possible.

Trace Nutrients, Micronutrients. A general term for any essential dietary component required in small quantities, like Trace elements and Vitamins deficiency leads to deficiency symptom, *e.g.* vitamin deficiency diseases.

D-Sedoheptulose 7-phosphate + D-Glyceraldehyde phosphate ⇌ (Transaldolase) D-Erythrose 4-phosphate + D-Fructose 6-phosphate

Sugar — Lysine — Enzyme protein

Transaldolase Reaction (above), and Binding of the Ketose to the ε-amino Group of Lysine Residue of the Enzyme (below)

Transacylation. Reversible transfer of acyl group (R-CO) from a donnor to an acceptor, *e.g.* transfer of the acyl residue CH_3-CO- by acetyl CoA to an acceptor Y :

$$CH_3\text{-CO-S-CoA} + Y \rightarrow CH_3\text{—CO—}Y + \text{CoA}$$

It is catalysed by transacylases, which are important in the synthesis and degradation of fatty acids, synthesis of conjugated bile acids via cholic acid-CoA compounds, and other reactions such as acetylation of amino acids and amines.

Transaldolation. A reaction of carbohydrate metabolism, in which a C_3-unit (equivalent to a dihydroxyacetone unit) is transferred from a ketose to an aldose. It is catalysed by transaldolase (EC 2.2.12). The C_3-unit does not exist in the free state, but remains bound to the ϵ-group of a lysine residue in the enzyme (Fig.).

Transamidation. Transfer of the amide nitrogen of Glutamine as an NH_2-group. T. is catalysed by transamidases. All glutamine transamidases so far investigated have a catalytically important thiol group in their active centres and are inhibited by the glutamine analogs, azaserine, 6-diazo-5-oxonorleucine (DON) and L-2-amino-4-oxo-2-chloropentanoic acid ("chloroketone"), *e.g.* anthranilate synthase (EC 4.1.3.27), carbamoyl phosphate synthetase (EC 6.3.5.5), transglutaminase (EC 2.3.2.13), 5′-phosphoribosyl-N-for-mylglycinamidine synthetase (EC 6.3.5.3), glutamate synthtase (EC 1.4.1.13).

Transamidinases, Amidinotransferases. Enzymes catalysing Transamidination catalyse transfer of the amidine group of arginine in the synthesis of creatine and other Phosphagens. They from Streptomyces griseus and S. baikiniensis catalyses amidine transfer in the biosynthesis of streptidine.

Transamidination. Reversible enzymatic transfer of the amidine group.

$$-\overset{\overset{\Large NH}{\|}}{C}-NH_2$$

, between guanidines. It is a group transfer reaction of nitrogen metabolism, which occurs in two stages and involves an intermediate enzyme-amidine complex :

$$R\text{-}NH\text{-}\overset{\overset{NH}{\|}}{C}-NH_2+\text{Enzyme-SH}\rightleftharpoons R\text{-}NH_2+\text{Enzyme-S-}\overset{\overset{NH_2}{\|}}{C}\text{-}NH_2$$

$$\text{Enzyme-S-}\overset{\overset{NH}{\|}}{C}\text{-}NH_2+R\text{-}NH_2\rightleftharpoons R\text{-}NH\text{-}\overset{\overset{NH}{\|}}{C}-NH_2+\text{Enzyme-SH}$$

Transaminases, Aminotransferases. (EC sub-group 2.6.1). Enz, catalysing transmination, *i.e.* the reversible transfer of the amino group of a specific amino acid to a specific oxoacid.

Coenzyme of transminases is pyridoxal 5′-phosphate, which becomes bound to the apoenzyme by condensation of its carboxyl group with the ε-amino group of a lysine residue, forming a Schiff's base or internal aldimine.

Transmination. Reversible transfer of amino groups, between two amino acids and their respective keto acids, catalysed by transaminases.

Transcarbamylation. Transfer of the carbamyl group of Carbamoyl phosphate.

Transcription. The DNA-dependent synthesis of RNA.

Transdeamination. Conversion of the amino group of an amino acid to ammonia by the combined action of a transaminase (TA) and L-glutamate dehydrogenese (GDH) (EC 1.4.1.2; 1.4.1.3) :

$$\text{Amino acid} + \text{2-oxoglutarate} \overset{\text{TA}}{\rightleftharpoons} \text{2 oxoacid} + \text{Glutamate;}$$

$$\text{Glutamate} + NAD^{+} + H_2O \xrightarrow{\text{GDH}} \text{2-oxoglutarate} + NADH + H^{+} + NH^{+}{}_{4}.$$

It is an important process in ureotelic organisms, where it accounts for most of the ammonium entering the Urea cycle via carbamoyl phosphate (the other nitrogen atom incorporated into urea is dervied directly from aspartate, formed by transamination of amino with oxaloacetate).

Transduction. Transfer of DNA from one bacterial cell to another by bacteriophage. There are two types. In generalized transduction the phage infects the bacterial cell (the donor) and enters a nonlysogenic cycle leading to lysis of the cell and release of progeny.

Transfer-RNA, tRNA, Soluble RNA, sRNA, Acceptor RNA, Transport RNA. The smallest known functional RNA, present in all living cells and essential for Protein biosyntheis. Different tRNAs contain between 70 and 85 nucleotide residues. There is at least one specific tRNA per cell for each of the 20 protein amino acids. There may be between 50 and 70 tRNA species within one cell; this multiplicity is the result of organelle specificity, and the fact that there may be two or more different but specific tRNAs for one amino acid. The source and specificity of a tRNA species is indicated by a code, *e.g.* $tRNA^{Val}{}_{yeast}$ is the valine-specific tRNA from yeast [^{14}C-Val] $tRNA^{Val}{}_{yeast}$ represents thenamed tRNA esterified with ^{14}C-labelled valine.

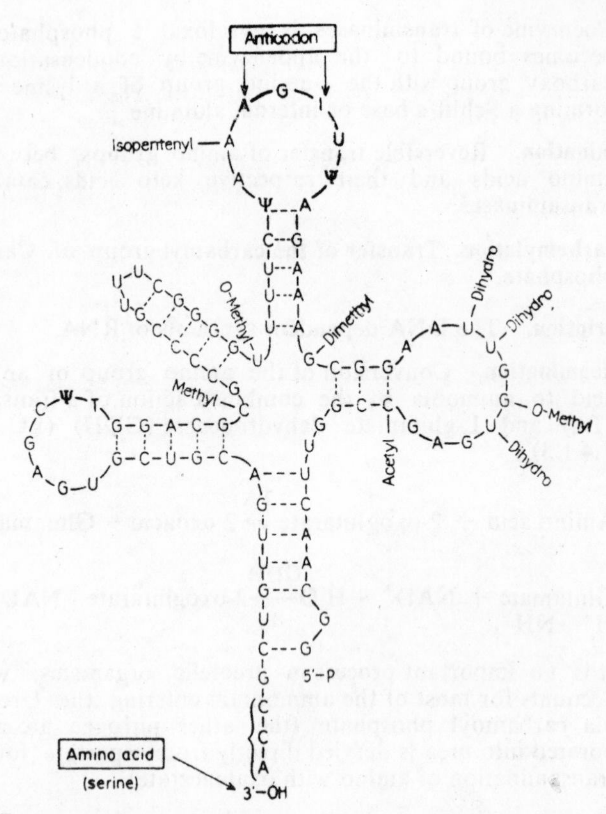

Clover Leaf Model of a tRNA Molecule (Serine-specific tRNA from yeast A, Anenine; C, Cytosine; G, guanine; I, inosine; U, uracil; T, thymine; ψ, pseudouracil.

Transformation. Conceptually the simplest form of genetic transfer. "Naked" DNA from a donor cell enters a recipient cell and is incorporated into the recipient DNA by genetic recombination. There is no other carrier substance or structure involved; small fragments of the donor DAN simply penetrate the membrane (and wall if it is present) of the recipient cell.

Transglycosidation. Transfer of a glycosidically bound sugar residue to another molecule with a suitable recipient OH-group. It is catalysed by transglycidases, *e.g.* galactosidase catalyses transfer of a galactose residue from lactose to the C6-hydroxyl group of glucose, or to a further lactose molecule to form the trisaccharide, 6-galactosidolactose.

Transition. Replacement of a purine by a different purine, or a pyrimidine by a different pyrimidine in the polynucleotide chain of DNA. It results in a gene mutation. It may occur spontaneously, or it may be promoted experimentally with mutagens.

Transketolase (EC 2.2.1.1). An enzyme that catalyses transketolation, an important process of carbohydrate metabolism, especially in the Pentose phosphate cycle and Calvin cycle. It has been found in a wide variety of cells and tissues, including mammalian liver, green plants and many bacterial species. The enzyme contains divalent metal cations and the coenzyme, thiamine pyrophate.

Translation. In the wider sense equivalent to Protein biosynthesis. In the narrower sense, it is the decoding process whereby each codon in mRNA is translated into one of 20 amino acids during protein synthesis on polysomes.

Transmethylation. Transfer of a methyl group($-CH_3$) from a physiological methyl donor to C-, O- and N-atoms of biomolecules. Transmethylation to oxygen produces the methoxy group ($-OCH_3$). The most important methyl donor is S-Adenosyl-L methionine.

Transplantation Antigens, Histocompatibility Antigens. Antigens on the surface of nucleated cells, particularly leucytes and to a lesser extent thrombocytes.

Transport. Passage of ions and certain molecules through biological membranes. Most polar molecules do not pass freely across biomembranes. Exchange of essential metabolites between a cell and its surroundings, or between cytoplasm and organelles, therefore depends on Transport mechanisms within the membranes.

Transsulfuration. Exchange of sulfur between L-homocysteine and L-cysteine, with L-cystathionine as an intermediate. Strictly speaking, it is not a group transfer reaction, because the sulfur bond formed in the synthesis of cystathionine is different from that broken in the formation of L-cysteine. It operates in the biosynthesis of L-cysteine from L-methionine, and in the biosynthesis of L-methionine. The methionine precusor, L-homocysteine, is formed by transsulfuration as follows :

L-Homoserine + Succinyl-CoA → CoA + O-Succinyl-L-homoserine

O-Succinyl-L-homoserine + L-Cysteine → L, Cystathionine + Succinate

L-Cystathionine + H_2O → L-Homocysteine + pyruvate + NH_3

Transversion. Replacement of a pyrimidine by a purine, or a purine by a pyrimidine in the polynucleotide chain of DNA. It results in a gene mutation.

Trehalose. A nonreducing disaccharide, consisting of two glucopyranoside residues. There are three forms of trehalose depending on the nature of the glycosidic linkage : α, α-T. α, β-T. or neotrehalose ($[\alpha]_D^{20}+95°$); β, β-T. or isotrehalose. It is present in algae, bacteria, numerous lower and higher fungi, and occurs sporadically in nonphotosynthetic tissues of higher plants. It is the "blood sugar" of insects. It is cleaved by many fungal enzymes, and it is fermented by certain yeasts.

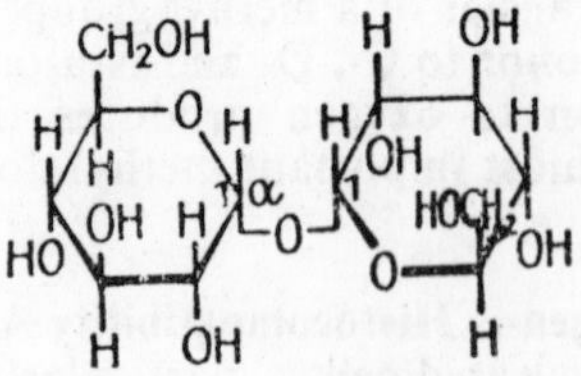

α, α-Trehalose

TRF. Abb. for thyrotropin releasing factor (Releasing hormones).

TRH. Abb. for thyrotropin releasing hormone.

Triamcinolone. 16α-hydroxy-9α-fluoroprednisolone; 9α-fluoro-11β, 16α, 17α, 21-terahydroxy-pregna-1,4-diene-3,20-dione. A synthetic steroid prepared from cortisol. Its antiinflammatory activity is 50 times greater than that of cortisone acetate and it does not cause undesirable salt retention. It is used in the treatment of arthritis, allergies, etc.

Tricetinidin. 3-deoxydelphinidin.

Trichochromes. Yellow-orange and violet natural pigments containing a substituted $\Delta^{2,2'}$-bis (1,4-benzothiazine) ring system Color is due to the conjugated chromophore system, S-C=C-C=N-. Biosynthetically, T. are related to the Melanins (see). Together with the phaeomelanins, T. are responsible for red and auburn colors of human hair and bird feathers. T.B and C are yellow-orange; T. E and F are violet.

Trichochrome B: $R_1 = H$; $R_2 = CH_2\text{-}CHNH_2\text{-}COOH$; $R_3 = COOH$

Trichochrome C: $R_1 = CH_2\text{-}CHNH_2\text{-}COOH$, $R_2 = H$; $R_3 = H$

Trichochrome E: $R_1 = H$; $R_2 = CH_2\text{-}CHNH_2\text{-}COOH$

Trichochrome F: $R_1 = CH_2\text{-}CHNH_2\text{-}COOH$; $R_2 = H$

Tricholomic Acid. Erythro-dihydroibotenic acid, a compound isolated from the Basidomycete (mushroom) Tricholoma muscarium. It possesses flavor promoting activity similar to, but much more active than, that of sodium glutamate. It also has a synergistic effect on the flavor improving property of inosinic acid and guanosine 5′-phosphate.

Tri(hydroxymethyl)methylamine, TRIS. $H_2N\text{-}C(CH_2OH)_3$. A widely used buffer substance, suitable for the pH range 7-9. The required pH is usually obtained by adding HCl to TRIS dissolved in water. TRIS buffers have a high temperatute/pH gradient.

Triose Phosphates. D-glyceraldehyde 3-phosphate ($PO_3H_2\text{-}OCH_2\text{-}CHOH\text{-}CHO$) and dihydroxyacetone phosphate ($PO_3H_2\text{-}OCH_2\text{-}CO\text{-}CH_2\ OH$). Important intermediates in Glycolysis and Alcoholic fermentation. The two triose phosphates are interconvertible via the ene-diol form, by the action of triose phosphates isomerase; the equilibrium mixture contains 96% ketotriose phosphate and 4% aldotriose phosphate.

Trioses. Glyceraldehyde and dihydroxyacetone phosphate. They contain 3 C-atoms and are the simplest monosaccdarides. Their phosphates are important metabolic intermediates.

Triterpenes. An extensive group of terpenes biosynthesized from six isoprene units. Apart from squalene, which is acyclic, most of this group are tetra or pentacyclic hydroarmatic compounds based on the parent hydrocarbon, sterane, *i.e.* they are steroids. Tail-to-tail condensation of two molecules of farnesylpyrophosphate produces squalene which serves as the precursor of the cyclic triterpenes.

tRNA Abb. for transfer-RNA.

Tropane Alkaloids. A group of Alkaloids. These are esters of various aminoalcohols based on a substituted tropane (N-methyl-8-azabicyclo-3,2,1-octane) ring system. This system is not optically active, because the two rings can only be linked in a cis configuration, thus resulting in a meso form.

Tropinone, Tropane-3-one. A possible biosynthetic precursor of the Tropane alkaloids present in Solanaceae spp. Robinson's laboratory synthesis of T. (Fig.) from succindialdehyde, methylamine and the calcium salt of acetone-dicarboxylic acid in aqueous solution at ordinary temperatures (yield 42%. 2 molecules of CO_2 are readily lost on subsequent treatment with acid), *i.e.* under mild or apparently physiological conditions, gave impetus to modern studies on alkaloid biosynthesis.

COOH
CH_2
$H_2C—CHO$
$H_2C—CHO$ + $H_2N—CH_3$ + C=O ⟶ N-CH_3 =O
CH_2
COOH

Succin-dialdehyde Methylamine Acetone-dicarboxylic acid Tropinone

Trp. Abb. for L-tryptophan.

Trypsin. (EC 3.4.21.4). A serine protease present as a zymogen (trypsinogen) in the pancreas of all vertebrates. Trypsinogen is released via the pancreatic duct into the duodenum. Conversion of trypsinogen into trypsin is initiated in the small intestine by enterokinase, and accelerated autocatalytically by traces of trypsin.

Tryptamine. β-indolyl-(3)-ethylamine. A biogenic amine, produced by decarboxylation of tryptophan. It stimulates the contraction of smooth muscle of blood vessels, uterus and central nervous system. It is found in both plants and animals, and as a bacterial degradation product of tryptophan.

L-Tryptophan, abb. **Trp, α-amino-β-indolepropionic Acid.** An aromatic, essential amino acid. Trp is nutritionally very important, although it is present in relatively small amounts in proteins. Acid hydrolysis of proteins completely destroys Trp. p-Dimethylaminobenzaldehyde or xanthydrol gives a violet coloration with Trp, which is used in its determination.

Tryptophan 2,3-oxygenase, Tryptophan Oxygenase, Tryptophan Pyrrolase, L-tryptophan : Oxygen 2,3-oxidoreductase (Decyclizing), (EC 1.13.11.11). An enzyme catalysing the oxidation of L-tryptophan to N. formylkynurenine, the first stage in the total degradation of L-Tryptophan in animals and microorganisms; this is also the first stage in the conversion of L-tryptophan into the nicotinamide moiety of NAD and NADP in molds and in some mammals, and in the conversion of L-tryptophan into Ommochromes in insects.

L-Tryptophan Synthase, Tryptophan Desmolase, L-serine Hydro-lyase (Adding Indoleglycerol-phosphate), (EC 4.2.1.20). The enzyme catalysing the synthesis of L-tryptophan from L-serine and indole 3-glycerol phosphate. This from Escherichia coli (Mr 149000) has $\alpha_2\ \beta_2$ subunit composition. The enzyme separates easily into monomeric subunit a (also called protein B) (Mr 29000) and dimeric subunit β_2 (also called protein B) (Mr of dimer 90 000) when eluted from DEAE cellulose with a sodium chloride gradient. The separated subunits catalyse partial reactions of L-tryptophan synthesis ;

$$\text{Indole 3-glycerol phosphate} \xrightarrow{\alpha} \text{indole} + \text{3-phosphoglyceraldehyde}$$

$$\text{Indole} + \text{L-serine} \xrightarrow{\beta_2} \text{L-tryptophan} + H_2O.$$

TSH. Abb. for thyroid stimulating hormone, or thyrotropin.

Tubercidin. 6-amino-9-β-D-ribofuranosyl-7-deazapurine. A purine antibiotic (Nucleoside antibiotics) from Streptomyces tubercidicus, and one of the group of 7-deaza-adenine-nucleoside analogs. The N7 of adenine is replaced by a methylene group. T. is biosynthesized from adenosine the C-atoms of the pyrrole ring are derived from a ribose moiety, which is introduced from 5-phosphoribosyl 1-pyrophosphate.

Biosynthesis of Tubercidin by Streptomyces Tubercidicus

Tubulins. Dimeric proteins of two closely related subunits (subunit Mr 60 000). They are the major components of microtubules, accompanied by smaller quantities of higher Mr proteins. Microtubules are composed of a series of parallel filaments, formed by end-to-end aggregation of T. molecules. Each molecule of tubulins strongly binds one molecule of GTP and loosely binds a second molecule. There is thus an analogy with actin, but the two proteins are structurally dissimilar. The alkaloid, Colchicine binds strongly to tubulins and causes disassembly of labile microtubules, such as the mitotic spindle.

Tumor Antigens. Carcinoembryonic antigens, which serve as an aid to the early recongnition of liver carcinoma and teratoblastomas (tumors of reproductive cells, especially in the testes and ovaries). They are embryonal plasma proteins produced in the placenta during pregnancy and in some organs of the embryo, but on longer detectable shortly after birth. They can be formed again later in life in response to malignant tumors.

Tunicamycin. A mixture of homologous, nucleoside antibiotics, produced by Streptomyces lysosuperificus, and active against viruses, Gram-positive bacteria, yeast and fungi. The structure of tunicamycin consists of one residue each of uracil, a C_{11}-aminodeoxy-dialdose (tunicamine), N-acetylglucosamine and a fatty acid. They differ from one another in the chain length of the fatty acid component. The major fatty acid components are trans α,β-unsaturated iso-acids.

Tunichrome. A green chromogen in the blood cells of tunicates, *e.g.* Ascidia nigra, Ciona intestinalis, Molgula manhattensis.

Turnover. The balance of synthesis and degradation of biomolecules in living organisms. All cell components are subject to continual degradation and resynthesis, *i.e.* they are subject to T.

Turpentine Oil. A volatile oil obtained by steam distillation of turpentine from various Pinus spp. It is a colorless liquid, most of which boils between 155 and 162°C. ρ 0.865—0.870, n_D^{25} 1.465—1.480. On exposure to air, it alters rapidly and finally resinfies. Its composition varies with origin, chief components being bicyclic monoterpenes of the carane and pinene type.

Tyr. Abb. for L-Tyrosine.

Tyramine. β-hydroxyphenylethylamine. A biogenic amine, found in plants (ergot, broom and other legumes) and animals (blood, urine, bile, liver), and as a bacterial degradation product of L-tyrosine. It is produced by the decarboxylation of L-tyrosine.

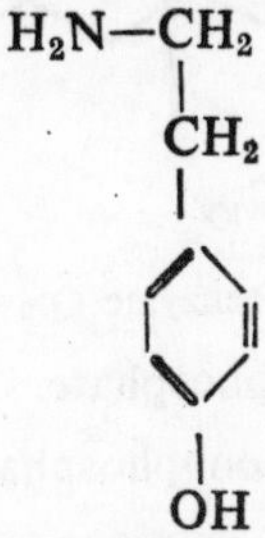

H_2N-CH_2
|
CH_2
|
(benzene ring)
|
OH

Tyramine

Tyrian Purple. 6,6′-dibromoindigotin. A red-violet pigment containing two brominated and oxidized indole ring systems. It is present in marine molluscs of the genera Murex and Nucella, and a few related whelks.

Tyrocidins. Homodetic, homomeric, cyclic peptide antibiotics active against Gram positive bacteria Like the Gramicidins they are produced by Bacillus brevis. The structure of tyrocidin **A**, confirmed by total synthesis, is cyclo-(-Val-Orn-Leu-D-Phe-Pro-Phe-D-Phe-Asn-Gln-Tyr-). In tyrocidin B, Phe_6 is replaced by Trp.

L-Tyrosine, abb. Tyr. L-α-amino-β-(p-hydroxy-phenyl)-propionic acid, an aromatic, ketogenic, proteogenic amino acid. It is nonessential in humans, since it can be synthesized by hydroxylation of L-phenylalanine. Millon's color reaction for proteins specific for L-T. residues.

U

U. Abb. for uracil.

UA Lase. Abb. for urea amidolyase.

Ubiquinone, Coenzyme Q, abb. **Q.** A low M_r electron transport component of the respiratory chain. Structurally, it is a 2, 3-dimethoxy-5-methylbenzoquinone carrying an isoprenoid side chain composed of dihydroisoprene units.

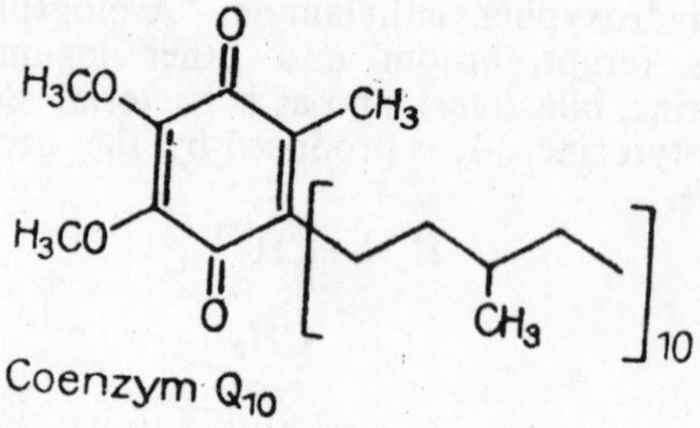

Coenzyme Q_{10}

UDP. Abb. for uridine 5′-diphosphate.

UMP. Abb. for uridine 5′-monophosphate.

Uncoupler. A chemical compound which prevents oxidative phosphorylation of ADP to ATP in the respiratory chain without affecting electron transport. It thus uncouples respiration and ATP formation; respiration continues or is even increased, but yields no energy. Some important U. are 2,4 dinitrophenol (abb. DNP), dicumarol, carbonyl cyanophenylhydrazone, salicylanilide, etc.

Ura. Abb. for uracil.

Uracil, abb. **U** or **Ura.** 2,4-dihydroxypyrimidine, a 1,3-diazine, which occurs as a pyrimidine base in all ribonucleic acids. It is the starting point for reductive and oxidative Pyrimidine degradation.

Urates. Salts of Uric acid.

Urd. Abb. for uridine.

Urea, Carbamide. $H_2N - CO - NH_2$, the diamide of carbonic acid. M_r 60.01, m.p. 132.7°C. Urea is the product of ammonia detoxification in the ureotelic animals.

Urea Amidolyase, ATP : Urea Amido-lyase, abb. **UALase, Urea Carboxylase (Hydrolysing)** (EC 6.3 4.6). A urea splitting enzyme present in some yeasts (Saccharomyces, Candida, etc.) and green algae (Chlorella, etc). where it replaces urease. It is a biotin enzyme and is inhibited by avidin. The catalysed reaction is an ATP-dependent cleavage of urea to CO_2 and NH_3 :

$$NH_2{}^-CO\text{-}NH_2 + HCO^-{}_3 + ATP \rightarrow 2HCO^-{}_3 + 2NH^+{}_4 + ADP + P_i$$

Urea Cycle, Arginine-urea Cycle, Ornithine Cycle, Krebs-Henseleit Cycle. A metabolic cycle present in mammals and other ureotelic animals (*e.g.* adult amphibians), which results in the synthesis of urea from carbon dioxide, ammonia and the α-amino nitrogen of L-aspartic acid. The process is energy-dependent.

Urease (EC 3.5.1.5). An enzyme of high catalytic activity that catalyses the hydrolysis of urea to CO_2 and NH_3 :

$$O = C \begin{matrix} \diagup NH_2 \\ \diagdown NH_2 \end{matrix} + 2H_2O \rightleftharpoons H_2CO_3 + 2NH_3.$$

It, is found especially in plant seeds and microorganisms, as well as invretebrates (crabs, mussels), and shows a high degree of substrate specificity; apart from urea, it only attacks urea derivatives like hydroxy—and dihydroxyurea, which also act as noncompetitive inhibitors of U.

Ureide Plants. Plant families that accumulate allantoin and/or allantoic acid, and use these compounds as nitrogen metabolism. It is excreted in the urine in certain animal families, particularly birds and reptiles. It has been isolated from bird excrements (guano). It is generated from xanthine by the enzyme xanthine oxidase in aerobic purine catabolism.

Uricase, Urate Oxidase (EC 1.7.3.3.). A copper-containing aerobic oxidase, which, in the presence of oxygen, catalyses the oxidation of poorly soluble uric acid or urates to soluble allantoin, with the formation of hydrogen peroxide.

Uricolysis. Oxidation and decarboxylation of uric acid to allantoin, catalysed by uricase as part of aerobic Purine degradation.

Uridine, abb. **Urd** 3-β-D-ribofuranosyluracil. A β-glycosidic Nucleoside of D-ribose and the pyrimidine base; uracil.

Uridine Phosphates. Nucleotides of uracil; phosphate esters of uridine. Uridine 5′-monophosphate, abb, UMP, uridylic acid, is produced de novo in Pyrimidine biosynthesis, or by degradation of nucleic acids. UMP is the starting point for the synthesis of other pyrimidine nucleotides.

Uronic Acids. Aldehyde carboxylic acids formed by oxidation of the terminal primary alcohol group of aldoses. They give the usual reactions for sugars and are widely distributed as components of glycosides, polyuronides, polysaccharides and mucopolysaccharides.

Ursodeoxycholic Acid: 3α, 7β-dihydroxy-5β-cholane-24-acid. A dihydroxylated steroid carboxylic acid, belonging to the bile acids.

Ursolic Acid. A simple unsaturated pentacylic triterpene carboxylic acid. It is a structural derivative of α-Amyrin in which the 28-methyl group is replaced by a carboxyl group.

UTP. Abb. for uridine 5′-triphosphate.

V

Vaccenic Acid. Δ^{11}-octadecanoic acid, CH_3-$(CH_2)_5$-$CH = CH$-$(CH_2)_9$-$COOH$. An unsaturated fatty acid. The trans form occurs in the glycerides of animal fats, *e.g.* breef, muton and butter fat, and in vegetable oils. The cis form is hemolytic, occurring in plasma and various animal tissues, and in Lactobacillus.

Vacuoles. Structures within plant cells composed of a three-layered membrane (tonoplast) enclosing the cell sap V. The cell sap within the vacuole is an aqueous solution of numerous substances in true or collidal solution; in addition to sugars and salts these include inner secretions, so that the V. is considered as an excretory organ. Vacuole is important in the maintenance of turgor (inner pressure) of the plant cell, by acting as an osmotic system.

Val. Abb. for L-valine.

Valepotriates. Iridoids from Valeriana and Kentranthus spp. Valeriana officinalis contains up to 5% V., which are responsible for the sedative properties of this drug. The most important representative is Valtratum.

Valerian Alkaloids. Terpene alkaloids containing a pyridine ring (therefore also considered as pyridine alkaloids) from valerian (Valeriana officinalis).

L-Valine, abb. Val. L-α-aminoisovaleric acid, $(CH_3)_2CH \cdot CH(NH_2)$-COOH, an aliphatic, neutral, essential, glucogenic, proteogenic, amino acid. It is degraded by deamination, followed by oxidative decarboxylation to isobutyryl-CoA. Partial β-oxidation gives 3-hydroxyisobutyryl-CoA, followed by loss of the CoA. The 3-hydroxyisobutyrate is then oxidized to methylmalonate semialdehyde, which is decarboxylated and oxidized to propionate. The intact molecule of L-valine is incorporated in the biosynthesis of Penicillin.

Valinomycin. Cyclo-(D-Val-Lac-Val-D-Hyv-)$_3$, an antibiotic, cyclic depsipeptide, especially active against Mycobacterium tuberculosis.

Valtratum, Valepotriatum, Valtrate, Valepotriate. The chief member of the Valeportriates from valerian root. Acid hydrolysis produces isovaleric acid and 4-acetoxymethyl-7-formylcyclopenta [c] pyran (baldrianal).

Vanadium V. A trace element required for normal growth by animals. Most dietary items contain less than 100ng V/g. It stimulates the oxidation of phospholipids and decreases cholesterol synthesis by inhibiting squalene synthase (a liver microsomal enzyme system); it also stimulates acetoacetyl-CoA deacylase in liver mitochondria. Its deficiency leads to abnormal bone growth.

Vasoactive Intestinal Peptide, VIP. An octacosapeptide from porcine small intestine, which causes vasodilation, lowers arterial blood pressure, increases cardiac output, enhances myocardial activity, increases glycogenolysis and relaxes the smooth muscle of trachea, stomach and gall bladder.

Vesopressin, Antidiuretic Hormone, abb. **ADH, Antidiuretin, Pitressin.** A neurohypophysial peptide hormone. ADH has a direct antidiuretic action on the kidneys. It also causes vasconstriction of the heart beat and increase of blood pressure. The amino acids in positions 3, 4 and 8 of the nonapeptide are variable The phylogenetic precursor of ADH and of Oxytocin is [8-arginine] vasotocin. ADH occurs as [8-arginine] ADH, M_r 1084 and [8-lysine] ADH, M_r 1056.

Veratramine. A Veratrum alkaloid of the jerveratrum type, with C-nor-D-homo structure, found in hellebores (Veratrum album, V. eschscholtzii, V. viride). It viride also contains veratrosine, a glycoalkaloid in which the 3-β-hydroxyl group of veratramine is linked glycosidically to D-glucose.

Veratrum Alkaloids. A group of steroid alkaloids found in the Solanaceae, and in the genera Veratrum (hellebores) and Fritillaria (fritillary). They are structural derivatives of the parent hydrocarbon cholestane (Steroids); in some members ring C is contracted and ring D is expanded (C-nor-D-homo type). These have a positive ionotropic action on the heart, and cause a decrease of blood pressure by reflex inhibition of the vasomotor centers. They were used for treatment of hypertension, but have been replaced by Rauwolfia alkaloids and other drugs.

Vernine. Obsolete name for guanosine.

Vertebrates Hormones. Hormones of vertebrate animals. Chemically, vertebrate hormones are a heterogeneous group, which can be subdivided into Steroid hormones derived from amino acids. Peptide hormones, Proteohormones and hormones derived from fatty acids (Prostaglandins).

Vinblastine, Vincaleucoblastine. A dimeric indole-indoline alkaloid. It is one of the most effective naturally occurring antitumor agents, and is used primarily in the treatment of Hodgkin's disease.

Vinca Alkaloids, Catharanthus Alkaloids. A group of about 60 iridoid indole alkaloids from Vinca (Catharanthus) spp Structurally, they are tetra or pentacyclic indole derivatives with an iridoid component. Tryptophan and mevalonic acid are biosynthetic precursors of these alkaloids. Vincamine has hypertensive activity and is used pharmaceutically, particularly in Hungary.

Vincristine, Leurocristine. A dimeric indole alkaloid closely related to vinblastine, from Vinca rosea (Vinca alkaloids). It is used mainly for the treatment of acute leukemia in children, and against various other neoplasmic growths.

Violacein. The major purple pigment of Chromobacterium violaceum, accompanied by smaller amounts of deoxyviolacein. Every C-atom is derived biosynthetically from tryptophan.

Violaxanthin. 3(S), 3′(S)-dihydroxy-β-carotene-(5R, 6S, 5′R, 6′ S)-5, 6,5′, 6′-diepoxide, a xanthophyll. It is one of the most important plant carotenoids, present as an orange or brown-yellow pigment in all green leaves, and especially plentiful in flowers and fruits of Viola tricolor, Taraxa cum, Tagetes, Tulipa, Citrus, Cytisus, etc.

Viomycin, Celiomycin, Florimycin, TuberactinomycinB. A polypeptide antibiotic from various Streptomyces spp., including S. floridae, S. puniceus and S. vinaceus, containing a 7-deazaadenine ring. It is active chiefly against Gram negative bacteria, and is used therapeutically against Mycobacterium tuberculosis.

Viridicatine. A quinoline alkaloid from moulds of the genius Penicillium. It is biosynthesized from anthranilic acid, phenylalanine and the methyl group of methionine. Biosynthesis is catalysed by the enzyme cyclopenase.

Virus Coat Proteins, Capsides. Proteins with the largest known M_r values (up to 40×10^6). They comprise many, usually identicals, subunits, called capsomeres (M_r 13000 – 60000). V.c.p. of tobacco mosaic virus consists of 2130 capsomers of M_r 17500. V.c.p. lie on the exterior of the virus particle, enclosing the DNA or RNA.

Viruses. Infectious particles composed of nucleic acid and protein. Viruses are various shapes (cubic, spherical or helical) and sizes (30 nm-1 μm). M_r of viruses lie in the range $<1-40\times 10^6$, and

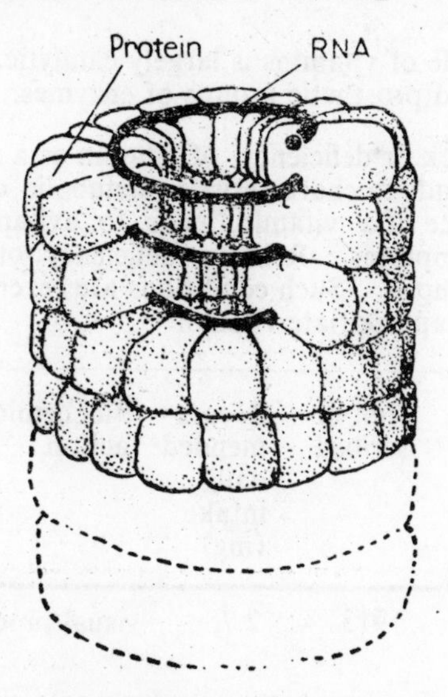

Schematic Representation of Part of Tobacco Mosaic Virus, Showing Spirally Arranged RNA and Protein Subunits

V. therefore pass through bacterial filters. They can be collected by ultracentrifugation and visualized under the electron microscope.

Viruses are parasites of animals, plants and microorganisms. They possess no metabolism of their own. The viruses of bacteria are called phages.

Vitamins [latin vita + amine). Substances present in the animal diet in only small quantities, and indispensable for the growth and maintenance of the organism. Most vitamins are essential for the metabolism of all living organisms, and they are synthesized by plants and microorganisms. The dietary requirement in the animal results from the evolutionary loss of this biosynthetic ability. In some animals the ability to synthesize a certain vitamin may not have been lost, *e.g.* most animals can synthesized ascorbic acid, for which it is therefore not a vitamin. Ascorbic acid is a vitamin C only for the primates and a few other animals (*e.g.* guinea pig). Some vitamins be synthesized from provitamins. A large proportion of the V. requirement of humans and higher animals is supplied by the intestinal flora, *e g.* most of the V.K. required by humans is provided in this way.

The role of vitamins is largely catalytic. They serve as coenzymes and prosthetic groups of enzymes.

The lack or deficiency of vitamin as a result of unbalanced nutrition leads to characteristic metabolic disturbances. Complete absence of a vitamin leads to avitaminosis, with typical clinical symptoms. Relative deficiency of a vitamin causes hypovitaminosis. Such conditions are reversible by administration of the appropriate vitamin.

Fat-soluble vitamins	First described	Recommended daily intake (mg)	Biochemical action	Clinical activity
A (retinal)	1913	2.7	visual process	epithelial protection vitamin; antixerophathalmic vitamin

D (calciferol)	1922	0.01—0.025	calcium and phosphate metabolism	antirachitic vitamin
E (tocopherol)	1922	5	electron transport	antisterility vitamin
K (phylloquinone menaquinone)	1935	1	electron transport	antihemorrhagic vitamin
Water soluble vitamins				
B_1 (thiamine)	1926	1.2	carbohydrate metabolism, transfer of active aldehyde	antineuritic vitamin
B_2—complex: Riboflavin	1932	1.7	respiration hydrogen transfer	antidermatitis vitamin
Folic acid	1941	1—2	one carbon unit transfer	theraphy of certain anemias
Niacin and niacinamide	1937	18	respiration, hydrogen transfer	pellagra preventative
Pantothenic acid	1933	3—5	transfer of acyl groups	chick antidermatitis factor; anti-gray hair factor
B_6 (pyrido xine)	1936	2	amino acid metabolism, transfer of amino groups	human deficiency symptoms not known
B_{12} (cobolamin especially cyanocobolamin)	1948	0.003	coenzyme of various metabolic reactions	antianemic vitamin, extrinsic factor
C (ascorbic acid)	1925	75	reducing agent for some mono oxygenases	antiscorbutic vitamin
H (biotin)	1935	0.25	coenzyme of various carboxylation reactions	the "skin" vitamin.

Vitamin A (retinol, axerophthol, xerophthol) (obsolete names; epithelial protection vitamin; growth vitamin). A group of fat-soluble vitamin with polyisoprenoid, structure. True Vitamin A is the alcohol retinol, also known as V.A1. 3-Dehydroretinol is VA_2, which is characterized by an additional double bond between C-3 and C-4 in the ring. V.A acid has also biological activity, but it is not converted into the alcohol or aldehyde in the organism.

Vitamin A occurs predominantly in animal products, such as milk, butter, egg yolk, cod liver oil, and in the body fat of many animals.

Deficiency of vitamin A in humans leads to night blindness, caused by the deficient regeneration of rhodopsin, and to hyperkeratosis of the epithelium of the eye (xerophthalmia).

Vitamin A can be stored in the organism for several months, chiefly in the liver in the form of its palmityl ester.

Deficiency of vitamin B_1 results in disturbances of carbohydrate metabolism, accompanied by an increase in the concentration of blood oxoacids (mostly pyruvate), which reflects the role of thiamine pyrophosphate as a coenzyme of pyruvate dehydrogenase. The typical deficiency disease, beriberi, results from a diet exclusively of polished rice.

Vitamin B_2 complex. A group of water-soluble vitamins consisting of folic acid, nicotinic acid and nicotinamide, pantothenic acid and riboflavin. The term Vitamin B_2 is now reserved for riboflavin alone.

Folic Acid (pteroylglutamic acid; the old name vitamin B_c is now obsolete). A pteridine derivative, especially plentiful in liver, yeast and green plants. The chemical structure of folic acid contains 3 moieties : 2-amino-4-hydroxypteridine, p-aminobenzoic acid and glutamic acid. Purines, especially guanine nucleotides, act as biosynthetic precursors of the pteridine moiety.

Folic acid is a growth factor for some bacteria. Micro-organisms also contain conjugates with several glutamic acid residues.

The biochemically active form of folic acid is Tetrahydrofolic acid, which is a coenzyme in the metabolism of Active one carbon units. In man, folic acid avitaminosis is more often caused by faulty uptake and/or utilization, then by a dietary deficiency. It usually results in abnormalities of the blood, *e.g.* megablastic anaemia, thrombocytopenia.

Nicotinic Acid (Niacin) and **Nicotinamide (Niacinamide)**, (pellagra preventative factor, Vitamin PP). These simple pyridine derivatives are especially plentiful in liver, fish, yeast and germinating cereal grains. Nicotinamide is a component of NAD and NADP.

Under certain nutritional conditions, *e.g.* when maize forms the bulk of the human diet, the deficiency of nicotinamide leads to Pellagra. This deficiency disease affects the skin (brown coloration), the digestive system (diarrhea) and the nervous system (dementia). Pellagra can be cured by feeding tryptophan; or nicotinamide may be administered therapeutically.

Pantothenic acid. Consists of 2, 4-dihydroxy 3, 3-dimethylbutyric acid (pantoic acid) linked to β-alanine by an amide bond. Pantoic acid is biosynthesized from valine. It is required as a precursor of Pantetheine (for the synthesis of Coenzyme A.

Experimental human deficiency results in burning sensations, muscle weakness, abdominal disorders, vasomotor instability and depression.

Riboflavin (lactoflavin; 6, 7-dimethyl-9 (D-1′-ribityl-isoalloxazine; Vitamin B_2). A yellow flavin derivative, which occurs chiefly in a bound form in flavin nucleotides or flavoproteins in yeasts, animal products and legume seeds. Milk contains free riboflavin. It is required as a precursor of Flavin mononucleotide and Flavin-adenine-dinucleotide which are coenzymes of Flavin enzymes.

In humans, riboflavin deficiency (ariboflavinosis) is characterized by lip lesions, a seborrheic dermatitis about the noise, ears and eyelids, and loss of hair.

Vitamin B_6 (pyridoxine; pyridoxol; adermine). Several naturally occurring compounds have vitamin B_6 activity in animal nutrition : pyridoxol (2-methyl-3-hydroxy-4, 5-di[hydroxymethyl]-pyridne) is the chief form of vitamin B_6 in vegetables; whereas, pyridoxal, pyridoxal phosphate, pyridoxamine and pyridoxamine phosphate are present in animal tissues. Vitamin B_6 is water soluble and occurs in liver, kidney, yeast, vegetables and cereals.

In humans, Vitamin B_6 deficiency is not a very specific condition, and symptoms are easily confused with those of most deficiencies of the Vitamin B group; occasionally there may be specific nervous disorders and anemia. Experimental Vitamin

B_6 deficiency in animals results in Vitamin B_6-pellagra, characterized by loss of hair, edema and red scaly skin. Vitamin B_6 deficiency inhibits the degradation of L-tryptophan, and the excretion of xanthurenic acid is used as an index of Vitamin B_6 deficiency.

Vitamin B_{12} (cobalamin; extrinsic factor; animal protein factor). A group of water-soluble compounds with extremely high biological activity. They all belong to a class of compounds known as corrinoids. The structure consists of a complex corrin ring system, with a centrally bound trivalent cobalt atom, a base-nucleotide moiety, and a monovalent group (called the cobalt ligand) bound to the cobalt.

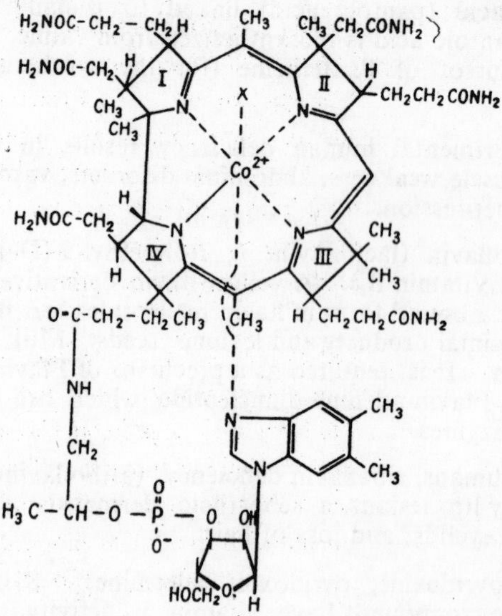

Cyanocobalamin is also known as antipernicious anemia factor. Pernicious anemia is characterized by a severely reduced production of red blood cells, deficient gastric secretion and disturbances of the nervous system. It is not usually caused by a dietary deficiency of V.B_{12}, but by poor absorption. Cure is effected by injection of small amounts (3–6 μg) of V.B_{12}. Excretion of methyl-malonic acid is used for the diagnosis of V.B_{12} deficiency.

Vitamin B_T. See Carnitine.

Vitamin C (ascorbic acid; antiscorbutic vitamin). A water-soluble vitamin with a wide natural distribution, especially in fresh vegetables and fruit. This vitamin is the γ-lactone of 2-oxo-L-gulonic acid, derived from carbohydrate metabolism. In most mammals it is synthesized from D glucuronate by reactions of the Glucuronate pathway. Man and the other primates, and the guinea pig cannot synthesize vitamin C.

It is a powerful reducing agent on account of its ene-diol grouping. Ascorbic acid oxidase (a copper enzyme) catalyzes the removal of hydrogen from ascorbic acid ($C_6H_8O_6$) to produce dehydroascorbic acid ($C_6H_6O_6$).

It is particularly important for the maintenance of the inner wall of blood vessels. Deficiency results in scurvy, a long-known avitaminosis, characterized by rupture of blood capillaries, hemorrhage of the skin and mucosas, inflammation of the gums, loosening of the teeth and painful swellings of the joints. Resistance of the organism to infectious diseases is also reduced.

Vitamin D (calciferol; antirachitic vitamin). A group of fat-soluble vitamin chemically related to the steroids. They are produced from provitamins, $\Delta^{5,7}$-unsaturated sterols, by UV-irradiation.

Vitamin D_1 is a molecular compound of lumisterol and ergocalciferol.

Vitamin D_4 is 22-dihydroergocalciferol, produced from 22-dihydroergosterol by UV-irradiation.

V D_3 is present in cod liver oil in particularly large quantities, and it is also formed in human skin from 7-dehydrocholesterol by the action of sunlight.

The V.D-complex is also present in *e.g.* herrings, egg yolk, butter, cheese, milk, pig liver and edible fungi.

V.D is important in calcium metabolism. It is required for calcium absorption and the mineralization of bone. The V.D deficiency disease, known as rickets, is characterized by a softening and malformation of the bones. It results from a poor absorption of calcium, coupled with deficient incorporation of calcium into bone tissue.

Vitamin E (tocopherol; antisterility factor). A group of fat-soluble V., containing a chromane ring with a polysioprenoid side-chain. Eight compounds with vitamin E activity are known:

α-, β-, γ-tocopherol etc., which differ in the number and positions of the methyl groups in the aromatic ring. Biologically, the most important member is α-tocopherol.

This vitamin occurs in wheat seedlings, and has been isolated from wheat seedling oil. It is also present in lettuce, celery, cabbage, maize, palm oil, ground nuts, soybeans, castor oil and butter. In animal experiments its deficiency results in death of the embryo in pregnant females. In the male, there is atrophy of the gonads and muscle dystrophy.

Vitamin F. The essential fatty acids, which cannot be synthesized in the body. These include the unsaturated fatty acids, in particular linoleic, linolenic and arachidonic acid. They occur in high concentrations in vegetable oils. Deficiency of essential fatty acids in rats leads to loss of hair, disturbances of water balance, and reproductive failure.

Vitamin G. An obsolete term for riboflavin, a component of the V.B_2-complex.

Vitamin H (biotin; bios II; coenzyme R). A sulfur-containing water soluble vitamin. Chemically it is a cyclic urea derivative: 2′-oxo-3, 4-imidazoline-2-tetrahydrothiophene-n-valeric acid. It is most commonly known as biotin. Biotin is biosynthesized from cysteine, pimelic acid and carbamyl phosphate. It acts as the prosthetic group of carboxylation enzymes the carbonyl group of biotin forms an amide bond with the ε-amino-group of a lysine residue in the enzyme protein.

Vitamin K (phylloquinone; antihemorrhagic vitamin; coagulation vitamin). A group of fat-soluble, naphthoquinone compounds with varying sizes of isoprenoid side chain. Vitamin K_1 is especially plentiful in green plants. Vitamin K_2 (farnoquinone; menaquinone-6; 2-methyl-3-difarnesyl-1, 4-naphthoquinone) is found chiefly in bacteria. Vitamin K_3 (menadione; 2-methyl-1, 4-naphthoquinone) is actually a provitamin.

Its deficiency causes the deficient production of blood coagulation factors, in particular prothrombin, leading to abnormally long clotting times. It serves as a cofactor in the carboxylation of glutamic acid residues during the post translational modification of prothrombin and other blood coagulation proteins.

Vitamin P (permeability factor). A term applied formerly to a group of plant Flavones *e.g.* hesperidin, eriodictin, and particularly quercetin. These flavones increase the permeability of blood vessels, and for this reason have been used pharmaceutically.

Vitamin PP (PP factor). An old name for nicotinamide, a component of the V B_2-complex.

Vitellin. A lipophosphoprotein present in egg yolk, together with Phosphovitin.

Vulgaxanthin. A yellow betaxanthin from sugar beet (Beta vulgaris).

W

Warburg's Atmungsferment. Cytochrome oxidase.

Warburg's Respiratory Enzyme. Cytochrome oxidase.

Water: H_2O, quantitatively the most important inorganic constituent of living cells A normal living cell contains about 80% W. Plants can contain up to 95%, jelly fish 98%, higher animals (60 75% (Tables 1 and 2). All life processes depend upon W.

Water is a reactant in enzyme (hydrolase)-catalysed hydrolytic cleavage of macromolecules (proteins, carbohydrates, fats), representing the first stage in the biological degradation of these substances. It is formed metabolically by the operation of the respiratory chain (respiratory W.), and it is a substrate of photosynthesis.

In animals, water is important in the regulation of body temperature. Heat is removed by evaporation of water from body and respiratory tract surfaces.

Wax Acids: CH_3-$(CH_2)_n$-COOH, long chain, even numbered monocarboxylic acids, which occur esterified in waxes, *e.g.* lauric acid (C_{12}), myristic acid (C_{14}), palmitic acid (C_{16}), carnaubic acid (C_{24}), cerotic acid (C_{26}), montanic acid (C_{28}), melissic acid (C_{30}) and other higher fatty acids.

WGA. Abb. for wheat germ agglutinin.

Withanolides. A group of C_{28} plant steroids based on the parent hydrocarbon, ergostane and containing a characteristic withanolide ring system with a δ-lactone side chain.

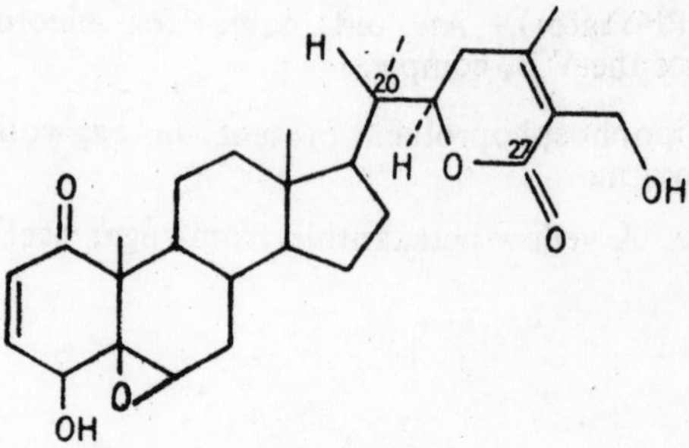

Withaferol A

Wood. A mixed polymer with lignin as a structural component. The deposition of lignin in the cellulose matrix of the cell wall is called lignification.

X

Xan. Abb. for xanthine.

Xanthine, abb. **Xan**: 2, 6-dihydroxypurine. A purine and the starting point for Purine degradation. It is found free, together with other purines. Some derivatives are physiologically important, especially xanthosine phosphates and the Methylated xanthines.

Xanthine Oxidase, Xanthine Dehydrogenase, Schardinger Enzyme. An enzyme of aerobic purine degradation, which catalyses the oxidation of hypoxanthine to xanthine, and xanthine to uric acid :

$$\text{Hypoxanthine} + H_2O + O_2 \rightarrow \text{Xanthine} + H_2O_2$$
$$\text{Xanthine} + H_2O + O_2 \rightarrow \text{Uric acid} + H_2O_2$$

It is a dimeric enzyme, containing 2 FAD, 2 Mo and 8 Fe (data for the enzyme from milk). The substrate specificity is low; it catalyses the oxidation of other purines (e g., adenine), aliphatic and aromatic aldehydes, pyrimidines, pteridines and other heterocyclic compounds. In animal tissues (e.g. calf liver) xanthine oxidase is in the Golgi apparatus; it is also a secretory enzyme present in milk, where its activity can be used to differentiate between fresh and heated or pasteurized milk.

Xanthocillin: 1, 4-di-(4-hydroxyphenyl)-2, 3-diisonitrilobutadiene (1, 3). A bacteriostic antibiotic used against local infections due to Gram positive and Gram negative organisms.

Xanthophylls. A group of Carotenoids.

Xanthoprotein Reaction. A qualitative test for protein, using concentrated nitric acid. The resulting yellow color is due to the nitration of aromatic amino acid residues.

Xanthopterin: 2-amino-4, 6-dioxotetrahydropteridine. It is the yellow pigment of bees, wasps and hornets. It is biosynthesized from guanine and two carbon atoms of a pentose.

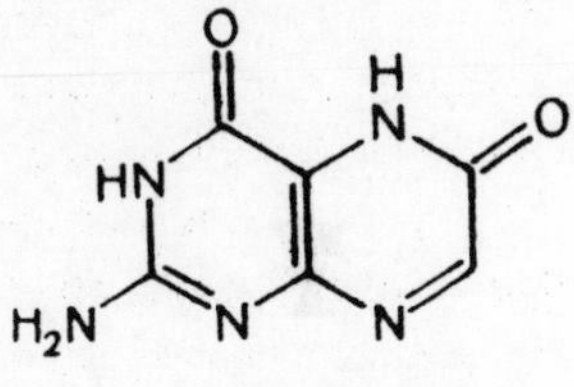

Xanthopterin

Xanthoxin. An endogenous growth regulator, occurring widely in higher plants possessing inhibitory properties similar to those of abscisic acid. It occurs in both the cis, trans, and in the biologically less active trans, trans form. Plant tissues can probably convert cis, trans-X, into (R)—(+)-abscisic acid.

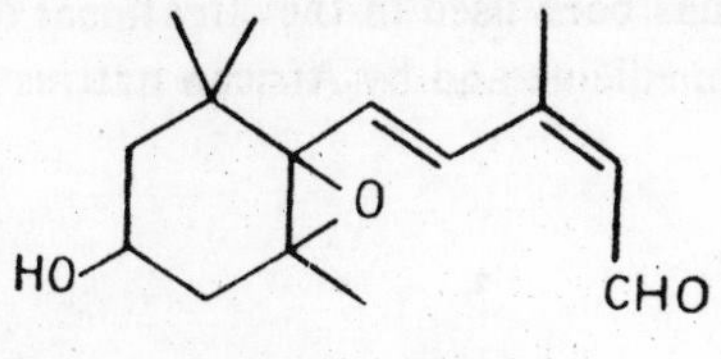

cis, trans-Xanthoxin

Xao. Abb. for xanthosine.

Xerophthol. Vitamin A.

XMP. Abb. for xanthosine 5′-monophosphate.

Xylitol: CH_2-OH-$(CHOH)_3$-CH_2OH, an optically inactive C_5-sugar alcohol, related to xylose. It is a byproduct of wood saccharification, and can also be prepared by the catalytic hydrogenation of xylose. It can be used as a sugar substitute in diabetic diets.

D-xylose, Wood Sugar. A monosaccharide pentose not fermented by yeast. Reduction of xylose gives xylitol, mild oxidation gives xylonic acid. It is produced by acid hydrolysis of Xylans. An important dietary component for herbivores, especially ruminants.

Xylosyl Nucleosides. Nucleosides in which the sugar component is xylose. They may acts as analogs of purine or pyrimidine ribosides.

Y

Yohimbine. A Rauwolfia alkaloid. It has five chiralc enters and therefore many stereoisomers; seven of these occur naturally, the most important being Corynanthine, obtained chiefly from the bark of the tropical tree, Corynanthe yohimbe. It is vasodilatory and has been used in the treatment of arteriosclerosis, in veterinary medicine and by African natives as an aphrodisiac.

Z

Zeatin: 6-(4-hydroxy-3-methyl-but-trans-2-enyl)-aminopurine. A naturally occurring Cytokinin. It occurs free in many plants, especially in immature maize kernels, and is identical with the previously described maize factor (abb. MF).

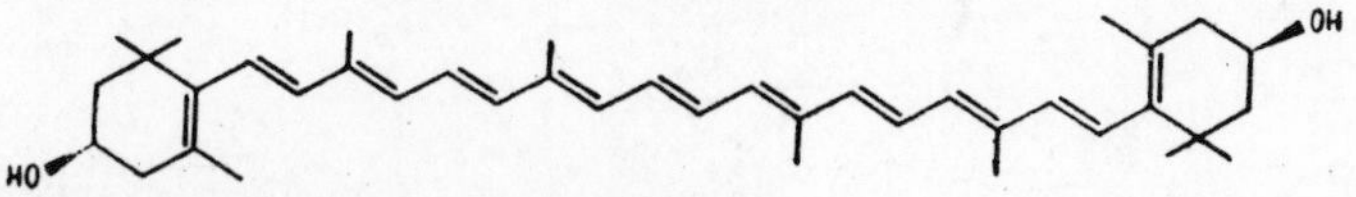

Zeaxanthin

Zeaxanthin: (3 R, 3′R)-β,β-carotene-3,3′-diol; 3, 3′-dihydroxy-β-carotene, a xanthophyll, isomeric with lutein. It occurs free and esterified as the dipalmitate, and shows no vitamin A activity.

Zinc Zn. An essential bioelement for the growth and development of plants, animals and microorganisms. It has a high affinity for nitrogen and sulfur ligands, and occurs in the cell in association with many different compounds, *e.g.* proteins (insulin), amino acids, nucleic acids.

Zwischenferment. Glucose 6-phosphate dehydrogenase.

Zymase. An old name for a mixture of 11 enzymes of glycolysis isolated from yeast after mechanical disruption of the cell wall. It catalyses alcoholic fermentation.

Zymogens. Inactive precursors of enzymes, usually proteolytic enzymes. They are converted into active enzymes by limited proteolysis.

Zymosterol, 5α-cholesta-8(9), 24-dien-3β-ol. A mycosterol which is present in yeast. It is an intermediate in the biosynthesis of cholesterol from lanosterol.